Minimal Surfaces, Geometric Analysis and Symplectic Geometry

ADVANCED STUDIES IN PURE MATHEMATICS 34

Chief Editor: Eiichi Bannai (Kyushu University)

Minimal Surfaces, Geometric Analysis and Symplectic Geometry

Edited by

Kenji Fukaya (Kyoto University)
Seiki Nishikawa (Tohoku University) and
Joel Spruck (Johns Hopkins University)

Mathematical Society of Japan

This book was typeset by $\mathcal{A}_{\mathcal{M}}\mathcal{S}$-TEX and $\mathcal{A}_{\mathcal{M}}\mathcal{S}$-LATEX, the TEX macro systems of the American Mathematical Society, together with the style files `aspm.sty` *and* `aspmfm.sty` *for $\mathcal{A}_{\mathcal{M}}\mathcal{S}$-TEX written by Dr. Chiaki Tsukamoto and* `aspmproc.sty` *for $\mathcal{A}_{\mathcal{M}}\mathcal{S}$-LATEX written by Dr. Akihiro Munemasa.*

Edited by the Mathematical Society of Japan.

Published by the Mathematical Society of Japan.

Distributed outside Japan by the American Mathematical Society.

ISBN 4-931469-18-3

2000 Mathematics Subject Classification.
Primary 53-06; *Secondary* 53A05, 53A10, 53B25, 53C21, 53C26, 53D05, 53D20, 53D40, 57R57, 58J50.

PRINTED IN JAPAN
by Tokyo Shoseki Printing Co., Ltd.

Preface

The eleventh year program of the Japan-U. S. Mathematics Institute, which was held during 1998-1999 at the Johns Hopkins University, was devoted to Minimal Surfaces, Geometric Analysis, and Symplectic Geometry. This program culminated in a week-long workshop and conference held at the Institute from March 16 to 21, 1999 under the co-sponsorship of the Japan Society for the Promotion of Science and the National Science Foundation as a part of the Japan-U. S. Cooperative Science Program. The workshop and conference were organized by K. Fukaya of Kyoto University, S. Nishikawa of Tohoku University, W. Minicozzi, J. Morava and J. Spruck of Johns Hopkins University, with the aim of facilitating interaction between the mathematicians working in these and related areas. About seventy mathematicians, from Japan, the U. S. and Europe, participated in the meeting and discussed recent developments.

This volume is a collection of articles contributed by speakers of the workshop and conference, which presents extended or modified versions of the lectures delivered at the meeting. It is the hope of the editors that this volume will prove to be a stimulus for further research into these rapidly developing areas.

The organizers and editors are grateful to the Japan Society for the Promotion of Science, the National Science Foundation, and the Department of Mathematics of the Johns Hopkins University for their financial support.

May, 2001
Kenji Fukaya
Seiki Nishikawa
Joel Spruck

All papers in this volume have been refereed and are in final form. No version of any of them will be submitted for publication elsewhere.

CONTENTS

Advanced Studies in Pure Mathematics 34, 2002
Minimal Surfaces, Geometric Analysis and Symplectic Geometry
pp. 1–7

Volume Minimizing Hypersurfaces in Manifolds of Nonnegative Scalar Curvature

Mingliang Cai

Abstract.

We prove that if a manifold of nonnegative scalar curvature contains a two-sided hypersurface which is locally of least area and admits no metric of positive scalar curvature, then it splits isometrically in a neighborhood of the hypersurface.

We report here on joint work with G. Galloway concerning the study of rigidity of manifolds with nonnegative scalar curvature. Let us first recall the following theorem of Schoen and Yau.

Theorem 1. *Let (M, g) be a smooth n-manifold with positive scalar curvature, $S > 0$. If Σ is a compact immersed two-sided stable minimal hypersurface in M, then Σ admits a metric of positive scalar curvature.*

The above theorem follows from the proof of Theorem 1 in [SY]. If M is merely assumed to have nonnegative scalar curvature, the conclusion of the above theorem may not hold. Consider, for example, $T^{n-1} \times S^1$, where T^{n-1} is an $n-1$ torus. It is known that T^{n-1} does not admit a metric of positive scalar curvature ([GL], [SY]). However, in this direction one has the following theorem (cf. [SY], [FCS]).

Theorem 2. *Let (M, g) be a smooth n-manifold with nonnegative scalar curvature, $S \geq 0$. Let Σ be a compact manifold which does not admit a metric of positive scalar curvature. If Σ is immersed as a two-sided stable minimal hypersurface in M, then Σ is totally geodesic. Furthermore, the ambient scalar curvature S, the intrinsic scalar curvature $\tilde{S}$ and the Ricci curvature in the normal direction Ric_{nn} along Σ all vanish.*

We outline here the proof of Theorem 2 for $n > 3$ (for $n = 3$, see [FCS]). Denote by Π the second fundamental form. The minimality and

---

2000 *Mathematics Subject Classification.* Primary 53C20; Secondary 53C21.

stability conditions of Σ imply that, for all smooth functions ϕ on Σ,

$$\int_\Sigma |\nabla\phi|^2 - \int_\Sigma (\mathrm{Ric}_{nn} + |\Pi|^2)\,\phi^2 \ge 0. \tag{1}$$

Using the Gauss curvature equation and relating the ambient and intrinsic scalar curvatures along Σ, one gets the following "rearrangement"

$$S = \tilde{S} + 2\,\mathrm{Ric}_{nn} + |\Pi|^2. \tag{2}$$

Substituting (2) into (1), we have

$$\int_\Sigma |\nabla\phi|^2 + \frac{1}{2}\int_\Sigma (\tilde{S} - S - |\Pi|^2)\,\phi^2 \ge 0. \tag{3}$$

Since $S \ge 0$, we conclude that

$$-\frac{1}{2}\int_\Sigma \tilde{S}|\phi|^2 \le \int_\Sigma |\nabla\phi|^2 \tag{4}$$

for any smooth function ϕ on Σ.

Now, consider the operator

$$L = \Delta - \frac{n-3}{4(n-2)}\tilde{S}.$$

We claim that all the eigenvalues of L are nonnegative. Suppose the contrary and let ϕ be a nonzero solution of

$$L\phi = -\lambda\phi$$

for some $\lambda < 0$. Multiplying the above equation by ϕ and integrating, we obtain

$$\frac{2(n-2)}{n-3}\int_\Sigma |\nabla\phi|^2 = -\frac{1}{2}\int_\Sigma \tilde{S}\phi^2 + \frac{2\lambda(n-2)}{n-3}\int_\Sigma \phi^2 < \int_\Sigma |\nabla\phi|^2 \tag{5}$$

where the inequality follows from (4). But this is not possible as $2(n-2)/(n-3) > 1$.

Now we show that the first eigenvalue, λ_0, is zero. We argue again by contradiction. Suppose the first eigenvalue $\lambda_0 > 0$ and let u be a first eigenfunction. It is well-known that the first eigenfunctions for operators of the form of L do not change sign, hence we may assume that $u > 0$. If we multiply the metric of Σ by $u^{4/(n-3)}$, the scalar curvature of Σ is transformed to

$$u^{-\frac{n+1}{n-3}}(\tilde{S}u - \frac{4(n-2)}{n-3}\Delta u) = \frac{4(n-2)}{n-3}u^{-\frac{n+1}{n-3}}\lambda_0 u > 0.$$

This contradicts our assumption that Σ does not admit a metric of positive scalar curvature.

Inequality (4) together with the equation in (5) implies that the eigenfunctions corresponding to the eigenvalue 0 must be constants and that $\tilde{S} = 0$. Substituting $\tilde{S} = 0$ and $\phi = 1$ into (3), we see that both S and Π vanish. Theorem 2 is thus proved.

Theorem 2 may be loosely paraphrased as: if Σ does not admit a metric of positive scalar curvature and if $\Sigma \subset M$ is infinitesimally of least area, then M infinitesimally splits along Σ. The aim of this paper is to establish a noninfinitesimal version of this result. Our main theorem is the following

Theorem 3. *Let (M, g) be a smooth n-manifold with nonnegative scalar curvature, $S \geq 0$. Let Σ be a compact manifold which does not admit a metric of positive scalar curvature. If Σ is immersed as a two-sided hypersurface in M which is locally of least area, then Σ has zero scalar curvature and a neighborhood of Σ in M splits isometrically as a product.*

By definition, a compact two-sided hypersurface Σ in a manifold M is locally of least area provided in some normal neighborhood V of Σ, $A(\Sigma) \leq A(\Sigma')$ for all Σ' isotopic to Σ in V, where A is the area functional. If the inequality is strict for $\Sigma' \neq \Sigma$, we say that Σ is locally strictly of least area. Note that "locally of least area" in the theorem cannot be replaced by "stable minimal". Take, for example, $S^2 \times S^1$, where S^2 is a modified sphere with an infinitesimally flattened equator E. Then $E \times S^1$ is a torus which does not admit a metric of positive curvature and which is stable minimal in $S^2 \times S^1$.

Theorem 3 was proved in [CG] for $n = 3$. We thank an anonymous referee for pointing out to us that ideas there also apply to higher dimensions.

The idea of the proof of Theorem 3 is as follows. We first show that Σ cannot be locally *strictly* of least area. If it were, then under a sufficiently small perturbation of the metric to a metric of (strictly) positive scalar curvature, Σ would be perturbed to a minimal hypersurface which would admit a metric of positive scalar curvature. But this would contradict our assumption. We then show that on each side of Σ there is a hypersurface which has the same volume as Σ. This implies that a neighborhood of Σ is foliated by local minimizers, which in turn implies that the neighborhood is a product.

The folowing lemma is proved in [CG] which shows that locally any metric of nonnegative scalar curvature can be perturbed to a nearby metric of positive scalar curvature.

Lemma 1. *Suppose Σ is a compact two-sided hypersurface in an n-manifold (M, g) with nonnegative scalar curvature, $S(g) \geq 0$. Then there exists a neighborhood U of Σ and a sequence of metrics $\{g_n\}$ on U such that $g_n \to g$ in C^∞ topology on U, and each g_n has strictly positive scalar curvature, $S(g_n) > 0$.*

The next lemma is proved in [CG] for $n = 3$. The arguments there probably do not extend beyond dimension 7. We adopt here an alternative approach suggested by the anonymous referee.

Lemma 2. *Let Σ be as in Theorem 3. Σ cannot be locally strictly of least area.*

Proof. Denote by X the set of C^∞ sections of the normal bundle of Σ with sufficiently small C^1 norm. For $u \in X$, let $H(u)$ be the mean curvature of $\mathrm{graph}_\Sigma\, u$ in normal coordinates. H is a Fredholm operator and has the linerization

$$H'(0) = -\Delta - (|\Pi|^2 + \mathrm{Ric}_{nn}).$$

Since both Π and Ric_{nn} vanish by Theorem 2, $H'(0) = -\Delta$ and hence the cokernel, as well as the kernel, of $H'(0)$ consists of constant functions on Σ. Denote by p the projection from $C^\infty(\Sigma)$ to Y, where $Y = \{u \mid \int_\Sigma u = 0\}$. The composition $p \circ H$ is then a submersion from X to Y (some shrinkage of the domain may be necessary) and $(p \circ H)^{-1}(0)$ is a one-dimensional submanifold of X whose graphs constitute a family of constant mean curvature hypersurfaces. The area functional A_g restricted to this submanifold has a strict minimum at the zero. Let $\tilde{g}$ be a small pertubation of g with positive scalar curvature, $\tilde{S} > 0$, and let $\tilde{H}$ be the corresponding mean curvature operator. The existence of $\tilde{g}$ is guaranteed by Lemma 1. When the perturbation is sufficiently small, $(p \circ \tilde{H})^{-1}(0)$ will be a one-dimensional submanifold whose graphs will be a family of constant mean curvature hypersurfaces in the metric $\tilde{g}$, and the area function $A_{\tilde{g}}$ has a local minimum in it close to 0. We first show that this local minimum is a minimal hypersurface.

To this end, let $u(t)$ be a parametrization of $(p \circ H)^{-1}(0)$ with $u(0) = 0$. Since $u'(0)$ is in the kernel of $p \circ H'(0)$, $u'(0)$ is a (non-zero) constant function. Without loss of generality, we assume $u'(0)$ is a positive constant. We then parametrize $(p \circ \tilde{H})^{-1}(0)$ by $\tilde{u}(t)$ in such a way that $\tilde{u}(t)$ is close to $u(t)$, $\tilde{u}(0)$ is the local minimum of $A_{\tilde{g}} \circ \tilde{u}$ and $\tilde{u}'(0)$ is a positive function.

For simplicity, denote $A_{\tilde{g}} \circ \tilde{u}$ by $\tilde{A}$, $\tilde{u}'(0)$ by ϕ and the graph corresponding to $\tilde{u}(0)$ by $\tilde{\Sigma}$.

Since 0 is an extremum of $\tilde{A}$, $\tilde{A}'(0) = 0$. On the other hand, the first variational formula shows that

$$\tilde{A}'(0) = \int_{\tilde{\Sigma}} \tilde{H}(0)\phi\,,$$

where $\tilde{H}(0)$ is the mean curvature of $\tilde{\Sigma}$. Since $\tilde{H}(0)$ is constant and ϕ is positive, the above shows that $\tilde{H}(0) = 0$, i.e., $\tilde{\Sigma}$ is an minimal hypersurface. Now we show that $\tilde{\Sigma}$ admits a metric of positive scalar curvature, contradicting our assumption on Σ as $\tilde{\Sigma}$ is diffeomorphic to Σ.

Since $(p \circ \tilde{H} \circ \tilde{u})'(0) = 0$, we have

$$-\tilde{\Delta}\phi - (\tilde{\Pi}^2 + \tilde{\mathrm{Ric}}_{nn})\phi = c\,, \tag{6}$$

where c is in the kernel of p and hence is a constant. We claim that $c \geq 0$. In fact, since $\tilde{A}''(0) \geq 0$ and

$$\tilde{A}''(0) = \int_{\tilde{\Sigma}} (-\tilde{\Delta}\phi - (\tilde{\Pi}^2 + \tilde{\mathrm{Ric}}_{nn})\phi)\phi\,,$$

it follows that $\int_{\tilde{\Sigma}} c\phi \geq 0$. This together with $\phi > 0$ implies that $c \geq 0$.

Applying the "rearrangement" to (6), we get

$$-\tilde{\Delta}\phi + \frac{1}{2}(\tilde{\tilde{S}} - \tilde{S} - \Pi^2)\phi = c \geq 0,$$

where $\tilde{\tilde{S}}$ is the intrinsic scalar curvature of $\tilde{\Sigma}$.

Similar to the proof of Theorem 2, we now multiply the metric on $\tilde{\Sigma}$ by $\phi^{2/(n-2)}$, the scalar curvature of the new conformed metric is then equal to

$$\begin{aligned} &\phi^{-\frac{n}{n-2}}(-2\tilde{\Delta}\phi + \tilde{\tilde{S}}\phi + \frac{n-1}{n-2}\frac{|\nabla\phi|^2}{\phi}) \\ &\qquad = \phi^{-\frac{n}{n-2}}(2c + (\tilde{S} + \Pi^2)\phi + \frac{n-1}{n-2}\frac{|\nabla\phi|^2}{\phi}). \end{aligned}$$

Since $c \geq 0$, $\phi > 0$ and $\tilde{S} > 0$, the above is positive. This is a contradiction and Lemma 2 is thus proved. □

Remark 1. It is clear from the proof that Lemma 2 holds for manifolds with $C^{2,\alpha}$ metrics, a fact which will be used later.

Remark 2. Since $u(0) = 0$ and $u'(0)$ is a positive constant, we know that $u(t)$ and t have the same sign when t is sufficiently small. This shows that when a constant mean curvature hypersurface is sufficiently close to Σ, it lies to one side of Σ and does not intersect with Σ unless it coincides with Σ.

We are now in a position to prove Theorem 3.

For simplicity, we assume that Σ is embedded. The general case can be reduced to this one by working in the normal bundle of Σ.

We denote by $\mathcal{F}$ the collection of minimal hypersurfaces which are C^1 close to Σ and have the same volume as Σ. Lemma 2 implies that each element in $\mathcal{F}$ is an accumulation point in $\mathcal{F}$. In fact, we can show that each element is a *two-sided* accumulation point. To see this, let us look at one of the two components of $M\backslash\Sigma$, say U. Taking two copies of U and gluing them along $\partial U = \Sigma$, we get a new manifold, N. Since Σ is totally geodesic, the induced metric on N is of class $C^{2,1}$. Moreover, Σ is locally of least area in the new metric. Applying Lemma 2 (see also Remark 1) to N, we obtain a sequence of mutually distinct hypersurfaces Σ_n in N such that Σ_n has the same volume as Σ and $\Sigma_n \to \Sigma$. It follows from Remark 2 that when n is sufficiently large, Σ_n lies to one side of Σ and does not intersect with Σ. This shows that U contains a sequence of hypersurfaces in $\mathcal{F}$ that is convergent to Σ. Since the choice of U is arbitrary, we conclude that Σ is a two-sided accumulation point in $\mathcal{F}$. The argument certainly applies to every element in $\mathcal{F}$.

We now show that when $|t|$ is sufficiently small, $\text{graph}_\Sigma\, u(t)$ is an element in $\mathcal{F}$, where $u(t)$ is as in the proof of Lemma 2. To do this, let us fix a point x_0 in Σ and consider $r(t) = \exp_{x_0} u(t)N$, where N is the normal vector to Σ. Since every element in $\mathcal{F}$ is a two-sided accumulation point, a continuity argument shows that for each t there is an element Σ_t in $\mathcal{F}$ passing through $r(t)$. Note that $(p \circ H)^{-1}(0)$ consists of all constant mean curvature hypersurfaces that are close to Σ and that Σ_t is a minimal hypersurface, hence, there is t' such that $\Sigma_t = \text{graph}_\Sigma\, u(t')$. Clearly, t' is uniquely determined, and thus we get a map $t \mapsto t'$. It is easy to see that this map is continuous and $0 \mapsto 0$. This implies that at least when $|t|$ is sufficiently small, $\text{graph}_\Sigma\, u(t)$ is a minimizer for the area functional. It then follows from the proof of Lemma 2 that $u(t)$ is a constant section for each t. We thus have obtained a smooth foliation of a neighborhood of Σ by totally geodesic hypersurfaces which are level surfaces of the distance function to Σ. A standard argument shows that the neighborhood is a product of Σ with an interval. This completes the proof of Theorem 3.

Remark 3. It would be interesting to extend Theorem 3 to non-compact hypersurfaces. In dimension 3, Fischer-Colbrie and Schoen ([FCS]) proved that a complete stable minimal surface in an orientable

3-manifold with nonnegative scalar curvature must be conformal to the complex plane or the cylinder A. In the latter case one can show that A is flat and totally geodesic (See [FCS] and [CM]). It seems reasonable to conjecture that if the cylinder A is actually area minimizing (in a suitable sense), then M is a product. (cf. Remark 4 in [CG]).

We would like to express our gratitude to a referee for valuable suggestions. We would also like to thank Bill Minicozzi for some helpful discussions.

References

[CG] M. Cai and G. J. Galloway, *Rigidity of 3-manifolds of nonnegative scalar curvature*, Comm. Anal. Geom., **8** (2000), 565–573.

[CM] T. H. Colding and W. P. Minicozzi, *Estimates for parametric elliptic integrals*, preprint (1999).

[FCS] D. Fischer-Colbrie and R. Schoen, *The structure of complete stable minimal surfaces in 3-manifolds of nonnegative scalar curvature*, Comm. Pure Appl. Math., **33** (1980), 199–211.

[GL] M. Gromov and H. B. Lawson, Jr., *The classification of simply connected manifolds of positive scalar curvature*, Ann. of Math., **111** (1980), 423–434.

[SY] R. Schoen and S.-T. Yau, *On the structure of manifolds with positive scalar curvature*, Manuscripta Math., **28** (1979), 159–183.

Department of Mathematics
University of Miami
Coral Gables, FL 33124
U. S. A.
mcai@math.miami.edu

Advanced Studies in Pure Mathematics 34, 2002
Minimal Surfaces, Geometric Analysis and Symplectic Geometry
pp. 9–14

The Gaussian Image of Mean Curvature One Surfaces in $\mathbb{H}^3$ of Finite Total Curvature

Pascal Collin, Laurent Hauswirth and Harold Rosenberg

Abstract.

The hyperbolic Gauss map G of a complete constant mean curvature one surface M in hyperbolic 3-space, is a holomorphic map from M to the Riemann sphere. When M has finite total curvature, we prove G can miss at most three points unless G is constant. We also prove that if M is a properly embedded mean curvature one surface of finite topology, then G is surjective unless M is a horosphere or catenoid cousin.

We consider complete surfaces M in hyperbolic 3-space $\mathbb{H}^3$ with mean curvature one and of finite total curvature. For a point $q \in M$, the Gauss map G sends q to the point at infinity obtained as the positive limit of the geodesic of $\mathbb{H}^3$ starting at q and having $\overrightarrow{H}(q)$ (the mean curvature vector of M at q) as its tangent at q. Bryant has shown that G is meromorphic on M and M admits a parametrization by meromorphic data analogous to the Weierstrass representation of minimal surfaces in Euclidean 3-space $\mathbb{R}^3$ [1], [4].

Yu [6] has shown that G can omit at most 4 points of the sphere at infinity S_∞, unless M is a horosphere and G is constant. For complete minimal surfaces in $\mathbb{R}^3$ of finite total curvature, Osserman had shown that the Gauss map omits at most 3 points of the sphere, unless M is a plane. In this paper we establish a result of this type in $\mathbb{H}^3$.

The conformal type of a complete surface of mean curvature one with finite total curvature in $\mathbb{H}^3$ is finite, i.e., M is conformally a compact Riemann surface $\overline{M}$ with a finite number of points removed (called the punctures), but G does not necessarily extend meromorphically to the punctures. M is called regular when G does extend meromorphically to the punctures.

Our first result is then:

2000 *Mathematics Subject Classification.* Primary 53A10; Secondary 53A35.

Theorem 1. *Let M be a complete surface immersed in $\mathbb{H}^3$ with mean curvature one and of finite total curvature. Then G can omit at most 3 points unless G is constant and M is a horosphere.*

Proof. If G is not regular, then G has an essential singularity at a puncture p_0. By Picard's theorem, G can omit at most two values in a neighborhood of this puncture. Thus in the following we can assume that G is meromorphic on $\overline{M}$, i.e., M is regular.

Let (g, ω) be local Weierstrass data of the minimal cousin of M in $\mathbb{R}^3$ (cf. [1], [4] for the details). The induced metric on M is given by $ds = (1 + |g|^2)|\omega|$, and the holomorphic quadratic differential

$$Q = \omega dg$$

is globally defined on M and meromorphic at each puncture of M, with a pole at each puncture which is at worst of order 2. Since dG is meromorphic on $\overline{M}$ (the conformal compactification of M), the 1-form $\omega^\# = -Q/dG$ is meromorphic on $\overline{M}$; in a local conformal coordinate, $\omega^\# = -(g'(z)/G'(z))\ \omega(z)$.

The Schwarzian quadratic differentials of g, G and Q are related on $\overline{M}$ ([1], [4]):

$$S(g) - S(G) = 2Q, \tag{1}$$

where $S(g)(z) = \big((g''/g')' - (1/2)(g''/g')^2\big)dz^2$. Writing $g(z) = a_0 + z^k(a_1 + a_2 z + ...)$, a calculation shows that $S(g)$ has at worst a pole of order 2 at z and the coefficient of dz^2/z^2 is $(1 - k^2)/2$.

Since Q is holomorphic on M, it follows from (1) that the branch points and non-simple poles of g and G on M coincide with each other and each of them has the same multiplicity (the branching order of g at z is defined to be $k - 1 = b_g(z)$). In particular, $\omega^\#$ has no poles on M.

We next observe that the zeros of $\omega^\#$ on M are the poles of G on M, and a pole of G of order k is a zero of $\omega^\#$ of order $2k$. First, suppose that $z \in M$ is a pole of G of order k. Then $k \geq 1$ and z may, or may not, be a pole of g. If it is a pole of g, then z is a pole of g of order k (by the Schwarzian derivative relation) and then is a zero of ω of order $2k$. Hence the order of a zero of $\omega^\#$ is of twice the order of the pole of G. If z is not a pole of g, then it is not a zero of ω but a zero of g' of order $k-1$ and a pole of G' of order $k+1$. Consequently $\omega^\#$ also has a zero whose order is twice the order of the pole of G. An analogous computation, in the case that G has no poles, implies that $\omega^\#$ is holomorphic and not zero.

Let $p_1, ..., p_r$ be the punctures, so $\overline{M} = M \cup \{p_1, ..., p_r\}$. After an isometry of $\mathbb{H}^3$, we can suppose that G has only simple poles on M and has no zeros or poles at the punctures. The metric

$$ds^{\#} = (1 + |G|^2)|\omega^{\#}|$$

is complete on $\overline{M}$, so $\omega^{\#}$ has a pole at each puncture [5]. The order of the pole of $\omega^{\#}$ at p_j is given by

$$P_{p_j}(\omega^{\#}) = \lambda_Q(p_j) + b_G(p_j),$$

where $Q(z) = (\gamma/(z-p_j)^{\lambda_Q(p_j)} + \cdots)dz^2$ is the Laurent expansion of Q at p_j. Then the total order of the poles of $\omega^{\#}$ is

$$P(\omega^{\#}) = \sum_{j=1}^{r} \lambda_Q(p_j) + \sum_{j=1}^{r} b_G(p_j). \tag{2}$$

By Riemann's relation for $\omega^{\#}$ on $\overline{M}$, we have

$$P(\omega^{\#}) - 2N = 2 - 2s, \tag{3}$$

where N is the degree of G (so $2N$ is the order of zeros of $\omega^{\#}$, since G has N simple poles on M) and s is the genus of M.

Let $q_1, ..., q_k$ be the points of S_∞ omitted by G, so that $G^{-1}\{q_1, ..., q_k\} \subset \{p_1, ..., p_r\}$ (we write G also for the meromorphic extension of G to $\overline{M}$). Then we have

$$kN \le \sum_{j=1}^{r} (1 + b_G(p_j)) \le r + b, \tag{4}$$

where b is the total branching order of G. Here $1 + b_G(p_j)$ is the total number of times that G takes its value at p_j, counted with multiplicity.

Riemann's relation applied to the 1-form dG on $\overline{M}$ yields:

$$2N - b = 2 - 2s. \tag{5}$$

Now by Lemma 3 of [5], we have at each puncture p_j:

$$\lambda_Q(p_j) + b_G(p_j) \ge 2.$$

Then equation (2) gives:

$$P(\omega^{\#}) \ge 2r. \tag{6}$$

This last inequality together with the equations (3) and (5) yields:

$$P(\omega^{\#}) = 4N - b \ge 2r.$$

Then the equation (4) implies:

$$4N - kN \geq r \geq 1, \tag{7}$$

and k is at most 3. □

Theorem 2. *Let M be a properly embedded surface in $\mathbb{H}^3$ with mean curvature one and of finite topology. If M is not a horosphere nor a catenoid cousin, then the Gauss map G of M is surjective.*

Proof. We know that M has finite total curvature and each end of M is regular [2]; also each end is asymptotic to an end of a horosphere or an end of a catenoid cousin. We also proved in [2] that the asymptotic boundary of an end is precisely the limiting value of G at the puncture. We can suppose M has at least two ends, since if M had only one end, the asymptotic boundary of M would be one point and M would be a horosphere [2].

We claim that each end of M is asymptotic to a catenoid cousin end. Suppose this were not true. Let E be an end of M asymptotic to a horosphere end. We work in the upper half-space model of $\mathbb{H}^3$, $\{x_3 > 0\}$, and assume E is asymptotic to a horosphere $x_3 = c > 0$. In particular, the mean curvature vector of E points up outside of some compact set of E. There are no ends of M above E. Indeed, their mean curvature vector would also point up (each such end is asymptotic to a horizontal horosphere or a catenoid cousin end whose limiting normal points vertically up) and M separates $\mathbb{H}^3$ into two connected components, so no such end is above E.

Then for $\varepsilon > 0$, the part A of M above $c + \varepsilon$ is compact. At the highest point of A (if A were not empty) the mean curvature vector of M points down. But this highest point can be joined by an arc in $\mathbb{H}^3 \setminus M$, to a point of E where the mean curvature vector points up. Thus M is completely below $x_3 = c$.

Let $\varepsilon > 0$, and let C be a small circle in the plane $x_3 = c - \varepsilon$ so that C is above M. Just as in the proof of the half-space theorem for properly immersed minimal surfaces in $\mathbb{R}^3$ [3], one can take a family of catenoid cousin ends $C(\lambda)$, $\partial C(1) = C$ with $C(1)$ above M, and $C(\lambda)$ converges to the plane $x_3 = c - \varepsilon$ as $\lambda \to 0$. Then some $C(\lambda)$ would touch M at a point $q \in M$, and the maximum principle would yield M equals this catenoid cousin. Thus each end of M is asymptotic to a catenoid cousin.

Next, observe that G is injective on the set of punctures; two distinct ends can not be asymptotic to the same point at infinity. This follows from the fact that each end is asymptotic to a catenoid cousin end and we know the direction of the mean curvature vector along the end. When M is embedded, M separates $\mathbb{H}^3$ and the mean curvature vector points into one of the components of the complement. Thus two ends can not be asymptotic to the same point at infinity.

Now, suppose that G is not surjective and omits a point q. Then there is exactly one catenoid cousin type end E of M asymptotic to q. Let $p \in \overline{M}$ be the puncture of E such that $G(p) = q$. We know G has local degree one at p. There is no other point $p' \in \overline{M}$ sent to q by G. For p' can not be a puncture of M, since G is injective on the punctures, and p' can not be a point of M because q is a value omitted. Hence the degree N of G on $\overline{M}$ is one.

We use the same notation as in Theorem 1. At each puncture p_j of M, $\omega^{\#}$ has a pole exactly of order 2. So, by equation (3), we have

$$2r - 2 = 2 - 2s \text{ and } r + s = 2.$$

Then M is the catenoid cousin ($r = 2$) and Theorem 2 is proved. □

References

[1] R. Bryant, *Surfaces of mean curvature one in hyperbolic space*, Astérisque, **154-155** (1987), Soc. Math. France, 1988, 321–347.

[2] P. Collin, L. Hauswirth and H. Rosenberg, *The geometry of finite topology Bryant surfaces*, Ann. of Math., **153** (2001), 623–659.

[3] D. Hoffman and W. Meeks, *The strong half-space theorem for minimal surfaces*, Invent. Math., **101** (1990), 373–377.

[4] M. Umehara and K. Yamada, *Complete surfaces of constant mean curvature-*1 *in the hyperbolic* 3*-space*, Ann. of Math., **137** (1993), 611–638.

[5] M. Umehara and K. Yamada, *A duality on CMC-*1 *surface in the hyperbolic* 3*-space and a hyperbolic analogue of the Osserman Inequality*, Tsukuba J. Math., **21** (1997), 229–237.

[6] Z.-H. Yu, *The value distribution of the hyperbolic Gauss map*, Proc. Amer. Math. Soc., **125** (1997), 2997–3001.

P. Collin
Université Paul Sabatier
118, route de Narbonne
31062 Toulouse
France
collin@picard.ups-tlse.fr

L. Hauswirth
Université Marne-la-Vallée
Cité Descartes
5, boulevard Descartes
77454 Champs-sur-Marne
Marne-la-vallée
France
hauswirth@math.univ-mlv.fr

H. Rosenberg
Université de Paris 7
2 Place Jussieu
75005 Paris
France
rosen@math.jussieu.fr

Advanced Studies in Pure Mathematics 34, 2002
Minimal Surfaces, Geometric Analysis and Symplectic Geometry
pp. 15–29

Behavior of Eigenfunctions near the Ideal Boundary of Hyperbolic Space

Harold Donnelly

Abstract.

The spectrum of the Laplacian on hyperbolic space is a proper subset of the positive reals. We study eigenfunctions, defined on the complements of compact sets, whose eigenvalues lie below the bottom of the spectrum. Such eigenfunctions may arise by perturbing the metric on compact subsets of the space. One divides the eigenfunctions by normalizing factors, so that the quotients have analytic boundary values on the ideal boundary at infinity. The renormalized eigenfunctions are approximated by special polynomials, in nontangential approach regions to the ideal boundary.

§1. Introduction

In [6] and [7], the authors studied asymptotic behavior of eigenfunctions, near infinity, for the Schrödinger operator $-\sum_{i=1}^{n} \partial^2/\partial x_i^2 + V$ in R^n. Let $\psi \in L^2(R^n)$ be a square integrable eigenfunction, of the Schrödinger operator, with eigenvalue $\lambda < 0$. If V decays rapidly, then a multiple, $\hat{\psi}(r,\theta) = \psi(r,\theta)/h(r)$, was shown to have analytic boundary values $A(\theta)$ on the sphere S^{n-1}, compactifying R^n at infinity. A detailed estimate was given for the asymptotic behavior of $\hat{\psi}$ near a zero of A. The two dimensional case was treated in [7], where applications were given to the structure of the nodal set of ψ. If $n > 2$, no such development seems feasible, as discussed in [6].

The present work gives analogous results for eigenfunctions of the Laplacian Δ on hyperbolic space H^n. One assumes that $\Delta\phi = \lambda\phi$, with $\lambda < (n-1)^2/4$ outside some compact subset. This reflects the fact [4], that the essential spectrum of Δ is now $[(n-1)^2/4, \infty)$. The case $n = 2$ was described in [3]. Here we proceed to generalize that work to arbitrary dimension n.

2000 *Mathematics Subject Classification.* Primary 58G25.

§2. Boundary values at infinity

Let H be the simply connected, complete, hyperbolic space of dimension $n \geq 2$. That is, H has the Poincaré metric of constant curvature -1. Fixing $p \in H$, the exponential map $\exp : T_pH \to H$ is a diffeomorphism. We endow this manifold H with the corresponding geodesic polar coordinates. The metric is then given by

$$\begin{aligned} (ds)^2 &= (dr)^2 + g^2(r)(d\theta)^2, \\ g(r) &= \sinh r. \end{aligned}$$

Suppose that $r_0 > 0$ and set $H(r_0) = \{x \in H \mid r(x) > r_0\}$. We consider eigenfunctions $\phi \in L^2(H(r_0))$ of the Laplacian Δ, associated to the given Riemannian metric. Thus, one has $\Delta\phi = \lambda\phi$. Our concern is with the behavior of ϕ as $r \to \infty$. Thus, we feel free to choose r_0 sufficiently large. The eigenfunction ϕ need not satisfy any constraints on the compact boundary of $H(r_0)$, where $r(x) = r_0$.

It seems natural to employ separation of variables. The spherical harmonics $Y_{k,j}(\theta)$, for $k \geq 0$ and $1 \leq j \leq q(k)$, form a complete orthonormal basis for $L^2(S^{n-1})$. Each $Y_{k,j}(\theta)$ belongs to a $q(k)$-dimensional eigenspace of the spherical Laplacian, with corresponding eigenvalue λ_k. One may expand

$$\phi(r,\theta) = \sum_{k=0}^{\infty}\sum_{j=1}^{q(k)} \phi_{k,j}(r)Y_{k,j}(\theta).$$

A computation using the local defining formula for Δ gives

$$\Delta\phi = \sum_{k=0}^{\infty}\sum_{j=1}^{q(k)} \Delta_k\phi_{k,j}(r)Y_{k,j}(\theta),$$

where

$$\Delta_k\phi_{k,j} = -\phi''_{k,j} - (n-1)\frac{g'}{g}\phi'_{k,j} + \lambda_k g^{-2}\phi_{k,j}.$$

Here the $'$ denotes differentiation in r. Thus $\phi_{k,j} \in L^2((r_0,\infty), g^{n-1}(r)dr)$. So Δ is decomposed as a direct sum of the operators Δ_k, with multiplicity $q(k)$.

Now Δ_k is unitarily equivalent to $D_k = g^{(n-1)/2}\Delta_k g^{(1-n)/2}$ acting on $L^2((r_0,\infty),\ dr)$. A calculation yields

$$D_k\psi = -\psi'' + [\gamma(r) + \lambda_k g^{-2}]\psi.$$

Here $\gamma(r) = ((n-1)/2)f'' + ((n-1)/2)^2(f')^2$, with $g = e^f$. In particular $\gamma(r) = ((n-1)/2)^2 + 0(e^{-2r})$.

Set $\overline{\phi}_{k,j} = g^{(n-1)/2}\phi_{k,j}$. Since $\Delta\phi = \lambda\phi$, one has the corresponding equation $D_k\overline{\phi}_{k,j} = \lambda\overline{\phi}_{k,j}$. Therefore

$$-\overline{\phi}_{k,j}{}'' + [\gamma(r) - \lambda + \lambda_k g^{-2}]\overline{\phi}_{k,j} = 0. \tag{2.1}$$

The potential term $\gamma(r) - \lambda + \lambda_k g^{-2}$ decays rapidly to $((n-1)/2)^2 - \lambda$. The hypothesis $\overline{\phi}_{k,j} \in L^2((r_0, \infty), dr)$ and the method of asymptotic integrations [5, pp. 370–384] give

Lemma 2.2. *The equation* (2.1) *has square integrable solutions on* (r_0, ∞) *if and only if* $E = (n-1)^2/4 - \lambda$ *is positive. When* $E > 0$*, there is a one-dimensional space of square integrable solutions. Moreover, any non-zero* L^2 *solution satisfies, for* r *large,*

$$\overline{\phi}_{k,j} \sim b_{k,j}e^{-\sqrt{E}r}, \ \overline{\phi}_{k,j}{}' \sim -\sqrt{E}\, b_{k,j}e^{-\sqrt{E}r}.$$

Here $\sim$ *means that the ratio approaches one as* $r \to \infty$*. The constant* $b_{k,j}$ *is not zero.*

Assume $E > 0$, and let $\overline{h}_k$ be a solution of (2.1). Suppose $\overline{h}_k$ lies inside the one-dimensional space of square integrable solutions, as specified in Lemma 2.2. If $\overline{h}_k(r_1) > 0$ for some $r_1 > r_0$, then $\overline{h}_k(r) > 0$ for all $r \geq r_1$. Otherwise, let $r_2 > r_1$ be the first zero of $\overline{h}_k$. Clearly, $\overline{h}_k'(r_2) \leq 0$. By the uniqueness theorem, for second order ordinary differential equations, this forces $\overline{h}_k'(r_2) < 0$. Since $\overline{h}_k \in L^2((r_0, \infty),\ dr)$, the function $\overline{h}_k$ must have a negative minimum $r_3 > r_2$. However, if r_0 is sufficiently large, then the potential term $\gamma(r) - \lambda + \lambda_k g^{-2} > 0$ for all $r \geq r_0$. Consequently, solutions to (2.1) cannot have negative local interior minimums. This contradiction shows that $\overline{h}_k(r) > 0$ for all $r \geq r_1$.

Now we fix $r_1 > r_0$. Define $\overline{h}_k \in L^2((r_0, \infty), dr)$ by requiring $\overline{h}_k$ to satisfy (2.1) and the normalization $\overline{h}_k(r_1) = 1$. The remarks above show that this defines $\overline{h}_k$ uniquely. Moreover $\overline{\phi}_{k,j}(r) = \overline{\phi}_{k,j}(r_1)\overline{h}_k(r)$, since both $\overline{\phi}_{k,j}$ and $\overline{h}_k$ lie in the one-dimensional space of solutions, specified by Lemma 2.2. If $\overline{\phi} = \phi g^{(n-1)/2}$, we may write

$$\overline{\phi}(r, \theta) = \sum_{k=0}^{\infty}\sum_{j=1}^{q(k)} \overline{\phi}_{k,j}(r_1)\overline{h}_k(r)Y_{k,j}(\theta). \tag{2.3}$$

The function $\phi(r, \theta)$ is analytic because $(\Delta - \lambda)\phi = 0$ and the elliptic operator $\Delta - \lambda$ has analytic coefficients [8, p. 178]. By Proposition 4.5

of the appendix, we may write

$$|\overline{\phi}_{k,j}(r_1)| \leq c_1 e^{-c_2\sqrt{\lambda_k}} \tag{2.4}$$

for positive constants c_1 and c_2. Recall that the eigenvalues of the spherical Laplacian satisfy $\lambda_k = 0(k^2)$.

It is also necessary to control the dependence of the $\overline{h}_k$ upon k. This is provided by

Lemma 2.5. *For all k, one has $0 < \overline{h}_k(r) \leq \overline{h}_0(r)$, whenever $r \geq r_1$.*

Proof. We have already shown that $\overline{h}_k(r) > 0$. The difference $a_k = \overline{h}_0 - \overline{h}_k$ satisfies, since $\lambda_0 = 0$,

$$\begin{aligned} -a_k'' + [\gamma(r) - \lambda]a_k - \lambda_k g^{-2}\overline{h}_k &= 0, \\ \overline{a}_k(r_1) &= 0. \end{aligned}$$

If a_k is ever negative, then, since $a_k \in L^2((r_0, \infty), dr)$, the function a_k must have a negative minimum. At such a local minimum, the differential equation for a_k cannot hold. This contradiction proves the lemma. □

One now has the necessary preparations to study the asymptotic behavior of eigenfunctions. Set $\hat{\phi}(r,\theta) = \overline{\phi}(r,\theta)\overline{h}_0^{-1}(r)$, or equivalently, $\hat{\phi}(r,\theta) = g^{(n-1)/2}(r)\phi(r,\theta)\overline{h}_0^{-1}(r)$. The central result of this section is

Theorem 2.6. *As $r \to \infty$, one has $\hat{\phi}(r,\theta) \to A(\theta)$, uniformly in θ. The function $A(\theta)$ is real analytic.*

Proof. By (2.4) and Lemma 2.5, we have $|\overline{\phi}_{k,j}(r)| = |\overline{\phi}_{k,j}(r_1)|\overline{h}_k(r) \leq c_1 e^{-c_2\sqrt{\lambda_k}}\, \overline{h}_0(r)$. Now

$$\hat{\phi}(r,\theta) = \sum_{k=0}^{\infty}\sum_{j=1}^{q(k)} \overline{\phi}_{k,j}(r)\overline{h}_0^{-1}(r)Y_{k,j}(\theta).$$

Lemma 2.2 guarantees that $A_{k,j} = \lim_{r\to\infty} \overline{\phi}_{k,j}\overline{h}_0^{-1}$ exists. The given bound on the functions $\overline{\phi}_{k,j}$ allows one to interchange the limit in r with the infinite sum in j. Note that $\|Y_{k,j}\|_2 = 1$, and thus $\|Y_{k,j}\|_\infty \leq c_3\lambda_k^{(n-2)/4}$, by standard elliptic theory. Moreover, the multiplicity $q(k) = 0(\lambda_k^{(n-2)/2})$. □

Thus, one has

$$A(\theta) = \lim_{r\to\infty} \hat{\phi}(r,\theta) = \sum_{k=0}^{\infty}\sum_{j=1}^{q(k)} A_{k,j}Y_{k,j}(\theta).$$

The estimate $|A_{k,j}| \leq c_1 e^{-c_2\sqrt{\lambda_k}}$ and Proposition 4.2 show that A is real analytic.

If $\phi_{k,j}$ is not identically zero, Lemma 2.2 guarantees that $A_{k,j} \neq 0$. In particular, $A(\theta)$ is not the zero function. We also have the expected

Corollary 2.7. *For any ℓ, $\lim_{r\to\infty} \nabla_\theta^\ell \hat{\phi}(r,\theta) = \nabla_\theta^\ell A$, where ∇_θ denotes the covariant derivative of S^{n-1} with its standard metric.*

Proof. The rapid decay $|\overline{\phi}_{k,j}(r)|\overline{h}_0^{-1}(r) \leq c_1 e^{-c_2\sqrt{\lambda_k}}$, and standard elliptic estimates for the derivatives of eigenfunctions, allow one to write

$$\nabla_\theta^\ell \hat{\phi}(r,\theta) = \sum_{k=0}^{\infty}\sum_{j=1}^{q(k)} \overline{\phi}_{k,j}(r)\overline{h}_0^{-1}(r)\nabla_\theta^\ell Y_{k,j}(\theta).$$

For the same reasons, one may interchange the sum in j and the limit in r, to get

$$\lim_{r\to\infty} \nabla_\theta^\ell \hat{\phi}(r,\theta) = \sum_{k=0}^{\infty}\sum_{j=1}^{q(k)} A_{k,j}\nabla_\theta^\ell Y_{k,j}(\theta) = \nabla_\theta^\ell A.$$

□

It is also useful to estimate the rate of convergence in Corollary 2.7. In this direction, there are constants B_ℓ, so that

Proposition 2.8. *For $r > r_1$, $|\nabla_\theta^\ell A - \nabla_\theta^\ell \hat{\phi}(r,\theta)| \leq B_\ell e^{-2r}$.*

Proof. Set $\hat{\phi}_{k,j} = \overline{\phi}_{k,j}\overline{h}_0^{-1}$, the coefficient of $Y_{k,j}(\theta)$ in the spherical harmonic expansion for $\hat{\phi}$. Then $\hat{\phi}'_{k,j} = [\overline{\phi}'_{k,j}\overline{h}_0 - \overline{\phi}_{k,j}\overline{h}'_0]\overline{h}_0^{-2}$. Define $w_{k,j} = \overline{h}_0^{2}\hat{\phi}'_{k,j} = \overline{\phi}'_{k,j}\overline{h}_0 - \overline{\phi}_{k,j}\overline{h}'_0$. By equation (2.1), we deduce that $w'_{k,j} = \lambda_k g^{-2}\overline{\phi}_{k,j}\overline{h}_0$. □

Now the functions $\overline{\phi}_{k,j}$ and $\overline{h}_0$ are both of order $e^{-\sqrt{E}r}$, according to Lemma 2.2. Moreover, $g^{-2} = 0(e^{-2r})$. So we may integrate up to infinity, yielding

$$w_{k,j}(x) = -\int_x^\infty \lambda_k g^{-2}(y)\overline{\phi}_{k,j}(y)\overline{h}_0(y)dy.$$

Note that $\lim_{x\to\infty} w_{k,j}(x) = 0$, by Lemma 2.2.

Recalling the definition of $w_{k,j}$ gives

$$\hat{\phi}'_{k,j}(x) = -\overline{h}_0^{-2}(x)\int_x^\infty \lambda_k g^{-2}(y)\overline{\phi}_{k,j}(y)\overline{h}_0(y)dy.$$

By Lemma 2.5, $|\overline{\phi}_{k,j}(y)| = |\overline{\phi}_{k,j}(r_1)\overline{h}_k(y)| \le \overline{\phi}_{k,j}(r_1)\overline{h}_0(y)$. So

$$|\hat{\phi}'_{k,j}(x)| \le \overline{h}_0^{-2}(x)\int_x^\infty \lambda_k g^{-2}(y)\overline{h}_0^{2}(y)dy|\overline{\phi}_{k,j}(r_1)|.$$

Using Lemma 2.2, $\overline{h}_0(y) \sim \overline{b}_0 e^{-\sqrt{E}r}$. Thus one has

$$|\hat{\phi}'_{k,j}(x)| \le c_4\int_x^\infty \lambda_k g^{-2}(y)dy|\overline{\phi}_{k,j}(r_1)|.$$

Now from (2.4)

$$|\hat{\phi}'_{k,j}(x)| \le c_5\int_x^\infty g^{-2}(y)dy\; e^{-c_6\sqrt{\lambda_k}}.$$

By definition $g(y) = \sinh y$. Therefore

$$|\hat{\phi}'_{k,j}(x)| \le c_7 e^{-c_6\sqrt{\lambda_k}}e^{-2x}. \tag{2.9}$$

The estimate (2.9) is quite appropriate for our present purpose. In fact

$$\nabla_\theta^\ell A(\theta) - \nabla_\theta^\ell\hat{\phi}(r,\theta) = \sum_{k=0}^{\infty}\sum_{j=1}^{q(k)}(A_{k,j} - \hat{\phi}_{k,j}(r))\nabla_\theta^\ell Y_{k,j}(\theta). \tag{2.10}$$

Since $A_{k,j} = \lim_{r\to\infty}\hat{\phi}_{k,j}(r)$, we have

$$A_{k,j} - \hat{\phi}_{k,j}(r) = \int_r^\infty \hat{\phi}'_{k,j}(x)dx.$$

The estimate (2.9) guarantees that the integral converges and also yields

$$|A_{k,j} - \hat{\phi}_{k,j}(r)| \le c_8 e^{-c_6\sqrt{\lambda_k}}e^{-2r}.$$

Returning to (2.10), one finds that

$$|\nabla_\theta^\ell A - \nabla_\theta^\ell\hat{\phi}(r,\theta)| \le \sum_{k=0}^{\infty}\sum_{j=1}^{q(k)} c_8 e^{-c_6\sqrt{\lambda_k}}|\nabla_\theta^\ell Y_{k,j}(\theta)|e^{-2r}.$$

Proposition 2.8 follows easily.

§3. Asymptotic estimate

We proceed to obtain more detailed information concerning the convergence of $\hat{\phi}(r,\theta)$ to $A(\theta)$ as $r \to \infty$. The first step is to derive a basic integral equation satisfied by $\hat{\phi}(r,\theta)$. This leads to an iterative scheme for developing $\hat{\phi}(r,\theta)$ in terms of $A(\theta)$. Near the zeroes of A, $\hat{\phi}$ may be approximated by certain polynomials. The order of these polynomials coincides with the order of vanishing of A.

Recall that $\overline{\phi}(r,\theta) = \phi(r,\theta)g^{(n-1)/2}(r)$ and ϕ is an eigenfunction of the hyperbolic Laplacian with eigenvalue λ. It follows that $\overline{\phi}$ satisfies the partial differential equation

$$-\frac{\partial^2 \overline{\phi}}{\partial r^2} + (\gamma(r) - \lambda)\overline{\phi} + g^{-2}\Delta_\theta \overline{\phi} = 0. \tag{3.1}$$

Here Δ_θ is the Laplacian on S^{n-1}. In fact, (3.1) follows by summing the equations (2.1). Alternatively, one derives (3.1) directly from the local coordinate formula for the Laplacian Δ.

The basic idea is to convert the partial differential equation for $\overline{\phi}$ into an integral equation for $\hat{\phi}(r,\theta) = \overline{\phi}(r,\theta)/\overline{h}_0(r)$. We may write

Proposition 3.2. *If $s > r > r_0$, then*

$$\hat{\phi}(s,\theta) = \hat{\phi}(r,\theta) - \int_r^s \overline{h}_0^{-2}(x) \int_x^\infty \overline{h}_0^2(y) g^{-2}(y) \Delta_\theta \hat{\phi}(y,\theta)\,dy\,dx.$$

Proof. One has $\partial\hat{\phi}/\partial r = \partial/\partial r(\overline{\phi}\,\overline{h}_0^{-1}) = (\partial\overline{\phi}/\partial r)\overline{h}_0 - \overline{\phi}(\partial\overline{h}_0/\partial r)\overline{h}_0^{-2}$. Set $H = \overline{h}_0^2 \partial\hat{\phi}/\partial r = (\partial\overline{\phi}/\partial r)\overline{h}_0 - \overline{\phi}(\partial\overline{h}_0/\partial r)$. Then, using equations (2.1) and (3.1), we find $\partial H/\partial r = \overline{h}_0 \partial^2\overline{\phi}/\partial r^2 - \overline{\phi}\partial^2\overline{h}_0/\partial r^2 = g^{-2}\Delta_\theta\overline{\phi}\,\overline{h}_0$. So $\partial H/\partial r = g^{-2}\overline{h}_0^2\Delta_\theta\hat{\phi}$. □

By Proposition 2.8, $|\Delta_\theta\hat{\phi}|$ is uniformly bounded in both r and θ. Also, $g^{-2} = 0(e^{-2r})$ and $\overline{h}_0 = 0(e^{-\sqrt{E}r})$, from Lemma 2.2. So we may integrate up to infinity, yielding

$$H(x) = -\int_x^\infty \overline{h}_0^2(y) g^{-2}(y) \Delta_\theta \hat{\phi}(y,\theta)\,dy.$$

Note that $H = (\partial\overline{\phi}/\partial r)\overline{h}_0 - \overline{\phi}(\partial\overline{h}_0/\partial r)$ approaches zero as $r \to \infty$. In fact, both $\overline{\phi}$ and $\overline{h}_0$ are of order $e^{-\sqrt{E}r}$, by Lemma 2.2 and Theorem 2.6. Equations (2.1) and (3.1) may be integrated to verify that $\partial\overline{\phi}/\partial r$ and $\partial\overline{h}_0/\partial r$ are bounded.

The definition of H now yields

$$\frac{\partial\hat{\phi}}{\partial r}(x,\theta) = -\overline{h}_0^{-2}(x)\int_x^\infty \overline{h}_0^2(y)g^{-2}(y)\Delta_\theta\hat{\phi}(y,\theta)dy.$$

Proposition 3.2 follows by integrating this equation between r and s.

We now let $s \to \infty$ in Proposition 3.2. Note that $\overline{h}_0(r) \sim \overline{b}_0 e^{-\sqrt{E}r}$ from Lemma 2.2. Moreover, the function $g^{-2}(r) = 0(e^{-2r})$ and $|\Delta_\theta\hat{\phi}|$ is bounded by Proposition 2.8. By Theorem 2.6 and the dominated convergence theorem,

$$A(\theta) = \hat{\phi}(r,\theta) - \int_r^\infty \overline{h}_0^{-2}(x)\int_x^\infty \overline{h}_0^2(y)g^{-2}(y)\Delta_\theta\hat{\phi}(y,\theta)dydx.$$

Let T denote the integral-differential operator defined by

$$Tf(r,\theta) = \int_r^\infty \overline{h}_0^{-2}(x)\int_x^\infty \overline{h}_0^2(y)g^{-2}(y)\Delta_\theta f(y,\theta)dydx.$$

The domain of T consists of functions where $|\Delta_\theta f|$ is bounded, uniformly in y and θ.

We may write

$$\hat{\phi}(r,\theta) = A(\theta) + T\hat{\phi}(r,\theta).$$

Substituting $\hat{\phi} = A + T\hat{\phi}$ in the right hand side gives

$$\hat{\phi} = A + TA + T^2\hat{\phi}.$$

Iterating any finite number of times yields, for any positive integer m,

$$\hat{\phi} = \sum_{j=0}^m T^j A + T^m(\hat{\phi} - A).$$

Proposition 2.8 and the dominated convergence theorem guarantee that we always remain within the domain of T.

An elementary calculation using Proposition 2.8 yields

$$T^m(\hat{\phi} - A) = 0(e^{-2(m+1)r}).$$

For this computation we use the familiar estimates $\overline{h}_0(r) \sim \overline{b}_0 e^{-\sqrt{E}r}$ and $g(r) = 0(e^{-2r})$, as noted repeatedly above.

So, with arbitrary m, we have

$$\hat{\phi}(r,\theta) = \sum_{j=0}^m T^j A(r,\theta) + 0(e^{-2(m+1)r}). \tag{3.3}$$

The function $A(\theta)$ is analytic and therefore has zeroes of finite order. We use (3.3) to investigate the behavior of $\hat{\phi}$ near an m'th order zero of A. Choosing a coordinate system centered at this zero, we have $A(\theta) = A_m(\theta) + 0(|\theta|^{m+1})$, where A_m is a non-zero polynomial of order m. We may denote $A_m(\theta) = \sum_{|L|=m} a_L \theta^L + 0(|\theta|^{m+1})$. Here $L = (\ell_1, \ell_2, \ldots, \ell_{n-1})$ is a multi-index of total length $|L| = \ell_1 + \ell_2 + \cdots + \ell_{n-1}$, with each ℓ_i being a non-negative integer, and furthermore $\theta^L = \theta_1^{\ell_1} \theta_2^{\ell_2} \cdots \theta_{n-1}^{\ell_{n-1}}$.

To compute the spherical Laplacian Δ_θ, it is convenient to employ normal coordinates for the standard metric on S^{n-1}. A result of Cartan [1] gives $g_{ij} = \delta_{ij} + 0(|\theta|^2)$, and thus

$$\Delta_\theta = -\sum_{i=1}^{n-1} \frac{\partial^2}{\partial\theta_i^2} + \sum_{i,j=1}^{n-1} a_{ij} \frac{\partial^2}{\partial\theta_i \partial\theta_j} + \sum_{i=1}^{n-1} b_i \frac{\partial}{\partial\theta_i},$$

where $a_{ij} = 0(|\theta|^2)$ and $b_i = 0(|\theta|)$.

To isolate the dominant contributions for the asymptotic expansion of $\hat{\phi}$, it is convenient to define $\overline{\Delta}_\theta = -\sum_{i=1}^{n-1} \partial^2/\partial\theta_i^2$. There is a corresponding integral-differential operator $\overline{T} f = \sum_{i=1}^{n-1} \overline{T}_i f$, where

$$\overline{T}_i f = -\int_r^\infty \overline{h}_0^{-2}(x) \int_x^\infty \overline{h}_0^2(y) g^{-2}(y) \frac{\partial^2 f}{\partial\theta_i^2}(y, \theta) dy dx.$$

Our asymptotic estimate of $\hat{\phi}(r,\theta)$ will be valid in the region $H_c = \{(r,\theta) \mid |\theta| < ce^{-r},\ r > r_2\}$, for any given $c > 0$ and $r_2 > r_0$, large enough. Suppose that B is an analytic function of θ satisfying $B(\theta) = 0(|\theta|^p)$ for some p. If $j \geq 0$, then the definition of T and Lemma 2.2 give $T^j B = 0(e^{-pr})$ in H_c. The point is that each application of T involves at most two θ-derivatives, but the double integration in r contributes a factor e^{-2r}. Similarly, one may bound $\overline{T}^j B = 0(e^{-pr})$. Since $\Delta_\theta - \overline{\Delta}_\theta = \Sigma a_{ij} \partial^2/\partial\theta_i\partial\theta_j + \Sigma b_i \partial/\partial\theta_i$, with $a_{ij} = 0(|\theta|^2)$ and $b_i = 0(|\theta|)$, we see that $(T - \overline{T})^j B = 0(e^{-(p+2j)r})$. Combining these observations with formula (3.3), where m is the order of vanishing of A, yields

$$\hat{\phi} = \sum_{j=0}^{m} \overline{T}^j A_m + 0(e^{-(m+1)r}).$$

Since A_m is a polynomial of total order m, in θ, and each $\overline{T}_i$ involves two derivatives in θ_i, we have

$$\begin{aligned}\hat{\phi} &= \sum_{j=0}^{\infty} \overline{T}^j A_m + 0(e^{-(m+1)r}) \\ &= \sum_{j=0}^{\infty} \left(\sum_{i=1}^{n-1} \overline{T}_i \right)^j A_m + 0(e^{-(m+1)r}) \\ &= \sum_{j=0}^{\infty} \sum_{j_1+j_2+\cdots+j_{n-1}=j} \frac{j!}{j_1! j_2! \cdots j_{n-1}!} \\ &\qquad \times \overline{T}_1^{j_1} \overline{T}_2^{j_2} \cdots \overline{T}_{n-1}^{j_{n-1}} A_m + 0(e^{-(m+1)r}).\end{aligned}$$

Thus

$$\text{(3.4)} \quad \hat{\phi} = \sum_{|L|=m} a_L \sum_{j=0}^{\infty} \sum_{j_1+j_2+\cdots+j_{n-1}=j} \frac{j!}{j_1! j_2! \cdots j_{n-1}!} \times (\overline{T}_1^{j_1} \theta_1^{\ell_1})(\overline{T}_2^{j_2} \theta_2^{\ell_2}) \cdots (\overline{T}_{n-1}^{j_{n-1}} \theta_{n-1}^{\ell_{n-1}}) + 0(e^{-(m+1)r}).$$

It remains to evaluate the individual expressions $\overline{T}_i^{j_i} \theta_i^{\ell_i}$, for fixed i. For this purpose, we need the following improvement of Lemma 2.2:

Lemma 3.5. $\overline{h}_0(r) = \overline{b}_0 e^{-\sqrt{E}r} + 0(e^{-(\sqrt{E}+2)r})$.

Proof. Lemma 2.2 gives $\overline{h}_0(r) \sim \overline{b}_0 e^{-\sqrt{E}r}$. Consider the ratio $u(r) = \overline{h}_0(r)/e^{-\sqrt{E}r} = \overline{h}_0(r)e^{\sqrt{E}r}$. Then $u' = (\overline{h}_0' + \sqrt{E}\,\overline{h}_0)e^{\sqrt{E}r}$. Set $w(r) = u'(r)e^{-2\sqrt{E}r} = e^{-\sqrt{E}r}(\overline{h}_0' + \sqrt{E}\,\overline{h}_0)$. Differentiating this gives the formula $w' = (\overline{h}_0'' - E\overline{h}_0)e^{-\sqrt{E}r} = (-E + \gamma - \lambda)\overline{h}_0 e^{-\sqrt{E}r} = (\gamma - (n-1)^2/4)\overline{h}_0 e^{-\sqrt{E}r}$. Here we used equation (2.1), with $k = 0$, and the definition $E = (n-1)^2/4 - \lambda$. Integrating up to infinity yields $w(x) = -\int_x^\infty (\gamma(y) - (n-1)^2/4)\overline{h}_0(y)e^{-\sqrt{E}y}dy$. Note that $w \to 0$ as $r \to \infty$, since $w(r) = (\overline{h}_0' + \sqrt{E}\,\overline{h}_0)e^{-\sqrt{E}r} = 0(e^{-2\sqrt{E}r})$, by Lemma 2.2. Moreover, the definition of w gives $u'(x) = w(x)e^{2\sqrt{E}x} = -e^{2\sqrt{E}x}\int_x^\infty (\gamma(y) - (n-1)^2/4)\overline{h}_0(y)e^{-\sqrt{E}y}dy$. Integrating up to infinity yields

$$u(r) - \overline{b}_0 = \int_r^\infty e^{2\sqrt{E}x} \int_x^\infty \left(\gamma(y) - \frac{(n-1)^2}{4} \right) \overline{h}_0(y) e^{-\sqrt{E}y} dy dx.$$

Now $\gamma(y) - (n-1)^2/4 = 0(e^{-2y})$. The integral then converges by Lemma 2.2. A calculation gives $u(r) - \overline{b}_0 = 0(e^{-2r})$. Lemma 3.5 follows after multiplying by $e^{-\sqrt{E}r}$. □

Suppose $\ell = \ell_i \geq 2$ is any integer. Clearly

$$\overline{T}_i\theta_i^\ell = \left(-\int_r^\infty \overline{h}_0^{-2}(x)\int_x^\infty \overline{h}_0^2(y)g^{-2}(y)dydx\right)\ell(\ell-1)\theta_i^{\ell-2}.$$

Using Lemma 3.5 and the elementary estimate $g(r) = e^r(1+0(e^{-2r}))/2$ implies

$$\overline{T}_i\theta_i^\ell(r,\theta) = -\ell(\ell-1)(\sqrt{E}+1)^{-1}\theta_i^{\ell-2}e^{-2r}(1+0(e^{-2r})).$$

An argument by mathematical induction gives

Lemma 3.6. *For $k \leq [\ell/2]$, the greatest integer in $\ell/2$,*

$$\overline{T}_i^k\theta_i^\ell(r,\theta) = \frac{(-1)^k}{k!}\prod_{s=1}^k(\sqrt{E}+s)^{-1}\frac{\ell!}{(\ell-2k)!}\theta_i^{\ell-2k}e^{-2kr}(1+0(e^{-2r})).$$

If $k > [\ell/2]$, then $\overline{T}_i^k\theta_i^\ell = 0$.

Proof. Suppose the required formula has been established for a given value $k-1$ with $k \leq [\ell/2]$. Then

$$\overline{T}_i^k\theta_i^\ell(r,\theta) = \overline{T}_i(\overline{T}_i^{k-1}\theta_i^\ell)(r,\theta) = \frac{(-1)^{k-1}}{(k-1)!}\prod_{s=1}^{k-1}(\sqrt{E}+s)^{-1}$$
$$\times\frac{\ell!}{(\ell-2k+2)!}\left(\int_r^\infty -\overline{h}_0^{-2}(x)\int_x^\infty \overline{h}_0^2(y)g^{-2}(y)e^{-2(k-1)y}\right.$$
$$\left.\times(1+0(e^{-2y}))dydx\right)(\ell-2k+2)(\ell-2k+1)\theta_i^{\ell-2k}.$$

Using Lemma 3.5 and $g(r) = (1/2)e^r(1+0(e^{-2r}))$ yields

$$\begin{aligned}\overline{T}_i^k\theta_i^\ell(r,\theta) \quad &= \quad \frac{(-1)^k}{(k-1)!}\prod_{s=1}^{k-1}\left(\sqrt{E}+s\right)^{-1}\frac{\ell!}{(\ell-2k)!}\theta_i^{\ell-2k}\\ &\quad\times\int_r^\infty e^{2\sqrt{E}x}(1+0(e^{-2x}))\int_x^\infty e^{-2\sqrt{E}y}\\ &\qquad\times 4e^{-2y}e^{-2(k-1)y}(1+0(e^{-2y}))dydx.\end{aligned}$$

The integral is easily calculated, which completes the induction. □

Define polynomials of $x = (x_1, x_2, \ldots, x_{n-1})$ by

$$P_L(x) = \sum_{0 \le j \le (1/2)|L|} \sum_{\substack{j_1+j_2+\cdots+j_{n-1}=j \\ 2j_i \le \ell_i}} \frac{(-1)^j j!}{(\ell_1 - 2j_1)!(\ell_2 - 2j_2)! \cdots (\ell_{n-1} - 2j_{n-1})!}$$

$$\times \frac{\ell_1!\ell_2! \ldots \ell_{n-1}!}{((j_1)!(j_2)! \cdots (j_{n-1})!)^2} \prod_{s_1=1}^{j_1} (\sqrt{E} + s_1)^{-1}$$

$$\cdots \prod_{s_{n-1}=1}^{j_{n-1}} (\sqrt{E} + s_{n-1})^{-1} \prod_{i=1}^{n-1} (x_i)^{\ell_i - 2j_i}.$$

Combining (3.4) and Lemma 3.6 gives our main result.

Theorem 3.7. *In the region $H_c = \{(r, \theta) \mid |\theta| < ce^{-r}, r > r_2\}$, for any given $c > 0$, one has*

$$\hat{\phi}(r, \theta) = e^{-mr} \sum_{|L|=m} a_L P_L(e^r \theta) + 0(e^{-(m+1)r}).$$

In the case of the hyperbolic plane, $n = 2$, Theorem 3.7 was proved in [3]. There one had a single polynomial $P_m(x)$, with m distinct real zeroes. This led to a detailed analysis of the nodal structure of $\hat{\phi}$, near $\theta = 0$, and as $r \to \infty$. For $n > 2$, a similar discussion does not appear feasible. The difficulty lies in the complicated structure of the zero set of polynomials of several variables and the related instability of the zero set under perturbation. Analogous problems arose in the earlier investigations [6] of Schrödinger operators in R^n, $n > 2$.

§4. Appendix — Analyticity and expansion in spherical harmonics

Let S^m denote the standard round m-dimensional sphere. The spherical harmonics $Y_{k,j}(\theta)$, for $k \ge 0$ and $1 \le j \le q(k)$, form a complete orthonormal basis for $L^2(S^m)$. Each $Y_{k,j}(\theta)$ is obtained by restriction, to $S^m \subset R^{m+1}$, of a homogeneous harmonic polynomial of degree k. The dimension of the space of degree k harmonic polynomials is $q(k) = 0(k^{m-1})$. Moreover, the spherical harmonics $Y_{k,j}(\theta)$ are eigenfunctions of the spherical Laplacian, with corresponding eigenvalue $\lambda_k = 0(k^2)$. The reader may consult [1] for detailed proofs of these elementary results.

Each $f \in L^2(S^m)$ has a generalized Fourier series, with coefficients $a_{k,j} = \int_{S^m} f(\theta) Y_{k,j}(\theta)$. If $f \in C^\infty(S^m)$, then $a_{k,j} = 0(\lambda_k^{-\ell})$ for any ℓ, according to partial integration. Moreover, one has a uniformly convergent expansion, as a consequence of standard elliptic theory,

$$f(\theta) = \sum_{k=0}^{\infty} \sum_{j=1}^{q(k)} a_{k,j} Y_{k,j}(\theta). \tag{4.1}$$

This expansion may be repeatedly differentiated, term by term, to yield the expansion of any higher order derivative of f.

The purpose of this appendix is to correlate the analyticity of f with the exponential decay of the $a_{k,j}$, in their dependence upon k. This result is implicit in the much more elaborate developments of [6]. However, it seems worthwhile to present a simple elementary proof. We begin with

Proposition 4.2. *If* $|a_{k,j}| \leq c_1 e^{-c_2 k}$, *then the series of* (4.1) *converges to a real analytic function* f.

Proof. Since we normalized $\|Y_{k,j}\|_2 = 1$, one has $\|Y_{k,j}\|_\infty \leq c_3 k^{(m-1)/2}$ by standard elliptic theory. The decay hypothesis, about $a_{k,j}$, allows the extension of f to a function

$$u(r, \theta) = \sum_{k=0}^{\infty} r^k \sum_{j=1}^{q(k)} a_{k,j} Y_{k,j}(\theta) \tag{4.3}$$

on some neighborhood of the closure of the unit ball B^{m+1} in R^{m+1}. Since u is the uniform limit of harmonic functions, u is harmonic. The standard Laplacian Δ of R^n is analytic hypoelliptic and thus u is analytic. It follows that the restriction f of u, to S^m, is also analytic. □

For the converse to Proposition 4.2, it is convenient to employ

Lemma 4.4. *Let* u *be a solution of the Dirichlet problem on* $B = B^{m+1}$, *with analytic boundary data* f. *Then* u *extends to a harmonic function on a neighborhood of the closure* $\overline{B}$.

Proof. The Cauchy-Kovalevskya theorem provides a harmonic extension h of f to a neighborhood of $S^m \subset R^{m+1}$. Let χ be a smooth cut-off function, equal to one near S^m, and with support contained within the domain of definition of h. Clearly, $u - \chi h = 0$ on S^m, and $\Delta(u - \chi h) = g \in C_0^\infty(B)$. If G is the Greens function of B, one consequently has $u - \chi h = Gg$. The explicit formula for G, obtained by the method of images [2, p. 264], now shows that u extends harmonically past the boundary of B. □

The converse to Proposition 4.2 is

Proposition 4.5. *If f is real analytic, then $|a_{k,j}| \leq c_1 e^{-c_2 k}$ in the expansion* (4.1).

Proof. Let u be the solution of the Dirichlet problem on B^{m+1}, with boundary data f. Since f is C^∞, the coefficients $a_{k,j}$ decay faster than any polynomial in k, as observed above. Thus the series in (4.3) converges for $r \leq 1$. By uniqueness in the Dirichlet problem, we have for $r \leq 1$,

$$u(r,\theta) = \sum_{k=0}^{\infty} r^k \sum_{j=1}^{q(k)} a_{k,j} Y_{k,j}(\theta). \tag{4.6}$$

By Lemma 4.4, we have for some $\delta > 0$ and $r \leq 1+\delta$, a uniformly convergent expansion

$$u(r,\theta) = \sum_{k=0}^{\infty} \sum_{j=1}^{q(k)} a_{k,j}(r) Y_{k,j}(\theta).$$

Since u is harmonic, separation of variables shows that each $a_{k,j}(r)$ satisfies a second order ordinary differential equation. By the uniqueness theory for ordinary differential equations, (4.6) holds when $r \leq 1+\delta$. Taking the L^2 norm gives

$$\sum_{k=0}^{\infty} (1+\delta)^{2k} \sum_{j=1}^{q(k)} |a_{k,j}|^2 = \int_{S^m} |u(1+\delta,\theta)|^2.$$

Therefore $|a_{k,j}|$ decays exponentially in k. $\square$

References

[1] M. Berger, P. Gauduchon and E. Mazet, Le Spectre d'une Variété Riemannienne, Lecture Notes, **194**, Springer-Verlag, Berlin-Heidelberg-New York, 1970.

[2] R. Courant and D. Hilbert, Methods of Mathematical Physics II, Interscience, New York, 1962.

[3] H. Donnelly, *Asymptotic properties of eigenfunctions—the hyperbolic plane*, J. d'analyse Math., **58** (1992), 135–151.

[4] H. Donnelly, *On the essential spectrum of a complete Riemannian manifold*, Topology, **20** (1981), 1–14.

[5] P. Hartman, Ordinary differential equations, Birkhäuser, Boston, 1982.

[6] M. Hoffman-Ostenhof, T. Hoffman-Ostenhof and J. Swetina, *Asymptotics and continuity properties near infinity of solutions of Schrödinger equations in exterior domains*, Ann. Inst. Henri Poincaré, **46** (1987), 247–280.

[7] M. Hoffman-Ostenhof, T. Hoffman-Ostenhof and J. Swetina, *Continuity and nodal properties near infinity of* 2*-dimensional Schrödinger equations*, Duke Math. J., **53** (1986), 271–306.

[8] L. Hörmander, Linear Partial Differential Operators, Springer-Verlag, Berlin-Heidelberg-New York, 1963.

Department of Mathematics
Purdue University
West Lafayette, IN 47907
U. S. A.

Advanced Studies in Pure Mathematics 34, 2002
Minimal Surfaces, Geometric Analysis and Symplectic Geometry
pp. 31–127

Floer Homology and Mirror Symmetry II

Kenji Fukaya

Abstract.

This is the second part of a series of articles explaining applications of Floer homology to mirror symmetry and D-brane. This article is independent of part I [Fu9]. We will associate an A_∞ category to a symplectic manifold. This is an improved version of previous ones [Fu1], [Fu4] in which there were some flaw. The correction is based on a book [FOOO] written jointly with Oh, Ohta, Ono. While correcting the flaw, we find various interesting new phenomena which are related to mirror symmetry.

We also discuss homological algebra of A_∞ category in this article.

This article is a survey article. So most of the material written here are minor modifications of the results which are already known to somebody. However it is rather hard to find a reference of them.

Table of contents

Key words and phrases. Symplectic geometry, Floer homology, mirror symmetry, D-brane, A_∞ algebra.

Partially supported by Grant in-Aid for Scientific Research 13852001.

§0. Introduction

This is the second part of a series of articles, describing a project in progress to study homological mirror symmetry [Ko1], [Ko2] and D-brane using Floer homology of Lagrangian submanifolds. See [Fu5], [Fu6], [Fu7], [FOOO], [Ko1], [Ko2], [KoS1], [Ot], [PZ], [Se2], [Se3], [ST], [SYZ] for related or different aspects of homological mirror symmetry. Though this is the second part, it is independent of the first part [Fu9]. In this second part, we focus the A_∞ category constructed by using Floer homology of Lagrangian submanifolds, and homological algebra of A_∞ categories. In this sense, this article is an updated version of author's previous papers [Fu1], [Fu4]. Time has passed after [Fu1], [Fu4] were written. During those period, we made several progress, some of which are explained in this article.

Among the points where the contents of this article overlaps with [Fu4], there are three points where we improve the constructions there. One is that we removed an assumption in [Fu4], that all Lagrangian submanifolds are monotone and have minimum Maslov index ≥ 3. (See §2 for the definition of Maslov index.) We assumed it in [Fu4] because we used results by Oh [Oh1] to define Floer homology of Lagrangian submanifolds. This assumption is not extremely restrictive for the purpose of [Fu4], that is to study relative Floer homology of 3 manifolds with boundary. However, for the purpose of this article, that is to study mirror symmetry, this assumption is rather restrictive. To study mirror symmetry, the case of Lagrangian submanifolds in Calabi-Yau 3 fold with Maslov index 0 is the most important. Such Lagrangian submanifolds are monotone but do not satisfy the condition that minimal Maslov index ≥ 3.

To define Floer homology of non monotone Lagrangian submanifolds or to remove the condition that minimal Maslov index ≥ 3, we need the obstruction theory developed in [FOOO]. The summary of a part of it is included in this article. (See [Ot] for another summary of [FOOO].) We generalize the construction of A_∞ category explained in [Fu1], [Fu4] to more general situation using the idea of [FOOO].

The second point is that we put precise sign in each formula. In the side of geometry, this requires us to describe orientations of the moduli spaces of pseudoholomorphic discs involved in the construction. Actually the argument we need to do so is already developed in detail in [FOOO] Chapter 6[1]. There, the case when only one or two Lagrangian submanifolds appear, are studied. But the method there can be generalized easily to the present situation where more than two Lagrangian submanifolds appear at the same time.

We need to study sign also in the algebraic construction involved. Various constructions on homological (homotopical) algebra of A_∞ category is developed in [Fu4] Part II. There we worked over $\mathbb{Z}_2$ coefficient. In this article, we give precise sign to the discussion there. The sign in the study of A_∞ category is rather hard and is not at all a trivial matter. Actually the author was unable to work over $\mathbb{Z}$ coefficient in [Fu4] Part II, because he could not find correct sign when he was writing [Fu4]. To fix the sign we develop more systematic way to write the formulas in [Fu4] Part II. As a consequence, the description of this article is considerably simplified compared to one in [Fu4] Part II.

Thirdly we solved in [FOOO] the trouble related to the existence of the identity element in the A_∞ category of Lagrangian submanifold, which was discussed in [Fu4] §13 in unsatisfactory way. In [FOOO] §20 we discussed the problem of identity in the A_∞ algebra we associate to a Lagrangian submanifold. As a consequence of [FOOO] §20 and §5 of present article, we have an A_∞ category with identity. The identity element plays a central role in the proof of Yoneda's lemma which we prove in §9 of present article, refining the proof given in [Fu4] §12.

Now the outline of each sections are in order. In §1, we introduce the notions of A_∞ category and filtered A_∞ category. The other sections of Chapter 1 are devoted to the construction of its example. Namely we associate a filtered A_∞ category to each symplectic manifold.

In §2, we describe an objects of this filtered A_∞ category. The object is roughly speaking a pair of a Lagrangian submanifold and a flat $U(1)$ bundle on it. But we need to add some additional topological data. The main point of this section is to describe precisely the additional data we add and explain the reason we need it. The additional data are relative spin structure (which is related to the orientation or sign), and the grading, (which is related to the degree). We follow [FOOO]

[1]When we quote [FOOO] in this article, we refer its preprint version which was completed in 2000 December and can be obtained from the author's home page http://www.kusm.kyoto-u.ac.jp/~fukaya/fukaya.html at the time of writing this article. We are now adding more materials to it and there will be some change of the order of the chapters in the final version.

Chapter 6 on the first point and follow Kontsevich, Seidel [Se1] on the second point.

§3, §4 are devoted to the definition of the module of morphisms and operations to define our filtered A_∞ category. To define an operator we need to consider two cases separately. The first case is when the Lagrangian submanifolds involved are mutually transversal. This case was discussed in [Fu1], [Fu4]. We discuss this case in §3.

In §4 we discuss the general case, namely the case when we do not assume transversality. Especially we need to study the case when two Lagrangian submanifolds involved coincide to each other. In a similar way, we can discuss more general case when they are of clean intersection. But we do not do it in this article. (See [Po] and [FOOO] §16.) The construction in §4 is a natural generalization of one in [FOOO] where we associate a filtered A_∞ algebra to a Lagrangian submanifold.

In §5 we discuss the problem of unit. Namely we define a notion, homotopy unit (which was first introduced in [FOOO] in the case of A_∞ algebra) and sketch the idea how to construct the homotopy unit in the case of the A_∞ category of Lagrangian submanifolds.

In Chapter 2 we discuss homological algebra of A_∞ category. After writing [Fu4], the author leaned that there are many works done in this direction, in the case of A_∞ algebra or differential graded category, especially by Russian mathematicians. However the reference on it is rather scattered and it is rather hard to find a good reference where we find an appropriate description (and its proofs) of the results we need, especially in the case of A_∞ category. So the author include it in this article. But he does not assert so much credit on it. Namely they are rather minor modification of the results already known to specialists in closely related situations, though some of the results in Chapter 2 are new in the sense they are not proved in the references. The author tries to quote appropriate reference in case he found it. However, since the author is far from being a specialist of homological algebra there should be many works which is closely related to Chapter 2 but is not known to the author.

In §6, we describe the notion of twisted complex and derived A_∞ category. Twisted complex is a natural analogue of chain complex. Namely in the case of abelian category we study chain complex and use it to construct derived category. In a similar way we use twisted complex in the case of A_∞ category. Twisted complex was introduced by Bondal-Kapranov [BoK] in the case of differential graded category. Kontsevitch [Ko1] proposed to use it in mirror symmetry. (It was also applied in [Fu7] to mirror symmetry.)

The author, in [Fu1], [Fu4], proposed to use the A_∞ category of A_∞ functors in the study of Floer homology of 3 manifolds with boundary. For each twisted complex of an A_∞ category $\mathcal{C}$, we can associate an A_∞ functor from $\mathcal{C}$ to $\mathcal{CH}$. Here $\mathcal{CH}$ is the A_∞ category whose object is a chain complex. In this sense, A_∞ functor to $\mathcal{CH}$ is a natural generalization of twisted complex. Also the idea of A_∞ functor is important to understand Floer homology of family of Lagrangian submanifolds (see [Fu5]) and its application to mirror symmetry. Moreover, to clarify the dependence (or independence) of the A_∞ category described in Chapter 1, we need to define a notion of homotopy equivalence of A_∞ categories and hence we use the notion of A_∞ functors. So we introduce the notion of A_∞ functors in section 7. There we construct a representable A_∞ functor and an A_∞ category whose objects are A_∞ functors.

In §8 we define homotopy equivalence of A_∞ category. We prove in §8 that an A_∞ functor which induces isomorphismes on cohomologies is a homotopy equivalence. This result is an algebraic analogue of J. H. C. Whitehead theorem in topology and is proved in [FOOO] §A5 in the case of A_∞ algebra. (A similar results should had been proved in various cases in the reference. The author was unable to locate the first place where this kinds of results appeared.)

In §9, we prove another main result of the homological algebra of A_∞ category, an A_∞ analogue of Yoneda's lemma. This result implies that we can embed any A_∞ category $\mathcal{C}$ to the A_∞ category of A_∞ functors from $\mathcal{C}$ to $\mathcal{CH}$. (Namely we identify an object of $\mathcal{C}$ to a functor represented by it.) As mentioned above, this point will be important to the further study of mirror symmetry and of Floer homology of 3 manifolds with boundary. An A_∞ analogue of Yoneda's lemma was proved in [Fu4] except signs. We give a proof with sign here and also we simplify the proof in [Fu4].

Those results on homological algebra of A_∞ category will be used in future to further study the filtered A_∞ category constructed in Chapter 1 of this article. For example we will prove that the filtered A_∞ category constructed in Chapter 1 of this article is independent of the various choices involved up to homotopy equivalence. This results together with other results are not included in this Part II and is postponed to Part III etc. So two chapters of this article are yet rather independent in this article but will be unified in future.

The author would like to thank Professors Yong-Geun Oh, Hiroshi Ohta, and Kaoru Ono, the joint authors of the book [FOOO]. He is happy to share many of the ideas described in this article with them.

The author also would like to thank Professor Seiki Nishikawa for his patience. The author apologize to other authors of this proceeding, since the delay of the completion of this article causes the delay of the publication of this proceeding.

Last but not the least, this article is dedicated to Professor Katsuhiro Shiohama for his 60th birth day. The author owes much to Professor Shiohama for his encouragement and support especially at the time when the author started his career as a mathematician.

Chapter 1: Floer homology and A_∞ category

§1. A_∞ category and filtered A_∞ category

To study the part of homological mirror symmetry conjecture we concern with in this article, we need to use the notion of A_∞ category appeared in [Fu1] in the study of topological field theory. (A_∞ structure had been studied in homotopy theory for a long time, especially in [St].) Actually we need to include the obstruction theory we developed in [FOOO], and to modify the definition of A_∞ category a bit, in order to correct the flaw pointed out in [Ko2], [Oh1]. We call this modified version *filtered A_∞ category.* We start with the usual definition of A_∞ category. Most of the constructions in this section are straight forward generalizations of the definition of filtered A_∞ algebra in [FOOO] Chapter 4.

We fix R, a commutative ring with unit. In our main application, R will be $\mathbb{C}$ or $\mathbb{Q}$.

Definition 1.1. An *A_∞ category*, $\mathcal{C}$, is a collection of a set $\mathfrak{Ob}(\mathcal{C})$, the set of objects, a graded free R module $\mathcal{C}(c_1, c_2)$ for each $c_1, c_2 \in \mathfrak{Ob}(\mathcal{C})$, the operations

$$\mathfrak{m}_k : \mathcal{C}[1](c_0, c_1) \otimes \cdots \otimes \mathcal{C}[1](c_{k-1}, c_k) \to \mathcal{C}[1](c_0, c_k),$$

of degree $+1$ for $k = 1, 2, \ldots$ and $c_i \in \mathfrak{Ob}(\mathcal{C})$. Here $\mathcal{C}[1](c_0, c_1)$ is $\mathcal{C}(c_0, c_1)$ with degree shifted. ($\mathcal{C}[1]^m(c_0, c_1) = \mathcal{C}^{m+1}(c_0, c_1)$). They are assumed to satisfy the A_∞ formula (1.3) described below.

To describe the A_∞ formula, we introduce notations. Let $a, b \in \mathfrak{Ob}(\mathcal{C})$, we put

$$(1.2)\quad B_k\mathcal{C}[1](a, b) = \bigoplus_{a=c_0, c_1, \ldots, c_{k-1}, c_k = b} \mathcal{C}[1](c_0, c_1) \otimes \cdots \otimes \mathcal{C}[1](c_{k-1}, c_k).$$

We define, in case $k = 0$,

$$B_0\mathcal{C}[1](a, a) = R, \qquad B_0\mathcal{C}[1](a, b) = 0 \quad \text{if } a \neq b.$$

We put

$$BC[1](a,b) = \bigoplus_k B_k\mathcal{C}[1](a,b), \quad B_k\mathcal{C}[1] = \bigoplus_{a,b} B_k\mathcal{C}[1](a,b),$$
$$BC[1] = \bigoplus_{a,b} BC[1](a,b).$$

We define a homomorphism

$$\Delta : B_k\mathcal{C}[1](a,b) \to \bigoplus_{k_1+k_2=k} \bigoplus_c B_{k_1}\mathcal{C}[1](a,c) \otimes B_{k_2}\mathcal{C}[1](c,b)$$

by

$$\Delta(x_1 \otimes \cdots \otimes x_k) = (x_1 \otimes \cdots \otimes x_{k_1}) \otimes (x_{k_1+1} \otimes \cdots \otimes x_k).$$

It induces maps $\Delta : B_k\mathcal{C}[1] \to \oplus_{k_1+k_2=k} B_{k_1}\mathcal{C}[1] \otimes B_{k_2}\mathcal{C}[1]$, $\Delta : BC[1] \to BC[1] \otimes BC[1]$. $(BC[1](a,a), \Delta)$ and $(BC[1], \Delta)$ are graded coalgebras. (They are coassociative but not cocomutative.)

Operations $\mathfrak{m}_k$ define homomorphisms: $B_k\mathcal{C}[1](a,b) \to \mathcal{C}[1](a,b)$. It can be extended uniquely to coderivations

$$\hat{d}_k : BC[1] \to BC[1], \quad \hat{d}_k : BC[1](a,b) \to BC[1](a,b)$$

by

$$\hat{d}_k(x_1 \otimes \cdots \otimes x_n) = \sum_\ell (-1)^{(\deg x_1+1)+\cdots+(\deg x_{\ell-1}+1)} x_1 \otimes \cdots \otimes \mathfrak{m}_k(x_\ell, \ldots, x_{\ell+k-1}) \otimes \cdots \otimes x_n.$$

We put

$$\hat{d} = \sum_k \hat{d}_k.$$

Now the A_∞ formula is

$$\hat{d} \circ \hat{d} = 0. \tag{1.3}$$

We can expand it and rewrite it using $\mathfrak{m}_k$. We thus obtain, for example,

$$0 = \mathfrak{m}_1\mathfrak{m}_1,$$
$$0 = \mathfrak{m}_1\mathfrak{m}_2(x \otimes y) + \mathfrak{m}_2(\mathfrak{m}_1(x) \otimes y) + (-1)^{\deg x+1}\mathfrak{m}_2(x \otimes \mathfrak{m}_1(y)).$$

(See [GJ], [FOOO] Chapter 4.) Namely $(\mathcal{C}(a,b), \mathfrak{m}_1)$ is a chain complex and $\mathfrak{m}_2$ is a derivation up to sign.

Remark 1.4. Here we follow the sign convention of [FOOO] and not of [Fu7].

Definition 1.5. Let $c \in \mathfrak{Ob}(\mathcal{C})$. We say an element $\mathbf{e}_c \in \mathcal{C}^0(c,c) = \mathcal{C}^1[1](c,c)$ is a *unit* if

$$(1.5.1) \qquad \mathfrak{m}_2(\mathbf{e}_c, x_1) = x_1, \quad \mathfrak{m}_2(x_2, \mathbf{e}_c) = (-1)^{\deg x_2} x_2$$

for $x_1 \in \mathcal{C}(c,c')$, $x_2 \in \mathcal{C}(c',c)$ and

$$(1.5.2) \qquad \mathfrak{m}_{k+\ell+1}(x_1, \dots, x_\ell, \mathbf{e}_c, y_1, \dots, y_k) = 0$$

for $k + \ell \neq 1$.

Definition 1.6. A_∞ category with one object is called an A_∞ *algebra.*

Example-Lemma 1.7. *Let $(A, d, \cdot)$ be a differential graded algebra. Namely $d : A^k \to A^{k+1}$, $\wedge : A^k \otimes A^\ell \to A^{k+\ell}$ with*

$$(1.8.1) \qquad d \circ d = 0,$$

$$(1.8.2) \qquad (x \cdot y) \cdot z = x \cdot (y \cdot z),$$

$$(1.8.3) \qquad d(x \cdot y) = dx \cdot y + (-1)^{\deg x} x \cdot dy.$$

We put

$$(1.9.1) \qquad \mathfrak{m}_1(x) = (-1)^{\deg x} dx,$$

$$(1.9.2) \qquad \mathfrak{m}_2(x, y) = (-1)^{\deg x (\deg y + 1)} x \cdot y,$$

$$(1.9.3) \qquad \mathfrak{m}_k = 0 \quad \textit{for } k > 2.$$

Then $(A, \mathfrak{m})$ is an A_∞ algebra.

Proof. $\mathfrak{m}_1\mathfrak{m}_1 = 0$ follows from (1.8.1). We calculate using (1.8.3) and (1.9.1), (1.9.2) that

$$\begin{aligned} \mathfrak{m}_1(\mathfrak{m}_2(x, y)) &= (-1)^{\deg x + \deg y + \deg x(\deg y + 1)} d(x \cdot y) \\ &= -\mathfrak{m}_2(\mathfrak{m}_1(x), y) - (-1)^{\deg x + 1} \mathfrak{m}_2(x, \mathfrak{m}_1(y)). \end{aligned}$$

This is A^∞ formula on $B_2A[1]$. We can also check

$$\begin{aligned} &\mathfrak{m}_2(\mathfrak{m}_2(x, y), z) + \mathfrak{m}_2(x, \mathfrak{m}_2(y, z)) \\ &= (-1)^{\deg x \deg y + \deg x \deg z + \deg x \deg z + \deg y}((x \cdot y) \cdot z - x \cdot (y \cdot z)), \end{aligned}$$

which is zero by (1.8.2). □

Definition-Example 1.10 ([BoK]). A *differential graded category* $\mathcal{C}$ is a collection of a set $\mathfrak{Ob}(\mathcal{C})$, the set of objects, a differential graded R module $\mathcal{C}(c_1, c_2)$ for each $c_1, c_2 \in \mathfrak{Ob}(\mathcal{C})$, the operations

$$\circ : \mathcal{C}(c_1, c_2) \otimes \mathcal{C}(c_2, c_3) \to \mathcal{C}(c_1, c_3),$$

which is a chain map and is associative in the sense of (1.8.2). We then define $\mathfrak{m}_k$ by (1.9). We obtain an A_∞ category.

For later use, we introduce some other notations. Let $\mathcal{C}$ be an A_∞ category and c be its object. We put

$$B\mathcal{C}[1](c) = \bigoplus_k B_k\mathcal{C}[1](c) = \bigoplus_k \mathcal{C}[1](c,c)^{\otimes k}.$$

(We remark that $B_k\mathcal{C}[1](c) \neq B_k\mathcal{C}[1](c,c)$.) $B\mathcal{C}[1](c)$ is a coalgebra and $\mathfrak{m}_k$ defines a structure of A_∞ algebra on it.

Actually our main example is a filtered A_∞ category rather than A_∞ category. We are going to define it. We first define a universal Novikov ring [No], [FOOO].

Definition 1.11. Let T be a formal parameter. We consider a formal power series

$$\sum_{i=1}^{\infty} a_i T^{\lambda_i} \tag{1.12}$$

where $a_i \in R$, $\lambda_i \in \mathbb{R}$. We assume $\lambda_i < \lambda_{i+1}$ and $\lim_{i\to\infty} \lambda_i = \infty$. We denote by $\Lambda_{R,\mathrm{nov}}$ the set of all such series. It has an obvious ring structure.

We consider its subring consisting of (1.12) such that $\lambda_i \geq 0$ in addition and denote it by $\Lambda_{R,0,\mathrm{nov}}$. The ring $\Lambda_{R,0,\mathrm{nov}}$ has a maximal ideal consisting of all series (1.12) such that $\lambda_i > 0$ in addition. We denote it by $\Lambda_{+,\mathrm{nov}}$.

We write Λ_{nov}, $\Lambda_{0,\mathrm{nov}}$, $\Lambda_{+,\mathrm{nov}}$ in place of $\Lambda_{R,\mathrm{nov}}$, $\Lambda_{R,0,\mathrm{nov}}$, $\Lambda_{R,+,\mathrm{nov}}$ in case no confusion can occur.

For each λ, we define $F^\lambda\Lambda_{\mathrm{nov}}$ so that it consists of the elements (1.12) satisfying $\lambda_i \geq \lambda$ in addition. It induces a filtration on Λ_{nov}. Namely each $F^\lambda\Lambda_{\mathrm{nov}}$ is an additive subgroup and $F^{\lambda_1}\Lambda_{\mathrm{nov}} \cdot F^{\lambda_2}\Lambda_{\mathrm{nov}} \subseteq F^{\lambda_1+\lambda_2}\Lambda_{\mathrm{nov}}$. Filtration on Λ_{nov} induces ones on $\Lambda_{0,\mathrm{nov}}$ and $\Lambda_{+,\mathrm{nov}}$.

Our filtration induces a uniform structure on Λ_{nov}, $\Lambda_{0,\mathrm{nov}}$ and $\Lambda_{+,\mathrm{nov}}$ in a usual way. Then these rings are complete with respect to it.

Remark 1.13. In [FOOO], we considered the set of series $\sum_{i=1}^{\infty} a_i T^{\lambda_i} e^{n_i}$ where n_i are integers, and denote by Λ_{nov} the set of all such series. However for the present purpose (that is to discuss Mirror symmetry where the case $c^1(M) = 0$ is important) it is more convenient to use Definition 1.11.

Now we define filtered A_∞ category.

Definition 1.14. A filtered A_∞ category $\mathcal{C}$ is a correction of a set $\mathfrak{Ob}(\mathcal{C})$, the set of objects, a graded torsion free filtered $\Lambda_{0,\mathrm{nov}}$ module $\mathcal{C}(c_1, c_2)$ for each $c_1, c_2 \in \mathfrak{Ob}(\mathcal{C})$, the operations

$$\mathfrak{m}_k : \mathcal{C}[1](c_0, c_1) \otimes \cdots \otimes \mathcal{C}[1](c_{k-1}, c_k) \to \mathcal{C}[1](c_0, c_k),$$

of degree $+1$ for $k = 0, 1, 2, \ldots$, and $c_i \in \mathfrak{Ob}(\mathcal{C})$. Note that $k = 0$ is included in the case of filtered A_∞ category. Here $\mathfrak{m}_0$ is a map

$$\mathfrak{m}_0 : \Lambda_{0,\mathrm{nov}} \to \bigoplus_c \mathcal{C}[1](c, c).$$

($\mathfrak{m}_0$ is not included in the case of A_∞ category.) We assume that $\mathfrak{m}_k$ respects the filtration in the sense that

$$\begin{aligned}\mathfrak{m}_k(F^{\lambda_1}\mathcal{C}[1](c_0, c_1) \otimes \cdots \otimes F^{\lambda_k}\mathcal{C}[1](c_{k-1}, c_k)) \\ \subseteq F^{\sum \lambda_i}\mathcal{C}[1](c_0, c_k).\end{aligned} \tag{1.15}$$

They induce coderivations

$$\hat{d}_k : B\mathcal{C}[1] \to B\mathcal{C}[1], \quad \hat{d}_k : B\mathcal{C}[1](a, b) \to B\mathcal{C}[1](a, b)$$

in the same way as before. Our filtrations on $\mathcal{C}$ induces one on $B\mathcal{C}[1]$ and we let $\hat{B}\mathcal{C}[1]$ be the completion. $\hat{d}_k$ induces a map: $\hat{B}\mathcal{C}[1] \to \hat{B}\mathcal{C}[1]$, which we denote by the same symbol. The sum

$$\hat{d} = \sum_{k=0}^{\infty} \hat{d}_k$$

converges by virtue of (1.15). Now we assume

$$\hat{d} \circ \hat{d} = 0. \tag{1.16}$$

We assume also

$$\mathfrak{m}_0 \equiv 0 \mod \Lambda_{+,\mathrm{nov}}. \tag{1.17}$$

We define a unit of filtered A_∞ category in the same way as Definition 1.5.

A filtered A_∞ category with one object is called a *filtered A_∞ algebra.*

For each filtered A_∞ category $\mathcal{C}$ and $c \in \mathfrak{Ob}(\mathcal{C})$, the operations $\mathfrak{m}_k : B_k\mathcal{C}(c) \to \mathcal{C}(c,c)$ define a structure of filtered A_∞ algebra on $\mathcal{C}(c,c)$.

We can construct an A_∞ category (of $\Lambda_{0,\mathrm{nov}}$ module) from a filtered A_∞ category $\mathcal{C}$ in the following way. Let $c \in \mathfrak{Ob}(\mathcal{C})$. We define

Definition 1.18. An element b of $F^+\mathcal{C}^1(c,c)$ is said to be a *bounding chain* if

$$\hat{d}(e^b) = 0.$$

Here $F^+\mathcal{C}^1(c,c) = \cup_{\lambda>0} F^\lambda\mathcal{C}^1(c,c)$ and

$$e^b = 1 + b + b \otimes b + b \otimes b \otimes b + \cdots \in \hat{B}\mathcal{C}(c,c).$$

We define $\tilde{\mathcal{M}}(c)$ be the set of all bounding chains of c.

Definition 1.19. Let $b_i \in \tilde{\mathcal{M}}(c_i)$, $i = 0, \ldots, k$, $k > 0$. We define

$$\mathfrak{m}_k^{(b_0,\ldots,b_k)} : \mathcal{C}[1](c_0,c_1) \otimes \cdots \otimes \mathcal{C}[1](c_{k-1},c_k) \to \mathcal{C}[1](c_0,c_k)$$

by

$$\begin{aligned} &\mathfrak{m}_k^{(b_0,\ldots,b_k)}(x_1,\ldots,x_k) \\ &= \sum_{\ell_0,\ldots,\ell_k} \mathfrak{m}_{k+\ell_0+\cdots+\ell_k}(b_0^{\ell_0}, x_1, b_1^{\ell_1}, \ldots, b_{k-1}^{\ell_{k-1}}, x_k, b_k^{\ell_k}). \end{aligned}$$

Here

$$b^\ell = \overbrace{b \otimes \cdots \otimes b}^{\ell \text{ times}}.$$

Proposition 1.20. *We put*

$$\mathfrak{Ob}(\mathcal{C}') = \bigcup_{c \in \mathfrak{Ob}(\mathcal{C})} \tilde{\mathcal{M}}(c) \times \{c\},$$

$\mathcal{C}'((c,b),(c',b')) = \mathcal{C}(c,c')$ *and let* $\mathfrak{m}_k^{(b_0,\ldots,b_k)}$ *be the operations. Then* $\mathcal{C}'$ *is an* A_∞ *category.*

The proof (which is easy) goes in exactly the same was as the proof of [FOOO] Lemma 13.37. Hence we omit it.

Actually $\tilde{\mathcal{M}}(c)$ is too big and it is more reasonable to define an equivalence relation $\sim$ on it and divide $\tilde{\mathcal{M}}(c)$ by $\sim$. See [FOOO], [Fu5], [Ot] on it.

We can continue and study A_∞ functor, natural transformations etc. We will do it later. We give our main example before continuing the discussion on algebraic formalism. If the reader is mainly interested in algebraic formalism then he can skip §2 $\sim$ §5 and proceed to Chapter 2.

§2. Floer homology and A_∞ category I – the set of objects –

The idea of this section is rather old. Inspired by S. Donaldson's lecture [Do], the author [Fu1] found that Floer homology and counting of holomorphic polygons will define an A_∞ category. However there was a trouble in defining Lagrangian intersection Floer homology as was mentioned in [Ko2], [Oh1]. We could overcome this trouble in [FOOO] and the construction is now presented here in a modified way. The notion of filtered A_∞ category introduced in the last section is defined for this purpose.

We explain, in this section, the definition of the objects of our filtered A_∞ category $\mathcal{LAG}$ and define a graded $\Lambda_{0,\mathrm{nov}}$ module $\mathcal{LAG}(c,c')$ for two objects when $c \neq c'$. Basically an object of $\mathcal{LAG}$ is a Lagrangian submanifold. However we need to add some topological data to it. One of the topological data to be added is related to the orientation problem of the moduli space of pseudoholomorhic discs. Another data to be added is related to the way to fix the degree of elements of $\mathcal{LAG}(c,c')$. To motivate those data we mention some comments on how they will be used. During those comments we assume that the reader is familiar to the Floer homology and pseudoholomorphic curves. (Please skip them otherwise.) Let (M,ω) be a symplectic manifolds. Let B be a closed 2 form which we call the B-field. We put $\Omega = \omega + 2\pi\sqrt{-1}B$.

Definition 2.1. Let $\mathfrak{Ob}_1(\mathcal{LAG}(M,\Omega))$ be the set of all pairs $(L,\mathcal{L})$ such that:

(2.1.1) L is a Lagrangian submanifold. Namely $\dim L = \dim M/2$ and $\omega|_L = 0$.
(2.1.2) $\mathcal{L}$ is a complex line bundle equipped with a unitary connection ∇ such that its curvature F_∇ coincides with the restriction of $2\pi\sqrt{-1}B$ to L. (Here we identify the Lie algebra of $U(1)$ with $\sqrt{-1}\mathbb{R}$.)

Remark 2.2. One may consider more general objects than $\mathfrak{Ob}_1(\mathcal{LAG})(M,\Omega)$. There are at least two generalizations.

(2.2.1) One may relax the condition on L so that $L \to M$ is a Lagrangian immersion.
(2.2.2) One may consider the vector bundle $\mathcal{L}$ together with its unitary connection ∇ whose curvature is $2\pi\sqrt{-1}B$ times the unit matrix.

The modification of the construction to include these cases will be discussed elsewhere. (See [Ak] on (2.2.1).)

It seems impossible to define an A_∞ category whose objects are all elements of $\mathfrak{Ob}_1(\mathcal{LAG}(M,\Omega))$ because of the transversality problem. (We use Bair's category theorem to achieve transversality quite frequently.) So instead we take and fix a countable set of Lagrangian submanifolds and let $\mathfrak{Ob}_2(\mathcal{LAG}(M,\Omega))$ be the set of all elements $(L,\mathfrak{L}) \in \mathfrak{Ob}_1(\mathcal{LAG}(M,\Omega))$ such that L is in this countable set.

The module of morphisms $\mathcal{LAG}((L_1,\mathfrak{L}_1),(L_2,\mathfrak{L}_2))$ of our filtered A_∞ category is Floer's chain complex, which is the free $\Lambda_{0,\mathrm{nov}}$ module generated by the intersection points $\in L_1 \cap L_2$. More precisely

$$\mathcal{LAG}((L_1,\mathfrak{L}_1),(L_2,\mathfrak{L}_2)) = \bigoplus_{p\in L_1\cap L_2} \mathrm{Hom}((\mathfrak{L}_1)_p,(\mathfrak{L}_2)_p)\otimes_{\mathbb{C}} \Lambda_{0,\mathrm{nov}}. \tag{2.3}$$

However there are two delicate points which will soon come to the story. The first of them is sign or orientation of the moduli space of pseudo-holomorphic discs which we will use to define operations $\mathfrak{m}_k$ (see §3, §4), and the other is the degree in the Floer homology.

We start with the first point. We refer [FOOO] Chapter 6 for the thorough argument on the orientation and present only a sketch of it in this article. We first fix an element $st \in H^2(M;\mathbb{Z}_2)$. We take a 3-skeleton $M^{(3)}$ of M. Then there exists a unique real rank 2 vector bundles $V(st)$ on $M^{(3)}$ such that $w^1(V(st)) = 0$, $w^2(V(st)) = st$, here w is the Stiefel-Whitney class.

Definition 2.4 ([FOOO]). L is said to be *relatively spin* in (M,st) if it is oriented and if the second Stiefel-Whitney class of (the tangent bundle of) L coincides with the restriction of st.

Let L be relatively spin in (M,st), and let $L^{(2)}$ be the two skeleton of L. Then $V\oplus TL$ is trivial on $L^{(2)}$.

A (M,st)*-relative spin structure* of L is by definition a spin structure of the restriction of $V\oplus TL$ to $L^{(2)}$.

We remark that the two spin structures are equivalent if it is equivalent on the first skeleton. Moreover oriented vector bundle is trivial on two skeleton if it is spin. Therefore the set of (M,st)-relative spin structures of L corresponds one to one to the set of trivializations of the restriction of $V\oplus TL$ to $L^{(1)}$ which can be extended to $L^{(2)}$. We use this remark to show the following:

Lemma 2.5. *The group $H^1(L;\mathbb{Z}_2)$ acts simple transitively on the set of all (M,st)-relative spin structures of L if it is nonempty.*

Proof. Let $\psi \in C^1(L;\mathbb{Z}_2)$ be a cocycle defining an element of $H^1(L;\mathbb{Z}_2)$. Let $\Psi : V\oplus TL|_{L^{(2)}} \to L^{(2)}\times\mathbb{R}^{n+2}$ be isomorphism of bundles whose restrictions to $V\oplus TL|_{L^{(1)}}$ define a relative spin structures of L. For each one cell Δ^1 of L we define a map

$$g_{\psi,\Delta^1} : (\Delta^1,\partial\Delta^1) \to (SO(n+2), I)$$

representing $\psi(\Delta^1) \in \pi_1(SO(n+2)) = \mathbb{Z}_2$. ($I \in SO(n+1)$ is the unit matrix.) We put

$$\Psi'(x,v) = g_{\psi,\Delta^1}(x)(x,v)$$

for $(x,v) \in V \oplus TL|_{L^{(1)}}$, $x \in \Delta^1$. Since ψ is a cocycle it follows that $\Psi' : V \oplus TL|_{L^{(1)}} \to L^{(1)} \times \mathbb{R}^{n+2}$ can be extended to $L^{(2)}$. It is easy to see that the relative spin structure determined by Ψ' depends only on the cohomology class of ψ and the relative spin structure Ψ. We put $[\psi] \cdot [\Psi] = [\Psi']$.

Conversely, let $\Psi_i : V \oplus TL|_{L^{(2)}} \to L^{(2)} \times \mathbb{R}^{n+2}$ be isomorphism of bundles whose restrictions to $V \oplus TL|_{L^{(1)}}$ define two relative spin structures of L. There exists a map $g : L^{(2)} \to SO(n+2)$ such that

$$\Psi_2(x,v) = g(x)(\Psi_1(x,v)).$$

Since $\pi_0(SO(n+2))$ is trivial we may modify Ψ_i so that $g(x) = 1$ for $x \in L^{(1)}$. Then for each 1 cell Δ^1 of L we have

$$[g|_{\Delta^1}] \in \pi_1(SO(n+2)) = \mathbb{Z}_2.$$

We regard it as a cochain. It is a cocycle since g can be extended to $L^{(2)}$. We thus obtain $\psi = g$ such that $[\psi] \cdot [\Psi_1] = [\Psi_2]$. □

We denote by $\mathfrak{Ob}_3(\mathcal{LAG}(M,\Omega,st))$ the set of all pairs of $(L,\mathcal{L}) \in \mathfrak{Ob}_2(\mathcal{LAG}(M,\Omega))$ and an (M,st)-relative spin structure on L. The reason we add relative spin structure is that it induces orientations of various moduli spaces we use in a canonical way. (See the next section and [FOOO] Chapter 6.)

Next we consider the degree problem. We use the notion of graded Lagrangian submanifold due to M. Kontsevich and P. Seidel [Se1] for this purpose. Let $(\mathbb{R}^{2n},\omega)$ be a symplectic vector space of dimension $2n$. We denote by $\mathrm{Lag}_n = \mathrm{Lag}(\mathbb{R}^{2n},\omega)$ the set of all oriented linear subspaces V of $\mathbb{R}^{2n}$ of dimension n such that $\omega|_V \equiv 0$. We call it the oriented Lagrangian Grassmanian. It is well-known that $\pi_1(\mathrm{Lag}_n) \cong \mathbb{Z}$ and it has a generator called the (universal) Maslov class. (See [AG]). Let $\widetilde{\mathrm{Lag}}_n$ be the universal covering space of Lag_n.

Let (M,ω) be a symplectic manifold. Then we have a fiber bundle $\mathrm{Lag}(M) \to M$ whose fiber at $p \in M$ is identified with $\mathrm{Lag}(T_pM)$. We remark that we can find an almost complex structure compatible with its symplectic structure, and it is unique up to homotopy. Hence the Chern classes of the tangent bundle of a symplectic manifolds are well defined.

Lemma 2.6. *The following two conditions are equivalent.*
(2.7.1) $c^1(M) = 0$ *where* $c^1(M)$ *is the first Chern class of* TM.
(2.7.2) *There exists a covering space* $\widetilde{\mathrm{Lag}}(M)$ *of* $\mathrm{Lag}(M)$ *such that its restriction to each fiber is identified with* $\widetilde{\mathrm{Lag}}_n \to \mathrm{Lag}_n$.

Proof. Since $\widetilde{\mathrm{Lag}} \to \mathrm{Lag}$ is a covering space, the obstruction to construct $\widetilde{\mathrm{Lag}}(M)$ lies in the second cohomology. Hence it is easy to see that (2.7.2) is equivalent to the triviality of the bundle $\mathrm{Lag}(M) \to M$ at the two skeleton $M^{(2)}$ of M.

Now let us assume (2.7.1). Then the complex vector bundle TM is trivial on two skeleton. Since Symplectic group $\mathrm{Symp}(n)$ is homotopy equivalent to $U(n)$, it follows that $\mathrm{Lag}(M) \to M$ is trivial on two skeleton. The proof of the converse is similar. □

From now on, we assume $c^1(M) = 0$. (In the case when the compatible almost complex structure is integrable, this condition implies that M has a Ricci flat Kähler metric, due to Yau's proof of Calabi conjecture.) We also *fix* a covering space $\widetilde{\mathrm{Lag}}(M)$ of $\mathrm{Lag}(M)$ as in (2.7.2).

Remark 2.8. In a way similar to the proof of Lemma 2.5, we can show that the set of such lifts $\widetilde{\mathrm{Lag}}(M)$ is an affine space over $H^1(M;\mathbb{Z})$ if it is nonempty.

Let L be an oriented Lagrangian submanifold. We have a canonical section s of the restriction of $\mathrm{Lag}(M)$ to L. Namely

$$s(p) = T_pL \subseteq T_pM.$$

Definition 2.9. A *graded Lagrangian submanifold* of $(M, \widetilde{\mathrm{Lag}}(M))$ is a pair of oriented Lagrangian submanifold L and a lift of s to $\tilde{s} : L \to \widetilde{\mathrm{Lag}}(M)$. We call $\tilde{s}$, the *grading* of L.

Definition 2.10. We denote by $\mathfrak{Ob}_4(\mathcal{LAG}(M, \Omega, st, \widetilde{\mathrm{Lag}}(M)))$ the set of all triples $(L, \mathcal{L}, \tilde{s})$ such that $(L, \mathcal{L}) \in \mathfrak{Ob}_3(\mathcal{LAG}(M, \Omega, st))$ and that $(L, \tilde{s})$ is a graded Lagrangian submanifold.

Example 2.11. Let $T^*N \to N$ be a cotangent bundle of an oriented manifold N with a canonical symplectic structure. The tangent bundle TT^*N is isomorphic to the complexification of the pull back π^*TN.

Hence its structure group is reduced to $U(n) \cap GL(n;\mathbb{R}) = O(n)$. Moreover, since N is oriented, it follows that the structure group is reduced to $SO(n)$. Since $\mathrm{Lag}_n = U(n)/SO(n)$ it follows that the bundle $\mathrm{Lag}_n(T^*N)$ is trivial. We thus obtain $\widetilde{\mathrm{Lag}}(T^*N)$ as in (2.5.2).

We remark that the zero section is a Lagrangian submanifold. Hence we have a section s_0 of $\mathrm{Lag}(T^*N)$ on the zero section. Since zero section $\cong N$ is homotopy equivalent to T^*N it induces a section s_0 of $\mathrm{Lag}(T^*N)$. It then induces a trivialization of $\mathrm{Lag}(T^*N)$ and hence a section $\tilde{s}_0$: $T^*N \to \widetilde{\mathrm{Lag}}(T^*N)$. We may choose $\tilde{s}_0$ such that $\tilde{s}_0(p)$ is transversal to the tangent space of the fibers of $T^*N \to N$.

Now let L be a Lagrangian submanifold of T^*N transversal to the fiber. Then, for each $p \in L$, there exists a path ℓ_p which joints T_pL to $s_0(p)$ in $\mathrm{Lag}(T_pT^*N)$ such that $\ell_p(t)$ is transversal to the tangent space of the fiber for each $t \neq 0$. The homotopy class of such ℓ_p is unique. We lift it so that $s_0(p)$ will be lifted to $\tilde{s}_0(p)$. In this way we obtain a lift $\tilde{s}(p) \in \widetilde{\mathrm{Lag}}(T_pT^*N)$. It is easy to see that this lift is independent of various choices.

We thus obtain a graded Lagrangian submanifold $(L, \tilde{s})$.

In a similar way, we can consider the case when (M, ω) has a singular Lagrangian fibration as follows. Let $\pi : M \to N$ be a smooth map with the following properties.

(2.12.1) There exists a subcomplex $X \subset M$ of codimension > 2 such that π is a submersion on $M - X$.
(2.12.2) The kernel of the differential of $d_p\pi$ at $p \in M - X$ is a Lagrangian vector subspace of T_pM.
(2.12.3) π is proper.

Now we define a section s' of $\mathrm{Lag}(M)$ on $M - X$ by putting $s'(p) = \mathrm{Ker}\, d_p\pi$. It defines a real vector bundle on $M - X$ whose fiber at p is $s'(p)$. This vector bundle tensored with $\mathbb{C}$ is isomorphic to the tangent bundle of $M - X$. Thus, in the same way as above, we have a trivialization of $\mathrm{Lag}(M)$ on $M - X$. It induces a bundle $\widetilde{\mathrm{Lag}}(M - X)$ on $M - X$. We can extend it to $\widetilde{\mathrm{Lag}}(M) \to M$ uniquely since the codimension of X in M is > 2. (Note the trivialization of the restriction of $\mathrm{Lag}(M)$ to $M - X$ may not be extended.)

Actually the condition for a Lagrangian submanifold to be graded is related to the (absolute) Maslov index $\eta : \pi_2(M, L) \to \mathbb{Z}$ as follows.

Let us first review Maslov index. Let $\varphi : (D^2, \partial D^2) \to (M, L)$ be a map representing an element of $\pi_2(M, L)$. The pullback bundle $\varphi^*(TM)$ has a trivialization since D^2 is contractible. We restrict this trivialization to ∂D^2. On the other hand, for each $t \in \partial D^2$ we have a Lagrangian subspace $T_{\varphi(p)}L \subset T_{\varphi(p)}M$. Hence we have $S^1 \to \mathrm{Lag}_n$. It determines an element of $\pi_1(\mathrm{Lag}_n) \cong \mathbb{Z}$. We call it *Maslov index* and write $\eta([\varphi])$.

We consider the composition $\pi_2(M) \to \pi_2(M,L) \to \mathbb{Z}$. One can verify easily that the composition coincides with the twice of the first Chern class $c^1 : \pi_2(M) \to \mathbb{Z}$.

Therefore, in the case when $c^1(M) = 0$, the homomorphism η induces a homomorphism : $\mathrm{Im}(\pi_2(M,L) \to \pi_1(L)) \to \mathbb{Z}$. We can extend it to $\pi_1(L)$ as follows. Using $c^1(M) = 0$ there exists a lift $\widetilde{\mathrm{Lag}}(M) \to M$. Let $\ell : S^1 \to L$ be a loop representing an element of $\pi_1(L)$. We define a map

$$\ell^+ : S^1 \to \mathrm{Lag}(M)$$

by

$$\ell^+(t) = T_{\ell(t)}L \in \mathrm{Lag}(T_pM).$$

Since $\widetilde{\mathrm{Lag}}(M) \to \mathrm{Lag}(M)$ is a covering space we have a lift

$$\tilde{\ell}^+(t) : [0,1] \to \widetilde{\mathrm{Lag}}(M)$$

of ℓ^+. Since $\widetilde{\mathrm{Lag}}/\mathbb{Z} = \mathrm{Lag}$ there exists $\overline{\eta}(\ell) \in \mathbb{Z}$ such that

$$\overline{\eta}(\ell) \cdot \tilde{\ell}^+(0) = \tilde{\ell}^+(1).$$

It is easy to see that $\overline{\eta}$ defines a homomorphism

$$\overline{\eta} : \pi_1(L) \to \mathbb{Z}.$$

Now it is easy to show the following two lemmata.

Lemma 2.13. *The composition of $\pi_2(M,L) \to \pi_1(L)$ with $\overline{\eta}$ is η.*

Lemma 2.14. *There exists a lift $\tilde{s}$ of $s : L \to \mathrm{Lag}(M)$ if and only if $\overline{\eta} : \pi_1(L) \to \mathbb{Z}$ is 0.*

We remark that $\overline{\eta}$ depends on the choice of $\widetilde{\mathrm{Lag}}(M)$, while η does not depend on it. Note the set of all choices of $\widetilde{\mathrm{Lag}}(M)$ is an affine space over $H^1(M;\mathbb{Z}) = \mathrm{Hom}(\pi_1(M),\mathbb{Z})$. In view of the exact sequence

$$\pi_2(M;L) \to \pi_1(L) \to \pi_1(M),$$

the group $\mathrm{Hom}(\pi_1(M),\mathbb{Z})$ controls the way to extend η (which is defined on the image of $\pi_2(M;L) \to \pi_1(L)$) to $\overline{\eta}$ which is defined on $\pi_1(L)$.

Now let $(L_1,\tilde{s}_1)$ and $(L_2,\tilde{s}_2)$ be graded Lagrangian submanifolds which intersect transversally each other. Let $p \in L_1 \cap L_2$. We are going to define an index $\eta_{L_1,L_2}(p)$.

We first consider a pair of family of Lagrangian submanifolds $\ell_0(\tau)$, $\ell_1(\tau) \in \mathrm{Lag}_n$, $\tau \in \mathbb{R}$ so that it is constant for $|\tau|$ large. We assume that

$$\ell_0(-\infty) \cap \ell_1(-\infty) = \ell_0(\infty) \cap \ell_1(\infty) = \{0\}. \tag{2.15}$$

We want to associate an integer, Maslov-Viterbo index $\eta(\ell_0, \ell_1)$, for such a pair. We need its relation to the index of Cauchy-Riemann operator also. Let us now describe it. We consider the product $\mathbb{R} \times [0,1] \subset \mathbb{C}$ and identify τ to the first coordinate. We consider the operator

$$\overline{\partial} : W^{1,p}(\mathbb{R} \times [0,1]; \mathbb{C}^n) \to W^{0,p}(\mathbb{R} \times [0,1]; \mathbb{C}^n \otimes \Lambda^{0,1}),$$

where $W^{k,p}$ denotes the set of sections of Sobolev (k,p) class, that is the linear space of sections whose derivatives up to order k are of L^p class. (We take $p > 2$. So $W^{1,p} \subseteq C^0$.) For an element $u \in W^{k+1,2}(\mathbb{R} \times [0,1]; \mathbb{C}^n)$, we put the boundary condition

$$u(\tau, 0) \in \ell_0(\tau), \quad u(\tau, 1) \in \ell_1(\tau). \tag{2.16}$$

We denote by $W^{1,p}(\mathbb{R}\times[0,1]; \mathbb{C}^n; \ell_0, \ell_1)$ the subset of $W^{1,p}(\mathbb{R}\times[0,1]; \mathbb{C}^n)$ satisfying the boundary condition (2.16). Then the $\overline{\partial}$ operator induces an operator:

$$\overline{\partial} : W^{1,p}(\mathbb{R} \times [0,1]; \mathbb{C}^n; \ell_0, \ell_1) \to W^{0,p}(\mathbb{R} \times [0,1]; \mathbb{C}^n \otimes \Lambda^{0,1}). \tag{2.17}$$

Lemma 2.18. *(2.17) is a Fredholm operator.*

The lemma is a consequence of (2.15). The proof is omitted.

We can calculate the index of (2.17) in the following way. We first take $g(\tau)$ such that $g(\tau)\ell_0(\tau) = \ell_0(-\infty)$. Then we consider the pair $\ell'_0(\tau) \equiv \ell_0(-\infty)$, $\ell'_1(\tau) \equiv g(\tau)\ell_1(\tau)$. It is easy to see that the index of (2.17) does not change if we replace $\ell_0(\tau)$, $\ell_1(\tau)$ by $\ell'_0(\tau), \ell'_1(\tau)$. Hence we may assume that $\ell_0(\tau)$ is independent of τ. We put $V = \ell_0(\tau)$.

For $V \in \mathrm{Lag}_n$, let $X(V)$ be the set of all $V' \in \mathrm{Lag}_n$ such that V' is not transversal to V. $X(V)$ is a (real) codimension one subcomplex such that it is a smooth submanifold outside a set of (real) codimension > 1 in $X(V)$.

Proposition 2.19. *There exists an orientation of the regular part of $X(V)$ such that the index of (2.17) is equal to the intersection number $\ell_1 \cdot X(V)$ when $V \equiv \ell_0(\tau)$.*

Sketch of a proof. We explain an idea of the proof of Proposition 2.19 by using the notion of spectral flow ([APS]). Let us consider the operator (2.17). It is an operator

$$\overline{\partial} = \frac{\partial}{\partial \tau} + J_{t,\tau} \circ \frac{\partial}{\partial t}. \tag{2.20}$$

$J_{t,\tau} \circ (\partial/\partial t)$ is a family of elliptic operators on $[0,1]$ and is parametrized by τ. Also the boundary condition (2.16) is τ dependent. We choose an appropriate bundle automorphism of φ^*TM on $\mathbb{R} \times [0,1]$ and use it to regard $J_{t,\tau} \circ (\partial/\partial t)$ as a family of elliptic operators whose boundary condition is τ *independent.* We thus obtain an operator

$$\frac{\partial}{\partial \tau} + P_\tau \tag{2.21}$$

where

$$P_\tau : W^{1,p}([0,1];\mathbb{C}^n;\Xi) \to W^{0,p}([0,1];\mathbb{C}^n).$$

Here Ξ is a boundary condition independent of τ. Then, as in [APS], the index of (2.21) is calculated by the spectral flow of the family of elliptic operators P_τ. We remark that the operator P_τ will have a kernel in a neighborhood of the points τ where $\ell_1(\tau)$ are not transversal to ℓ_0.

Thus the index of the spectral flow coincides with the intersection number $\ell_1 \cdot X(V)$. □

For our purpose, we need to relax the condition (2.15) and consider the case

$$\ell_0(-\infty) = \ell_1(-\infty), \quad \ell_0(\infty) \cap \ell_1(\infty) = \{0\}. \tag{2.22}$$

(We still assume $\ell_0(\tau) \equiv \ell_0(0)$.) We perturb ℓ so that it is transversal to $X(\ell(0))$ on $(0,1]$. Then we will put

$$\eta(\ell_0,\ell_1) \text{ “ = ” } \ell_1|_{(0,1]} \cdot X(\ell_0(0)) \tag{2.23}$$

where $\cdot$ is the intersection number. However, since $\ell_0(0) \in X(\ell_0(0))$, we need to be very careful to define the intersection pairing in (2.23). Actually $\ell_0(0) \in X(\ell_0(0))$ is a singular point of $X(\ell_0(0))$ where n components of codimension one meet each other. To define right hand side of (2.23) rigorously, we modify ℓ_1 in a neighborhood of 0 as follows. Let $U(\ell_0(0))$ be a small neighborhood of $\ell_0(0)$ in Lag_n. We will define components $U_i(\ell_0(0))$ such that

$$U(\ell_0(0)) - X(\ell_0(0)) = \bigcup_{i=0}^{n} U_i(\ell_0(0)), \tag{2.24}$$

as follows.

We take a symplectic isomorphism $\mathbb{C}^n \simeq T^*\ell_0(0)$. For each

$$V \in U(\ell_0(0)) - X(\ell_0(0)),$$

there exists a non degenerate quadratic function f on $\ell_0(0)$ such that V is a graph of df. We define $U_i(\ell_0(0))$ so that $V \in U_i(\ell_0(0))$ if f is a quadratic function of index i.

We may choose $U(\ell_0(0))$ so that $U_i(\ell_0(0))$ is contractible. Note that $\overline{U}_i(\ell_0(0)) \cap \overline{U}_{i+1}(\ell_0(0))$ is of codimension one in $X(\ell(0))$ and they are all of the codimension 1 components of $X(\ell(0)) \cap U(\ell_0(0))$.

Now we modify ℓ_1 so that $\ell_1(\epsilon)$ is in $U_n(\ell_0(0))$ for small ϵ. Now we define $\eta(\ell_0, \ell_1)$ by

$$\eta(\ell_0, \ell_1) = \ell_1|_{[\epsilon,1]} \cdot X(\ell_0(0)). \tag{2.25}$$

If we take another perturbation ℓ such that $\ell_1'(\epsilon)$ is in $U_k(\ell_0(0))$ for small k, then

$$\ell_1|_{[\epsilon,1]} \cdot X(\ell_0(0)) + n = \ell_1'|_{[\epsilon,1]} \cdot X(\ell_0(0)) + k. \tag{2.26}$$

Let $-\ell$ be the path such that $-\ell(t) = \ell(1-t)$ then we have

Lemma 2.27. *$\eta(\ell_0, \ell_1) + \eta(\ell_1, \ell_0) = -n$, if (2.22) holds.*

Proof. Let us perturb ℓ_1 so that $\ell_1(\epsilon)$ is in $U_n(\ell_0(0))$ for small ϵ. Then we have

$$\ell_1|_{(0,1)} \cdot X(\ell_0(0)) + \ell_0|_{(0,1)} \cdot X(\ell_1(0)) = 0.$$

Let us change the pair ℓ_1, ℓ_0 by the one parameter family of authomorphisms of $\mathbb{C}^n$ as follows: Let us denote the modified pair by ℓ_1', ℓ_0': we take it so that ℓ_1' is constant.

We then find that $\ell_0'(\epsilon) \in U_0(\ell_1'(0))$ for ϵ small. Lemma 2.27 now follows from (2.26). □

Now we are going to define an index of the intersection point of two graded Lagrangian submanifolds $(L_1, \tilde{s}_1)$, $(L_2, \tilde{s}_2)$. Let $p \in L_1 \cap L_2$. We take a pair $(\tilde{\ell}_0, \tilde{\ell}_1)$ of path in $\widetilde{\mathrm{Lag}}(T_pM)$ such that $\tilde{\ell}_0(0) = \tilde{\ell}_1(0)$, and $\tilde{\ell}_0(1) = \tilde{s}_1(p)$, $\tilde{\ell}_1(1) = \tilde{s}_2(p)$. We put $\ell_i = \pi \circ \tilde{\ell}_i$. We then put

$$\eta_{(L_1,\tilde{s}_1),(L_2,\tilde{s}_2)}(p) = \eta(\ell_0, \ell_1). \tag{2.28}$$

It is easy to see that the right hand side is independent of the path $(\tilde{\ell}_0, \tilde{\ell}_1)$.

Definition 2.29. Let

$$(L_1, \tilde{s}_1, \mathfrak{L}_1),\ (L_2, \tilde{s}_2, \mathfrak{L}_2) \in \mathfrak{Ob}_4(\mathcal{LAG}(M, \Omega, st, \widetilde{\mathrm{Lag}}(M))).$$

We assume $L_1 \ne L_2$. We define:

$$\begin{aligned}(2.30)\quad &\mathcal{LAG}^k((L_1,\tilde s_1,\mathfrak{L}_1),(L_2,\tilde s_2,\mathfrak{L}_2)) \\ &= \bigoplus_{\substack{p\in L_1\cap L_2 \\ \eta_{(L_1,\tilde s_1),(L_2,\tilde s_2)}(p)=k}} \mathrm{Hom}((\mathfrak{L}_1)_p,(\mathfrak{L}_2)_p)\otimes_{\mathbb{C}} \Lambda_{0,\mathrm{nov}}.\end{aligned}$$

We discuss the case $L_1 = L_2$ later. (We actually need one more small modification to (2.30) for the orientation problem. We will discuss it later. See (3.16).)

Lemma 2.27 implies:

$$(2.31)\qquad \eta_{(L_1,\tilde s_1),(L_2,\tilde s_2)}(p) = -n - \eta_{(L_2,\tilde s_2),(L_1,\tilde s_1)}(p)$$

Example 2.32. Let $u = df$ be an exact 1 form on N and $L_2 \subset T^*N$ be its graph. Let L_1 be a zero section. We assume that f is a Morse function. If $p \in L_1 \cap L_2$ then $p = (x, 0)$ and $df(x) = 0$. We define $\tilde s_1$, $\tilde s_2$ as in Example 2.11. We find

$$\eta_{(L_1,\tilde s_1),(L_2,\tilde s_2)}(p) = \mathrm{Index}(\nabla_x^2 f) - n.$$

Here $\nabla_x^2 f$ is the Hessian of f at x. Namely in this case the Maslov index reduces to the Morse index.

Remark 2.33. Let $(L_1, \tilde s_1)$ be a graded Lagrangian submanifold. We assume that L_1 is connected. When we change the lift $\tilde s_1$, then the degree η changes only by an overall integer. Let us state it more precisely. We recall that $\widetilde{\mathrm{Lag}}(M) \to \mathrm{Lag}(M)$ is a normal covering whose deck transformation group is $\mathbb{Z}$. Let $\tilde s_1'$ be another lift of $s_1(p) = T_p(L)$. Then there exists an integer k such that $\tilde s_1 = k \cdot \tilde s_1'$. We have:

$$(2.34.1)\qquad \eta_{(L_1,\tilde s_1),(L_2,\tilde s_2)}(p) = \eta_{(L_1,\tilde s_1'),(L_2,\tilde s_2)}(p) + k$$

and

$$(2.34.2)\qquad \eta_{(L_1,\tilde s_2),(L_2,\tilde s_1)}(p) = \eta_{(L_1,\tilde s_2),(L_2,\tilde s_1')}(p) - k$$

for every $(L_1, \tilde s_2)$ and $p \in L_1 \cap L_2$. More precisely the identification of the deck transformation group of $\widetilde{\mathrm{Lag}}(M) \to \mathrm{Lag}(M)$ and $\mathbb{Z}$ has an ambiguity $\mathrm{Aut}(\mathbb{Z}) = \{\pm 1\}$. We can fix it by requiring (2.34).

So far, we restricted ourselves to a symplectic manifold with $c^1 = 0$ and Lagrangian submanifold with Maslov index 0. This allows us to define Floer homology with $\mathbb{Z}$ grading. However it is useful for some other purposes to relax the condition c^1 = Maslov index = 0. Let us discuss the general case here, following [Se1]. Let us define $N \in \mathbb{Z}_{\geq 0}$ by

$$\mathrm{Im}(c^1 : \pi_1(M) \to \mathbb{Z}) = N\mathbb{Z}.$$

N is called the *minimal Chern number* of (M, ω). In a way similar to the proof of Lemma 2.6, we can construct a covering space

$$\pi : \widetilde{\mathrm{Lag}}^N(M) \to \mathrm{Lag}(M) \tag{2.35}$$

such that π induces $2N$ hold covering on each fiber. (We remark that two hold fiberwise covering of $\mathrm{Lag}(M)$ always exists because M is orientable.)

For any N' dividing N, we put

$$\widetilde{\mathrm{Lag}}^{N'}(M) = \widetilde{\mathrm{Lag}}^N(M)/\mathbb{Z}_{N/N'}$$

which is a $2N'$ hold fiberwise covering of $\mathrm{Lag}(M)$.

Let $L \subset M$ be a Lagrangian submanifold. We can define

$$\overline{\eta} : \pi_1(L) \to \mathbb{Z}_{2N}$$

using (2.35). Moreover the composition of $\pi_2(M; L) \to \pi_1(L)$ and $\overline{\eta}$ coincides with the mod $2N$ reduction of $\eta : \pi_2(M; L) \to \mathbb{Z}$. (Note that the Maslov index $\eta : \pi_2(M; L) \to \mathbb{Z}$ is defined for any pair of symplectic manifold M and its Lagrangian submanifold L.) Now we have:

Lemma 2.36. *A lift*

$$\tilde{s} : L \to \tilde{\mathrm{Lag}}^{N'}(M)|_L$$

of the map $s : L \to \mathrm{Lag}(M)|_L$ exists if and only if the image of $\overline{\eta} : \pi_1(L) \to \mathbb{Z}_{2N}$ is contained in $2N'\mathbb{Z}_{2N}$.

Such a lift $\tilde{s}$ is called an N' *grading* of L'. Let L_i $(i = 1, 2)$ be a pair of Lagrangian submanifolds with N' grading. We assume they are transversal to each other. Let $p \in L_1 \cap L_2$. Then we can define its index $\eta(p) \in \mathbb{Z}_{2N'}$ in a similar way to Definition 2.25. Thus we have a $\mathbb{Z}_{2N'}$ graded Floer homology.

§3. Floer homology and A_∞ category II – the operator $\mathfrak{m}_k$ – the transversal case –

We next define

$$\mathfrak{m}_k : \mathcal{LAG}[1](c_0, c_1) \otimes \cdots \otimes \mathcal{LAG}[1](c_{k-1}, c_k) \to \mathcal{LAG}[1](c_0, c_k)$$

where $c_i = (L_i, \tilde{s}_i, \mathfrak{L}_i) \in \mathfrak{Ob}_4(\mathcal{LAG}(M, \Omega, st, \widetilde{\mathrm{Lag}}(M)))$, under the additional asumption that $L_i \neq L_{i+1}$. (Here $L_{k+1} = L_0$ by convention.) It follows from our assumption that L_i is transversal to L_{i+1}. Again the idea presented here is not new and is discussed already in [Fu1], [Fu4]. However we here give a more precise argument especially on degree and orientation. In fact there were several errors on those points in [Fu1].

We use moduli spaces of pseudoholomorphic discs. Let us consider a pair $(D^2, \vec{z})$ of 2 disc D^2 with the canonical complex structure and $\vec{z} = (z_0, \dots, z_k)$, the ordered set of $k+1$ points on its boundary. We assume that $(z_0, \dots, z_k)$ respects the cyclic order. We denote by $\tilde{\mathcal{M}}_{k+1}$ the space of all such pairs. The group $PSL(2;\mathbb{R}) = \mathrm{Aut}(D^2, J)$ acts on it, and let $\mathcal{M}_{k+1}$ be the quotient space. It is well known that $\mathcal{M}_{k+1}$ is diffeomorphic to $\mathbb{R}^{k-2}$ and carries a natural orientation. We can compactify it to $\mathcal{CM}_{k+1}$ in a way similar to the Deligne-Mumford compactification of the moduli space of marked closed Riemann surfaces. (See [FOh], [FOOO] §3.)

Let $(D^2, \vec{z}) \in \tilde{\mathcal{M}}_{k+1}$. Let $\partial_i D^2$ be the part of ∂D^2 between z_{i-1} and z_i. (Here $z_{-1} = z_{k+1}$ by notation.)

Let $p_i \in L_i \cap L_{i+1}$. ($L_{k+1} = L_0$ by notation.) We fix a compatible almost complex structure J on M. Namely we assume

$$\omega(JX, JY) = \omega(X, Y), \quad \omega(X, JX) \geq 0.$$

We consider the set of all $((D^2, \vec{z}), \varphi)$ such that

(3.1.1) $(D^2, \vec{z}) \in \mathcal{M}_{k+1}$, $\varphi : D^2 \to M$,
(3.1.2) φ is pseudoholomorphic,
(3.1.3) $\varphi(\partial_i D^2) \subset L_i$,
(3.1.4) $\varphi(z_i) = p_i$.

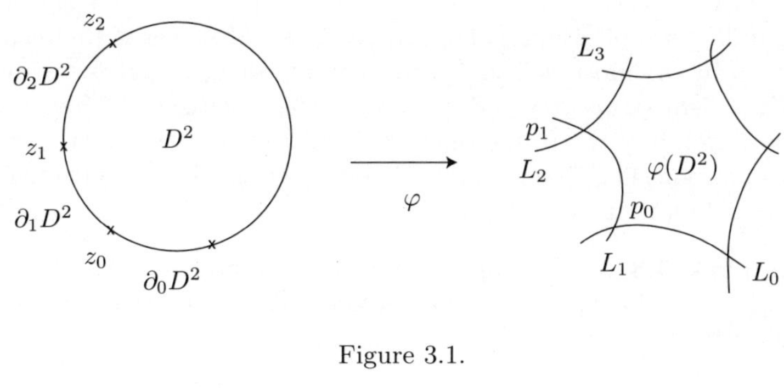

Figure 3.1.

We denote by $\tilde{\mathcal{M}}_{k+1}(L_0, \dots, L_k; p_1, \dots, p_k)$ the set of all such

$((D^2, \vec{z}), \varphi)$.

The group $PSL(2;\mathbb{R})$ acts on $\tilde{\mathcal{M}}_{k+1}(L_0, \dots, L_k; p_1, \dots, p_k)$ by

$$u \cdot ((D^2, \vec{z}), \varphi) = ((D^2, u(\vec{z})), \varphi \circ u^{-1}).$$

Let $\mathcal{M}_{k+1}(L_0, \dots, L_k; p_1, \dots, p_k)$ be the quotient space. We can compactify our moduli space $\mathcal{M}_{k+1}(L_0, \dots, L_k; p_1, \dots, p_k)$ by using a notion of stable map from open Riemann surface. (See [FOOO] §3.) Let $\mathcal{CM}_{k+1}(L_0, \dots, L_k; p_1, \dots, p_k)$ be the compactification.

Theorem 3.2. *There exists a Kuranishi structure (with corners) of dimension* $n + (k+1) - \sum \eta_{(L_i, \tilde{s}_i),(L_{i+1}, \tilde{s}_{i+1})}(p_i) - 3$ *on* $\mathcal{CM}_{k+1}(L_0, \dots, L_k; p_1, \dots, p_k)$.

The notion of Kuranishi structure is defined in [FOn2], to handle transversality problem appeared in the construction of fundamental chains of various moduli spaces in a uniform way. We do not try to define it here. Roughly speaking it is a way to restate the following imprecise statement in a rigorous way:

"Theorem 3.2′"[2]**.** *By a " generic perturbation ",* $\mathcal{CM}_{k+1}(L_0, \dots, L_k; p_1, \dots, p_k)$ *will become a manifold with corner. Its dimension is* $n + (k+1) - \sum \eta_{(L_i, \tilde{s}_i),(L_{i+1}, \tilde{s}_{i+1})}(p_i) - 3$.

The word "generic perturbation" in "Theorem3.2′" should be made precise. Kuranishi structure is a way to include as the most general perturbation as possible. Usually the reader do not have to be bothered with the detail of the study of a space with Kuranishi structure. The frame work of Kuranishi structure is designed so that desired fundamental chain (usually over $\mathbb{Q}$) which has the properties expected from naive guess can be constructed in the situation we are interested in (that is the moduli space of pseudoholomorphic discs). However if the reader is interested in the most delicate parts of the proofs, he needs to investigate Kuranishi structures. (For example the discussion of Remark 3.3 requires a detail of the frame work of Kuranishi structure.) We pretend as if a statement like "Theorem 3.2′" is correct usually and give remarks on Kuranishi structure when necessary.

Remark 3.3. In our case, the fundamental chain of $\mathcal{CM}_{k+1}(L_0, \dots, L_k; p_1, \dots, p_k)$ can be defined over integer. In fact, by Lemma 2.10, the Maslov index is 0 in our case. Moreover by Lemma

[2]We write Theorem in the quote since the statement as is stated is not correct.

2.6, $c^1(M) = 0$. It follows that our Lagrangian submanifold is semi-positive in the sense of [FOOO] Chapter 3. Therefore, we can apply the method of [FOOO] §9, §A3 to obtain a perturbation of our moduli space so that it is a simplicial complex with fundamental chain over $\mathbb{Z}$. (It is actually a space with Whitney stratification.) Using this remark, our filtered A_∞ category is defined over $\Lambda_{\mathbb{Z},0,\mathrm{nov}}$. Also we can work over $\Lambda_{\mathbb{Z}_2,0,\mathrm{nov}}$ and forget all the orientation problems (as we did in [Fu4]). However to work over $\mathbb{C}$ (or $\Lambda_{\mathbb{C},\mathrm{nov}}$) seems to be natural when our main interest is in mirror symmetry and not in applications of Floer homology to symplectic topology.

The proof of Theorem 3.2 consists of two parts. One is to use implicit function theorem and Taubes' type gluing argument to construct Kuranishi structure. This part is in fact a straightforward analogue of the argument presented in [FOn2] Chapter 3, [FOh] and [FOOO] Chapter 5. So we do not repeat it. The second part is a calculation of the (virtual) dimension, which is related to the study of Maslov index in the last section. We discuss this second point now. The argument here is also an analogy of one in [FOOO] Chapter 6.

First we remark that the number $k+1-3$ is the dimension of the moduli space $\mathcal{CM}_{k+1}$. There is a natural projection

$$\mathcal{CM}_{k+1}(L_0, \ldots, L_k; p_1, \ldots, p_k) \to \mathcal{CM}_{k+1}.$$

Hence we are only to show that the virtual dimension of the fiber of the projection is $n - \sum \eta_{(L_i,\tilde{s}_i),(L_{i+1},\tilde{s}_{i+1})}(p_i)$. (In the case of the space with Kuranishi structure, its virtual dimension is its dimension by definition. So it suffices to calculate the virtual dimension.)

Namely we fix $(D^2, \vec{z}) \in \mathcal{M}_{k+1}$ and study the moduli space of pseudoholomorphic maps φ. (Actually we need to study the case when the domain of φ is singular also. However since the study of it is similar to and is written in detail in [FOOO] Chapter 5, we omit it.)

The study of virtual dimension is a problem calculating the index. We first choose a metric on the domain and fix a function space to work with. For our purpose, it is convenient to use an alternative representative of $(D^2, \vec{z})$. Namely we take a one dimensional Kähler manifold Σ with the following properties:

(3.4.1) There exists a compact subset Σ_0 such that the complement $\Sigma - \Sigma_0$ is isometric to the disjoint union of $k+1$ copies of $(-\infty, 0] \times [0,1]$.
(3.4.2) Σ is conformaly equivalent to $D^2 - \{z_1, \ldots, z_{k+1}\}$.

By using [FOh] Theorem 10.4, such Σ with a singular Riemannian metric is given in a canonical way. Also Σ and metric in [FOh] Theorem 10.4 depends smoothly on $(D^2, \vec{z})$. This fact is essential to work out the

Taubes' type gluing construction and the analytic detail of the proof of Theorem 3.2, as in [FOOO] Chapter 5. For our purpose, that is to calculate the index, we do not need it.

Note that $k+1$ copies of $(-\infty, 0] \times [0,1]$ in (3.4.1) corresponds to the marked points $z_0, \dots, z_{k+1}$ by (3.4.2). We let $\mathrm{End}_i\,\Sigma \equiv (-\infty, 0] \times [0,1]$ be the copy corresponding to z_i.

Let $\varphi : D^2 \to M$ be a smooth map satisfying (3.1.1), (3.1.3), (3.1.4). It induces a map $\Sigma \to M$, which we denote by the same symbol. We consider the operator

$$(3.5) \qquad D_\varphi \overline{\partial} : W^{1,p}(\Sigma, \varphi^*TM; \varphi^*TL) \to W^{0,p}(\Sigma, \varphi^*TM \otimes \Lambda^{0,1}\Sigma).$$

Here (3.5) is a linearization operator of the pseudoholomorphic curve equation:

$$\overline{\partial}\varphi = 0,$$

$W^{k,p}$ in (3.5) denotes the space of all sections whose derivative up to order k is of L^p class, and

$$\begin{aligned} &W^{1,p}(\Sigma, \varphi^*TM; \varphi^*TL) \\ &= \{u \in W^{1,p}(\Sigma, \varphi^*TM) \mid u(x) \in T_{\varphi(x)}L_i \text{ if } x \in \partial_i\Sigma\} \end{aligned}$$

where $\partial_i\Sigma$ is a part of $\partial\Sigma$ which corresponds to $\partial_i D^2$. (We remark that elements of $W^{1,p}$ is continuous since we assumed $p > 2$.)

Note that ends of Σ are of product type. We also assumed that L_i is transversal to L_{i+1} at p_i. Hence the operator (3.5) is nondegenerate at the end. Therefore, by a standard argument, we can prove that (3.5) is Fredholm. We are going to calculate its index.

We first trivialize φ^*TM on Σ. The trivialization is unique up to homotoy since Σ is contractible. We next take a path $s_i : \partial_i\Sigma \to \mathrm{Lag}_n$. s_i is defied by $s_i(x) = T_{\varphi(x)}TL \subset T_{\varphi(x)}TM = (\varphi^*TM)_x \cong \mathbb{C}^n$.

We thus reduced the problem calculating the virtual dimension of the moduli space $\mathcal{CM}_{k+1}(L_0, \dots, L_k; p_1, \dots, p_k)$ to the problem calculating the index of the operator

$$(3.6) \qquad \overline{\partial} : W^{1,p}(\Sigma, \mathbb{C}^n; s_0, \dots, s_k) \to W^{0,p}(\Sigma, \mathbb{C}^n \otimes \Lambda^{0,1}\Sigma),$$

where

$$W^{1,p}(\Sigma, \mathbb{C}^n; s_0, \dots, s_k) = \{u \in W^{1,p}(\Sigma, \mathbb{C}^n) \mid u(x) \in s_i(x) \text{ if } x \in \partial_i\Sigma\}.$$

The index of (3.6) does not change if we change s_i in a homotopy class. Hence we may assume:

$$(3.7) \qquad \begin{cases} s_i(\tau, 1) = T_{p_i}L & \text{if } \tau < -T,\ (\tau, 1) \in \mathrm{End}_{i+1}\,\Sigma, \\ s_i(\tau, 0) = T_{p_i}L & \text{if } \tau < -T,\ (\tau, 0) \in \mathrm{End}_i\,\Sigma. \end{cases}$$

We next construct an elliptic complex whose index is $\eta_{(L_i,\tilde{s}_i),(L_{i+1},\tilde{s}_{i+1})}(p)$. We put:

$$Y = D^2 \cup \{x + \sqrt{-1}y \mid x \geq 0, y \in [-1,1]\} \subset \mathbb{C}.$$

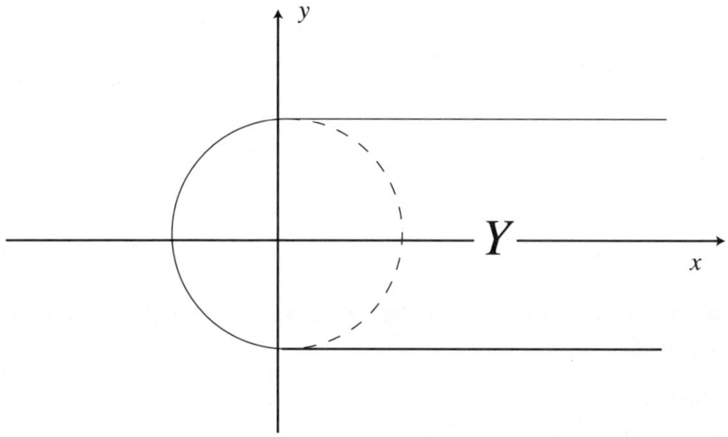

Figure 3.2.

We remark $\partial Y \cong \mathbb{R}$ where $-\infty$ corresponds to $\infty - \sqrt{-1}$ and ∞ corresponds $\infty + \sqrt{-1}$. We also define a path $\tilde{\ell}_i : \mathbb{R} \to \widetilde{\mathrm{Lag}}(T_{p_i}M)$ such that $\tilde{\ell}_i(-\infty) = \tilde{s}_i(p)$ and $\tilde{\ell}_i(\infty) = \tilde{s}_{i+1}(p)$, where $\tilde{s}_i$ is the grading of the Lagrangian submanifold of L_i we have chosen. We may also assume that $\ell_i(t)$ is locally constant if $|t| > T$. We thus obtain a Fredholm operator

$$(3.8.i) \quad \overline{\partial} : W^{1,p}(Y, T_{p_i}M; \ell_i) \to W^{0,p}(Y, T_{p_i}M \otimes \Lambda^{0,1}\Sigma).$$

Here $\ell_i = \pi \circ \tilde{\ell}_i$ and $W^{1,p}(Y, T_{p_i}M; \ell_i)$ is defined by

$$W^{1,p}(Y, T_{p_i}M; \ell_i) = \{u \in W^{1,p}(Y, T_{p_i}M) \mid u(x) \in \ell_i(x) \text{ if } x \in \partial Y \cong \mathbb{R}\}.$$

Now we have the following two Lemmata 3.9 and 3.13.

Lemma 3.9. *The index of* (3.8.i) *is* $\eta_{(L_i,\tilde{s}_i),(L_{i+1},\tilde{s}_{i+1})}(p)$.

Proof. Take a biholomorphic map $\psi : \mathbb{R} \times [0,1] \to Y - \{-1\}$ such that (∞, t) corresponds $\infty + t\sqrt{-1}$ and $(-\infty, t)$ corresponds $-1 \in \partial Y$. We pull back the boundary condition ℓ_i to $\mathbb{R} \times \{0,1\}$ and obtain $(\ell_{i,0}, \ell_{i,1})$. We then consider the operator

$$(3.10) \ \overline{\partial} : W^{1,p}(\mathbb{R} \times [0,1]; \mathbb{C}^n; \ell_{i,0}, \ell_{i,1}) \to W^{0,p}(\mathbb{R} \times [0,1]; \mathbb{C}^n \otimes \Lambda^{0,1}).$$

The operator (3.10) is similar to (2.17). However (3.10) is not a Fredholm operator since the operator is degenerate when $\tau \in \mathbb{R}$ goes to $-\infty$. In fact $\ell_{i,0}(-\infty) = \ell_{i,1}(-\infty)$. So to obtain a Fredholm operator, we need to use a weighted Sobolev norm. We consider the weighted Sobolev norm

$$\|u\|_{1,p;\delta}^p = \|e^{\delta|\tau|}u\|_p^p + \|e^{\delta|\tau|}\nabla u\|_p^p$$

where $\| \ \|_p$ is the usual L^p norm and ∇ is a covariant derivative. We define $W^{1,p;\delta}(\mathbb{R} \times [0,1]; \mathbb{C}^n; \ell_{i,0}, \ell_{i,1})$ using this weighted Sobolev norm and the same boundary condition as $W^{1,p}(\mathbb{R} \times [0,1]; \mathbb{C}^n; \ell_{i,0}, \ell_{i,1})$. We then consider

$$(3.11.\delta) \qquad \begin{aligned} \overline{\partial} : W^{1,p;\delta}&(\mathbb{R} \times [0,1]; \mathbb{C}^n; \ell_{i,0}, \ell_{i,1}) \\ &\to W^{0,p;\delta}(\mathbb{R} \times [0,1]; \mathbb{C}^n \otimes \Lambda^{0,1}). \end{aligned}$$

(3.11.δ) is a Fredholm operator for nonzero small δ. We can prove the following easily.

Sublemma 3.12. *The index of* (3.11.δ) *is* $\eta(\ell_{i,0}, \ell_{i,1}) - n$ *if* $\delta > 0$ *and if* $|\delta|$ *is small, and is* $\eta(\ell_{i,0}, \ell_{i,1})$ *if* $\delta < 0$ *and if* $|\delta|$ *is small.*

In fact, if we study (3.10) using spectral flow as in the sketch of the proof of Proposition 2.19, then we find that n is the number of eigenvalues converging to 0 as $\tau \to -\infty$. Hence the index changes by n when we move δ from negative to positive. This difference corresponds to the way to perturb $\ell_{i,1}(\tau)$ for τ close to $-\infty$. Namely if we perturb so that $\ell_{i,0}(-\infty)$ is transversal to $\ell_{i,1}(-\infty)$. (Then the operator (3.11.δ) will become Fredholm.) We can prove the sublemma using this observation.

Now we consider the case when $\delta > 0$. This means that we consider the solution of $\overline{\partial} u = 0$ with u converges to 0 in the exponential order as $\tau \to -\infty$. Then when we transform u by ψ, it will corresponds to an element of the kernel of (3.8.i) such that its value at -1 is zero. Therefore we find that the index of (3.11.δ) for $\delta > 0$ is the index of (3.8.i) minus n. Lemma 3.9 follows from Sublemma 3.12. □

Lemma 3.13. *The index of* (3.6) *plus the sum of the indices of* (3.8.i) *for* $i = 0, \ldots, k$ *is* n.

Proof. We glue the elliptic operator (3.6) with the elliptic operators (3.8.i) on their boundaries. Then we obtain an elliptic operator

$$(3.14) \qquad \overline{\partial} : W^{1,p}(D^2; \mathbb{C}^n; \ell) \to W^{0,p}(D^2; \mathbb{C}^n \otimes \Lambda^{0,1}).$$

Here notations in (3.14) are defined as follows.

$\ell : S^1 \to \mathbb{C}^n$ is a path obtained by joining $s_0, \ell_0, s_1, \ell_1, s_2, \ell_2, \dots, \ell_k, s_k$ in this order. $W^{1,p}(D^2; \mathbb{C}^n; \ell)$ is the set of $W^{1,p}$ sections u of $\mathbb{C}^n$ on D^2 such that $u(t) \in \ell(t)$ for $t \in S^1 = \partial D^2$.

Now the index of (3.14) is equal to the index of (3.6) plus the sum of the indices of (3.8.i) for $i = 0, \dots, k$ by excision property of indices.

On the other hand, the homotopy invariance of index implies that the index of (3.14) depends only on the homotopy type of ℓ. Moreover, in fact, ℓ is lifted to $\widetilde{\mathrm{Lag}}_n$ since $s_0, \ell_0, s_1, \ell_1, s_2, \ell_2, \dots, \ell_k, s_k$ are all lifted. Hence we may assume that ℓ is constant. Then it is easy to see that the index of (3.14) is n. The proof of Lemma 3.13 is complete. □

From Lemma 3.9 and Lemma 3.13 we find that the virtual dimension of the moduli space $\mathcal{CM}_{k+1}(L_0, \dots, L_k; p_0, \dots, p_k)$ is as asserted in Theorem 3.2. □

We next discuss the orientation. To define an orientation of the moduli space $\mathcal{CM}_{k+1}(L_0, \dots, L_k; p_0, \dots, p_k)$, we need one extra data, which we discuss now. We consider the complex (3.8.i). Its index as a virtual vector space depends only on the homotopy class of the path ℓ_i. Since $\ell_i = \pi \circ \tilde{\ell}_i$, and $\tilde{\ell}_i$ is a path joining $\tilde{s}_i(p)$ and $\tilde{s}_{i+1}(p)$ which is unique up to homotopy, it follows that the homotopy class of ℓ_i is defined in a canonical way for *graded* Lagrangian submanifold. Therefore the virtual vector space, which is an index of (3.8.i), is well-defined. Now we *choose* the orientation of this virtual vector space, the index of (3.8.i).

Remark 3.15. In a similar context of finite dimensional Morse theory, we need to fix an orientation of the stable (or unstable) manifold of each critical point in order to fix an orientation of the moduli space of connecting orbit. (See for example [FOn2] §21.)

The orientation of the index virtual vector space of (3.8.i) corresponds to the choice of orientation of stable manifold in the finite dimensional Morse theory.

To describe the choice of the orientation of the index of (3.8.i) in more canonical way, we proceed as follows. We modify the definition in (2.30) and put

$$\begin{aligned} &\mathcal{LAG}^k((L_1, \tilde{s}_1, \mathfrak{L}_1), (L_2, \tilde{s}_2, \mathfrak{L}_2)) \\ &= \bigoplus_{\substack{p \in L_1 \cap L_2 \\ \eta_{(L_1,\tilde{s}_1),(L_2,\tilde{s}_2)}(p) = k}} \mathrm{Hom}((\mathfrak{L}_1)_p, (\mathfrak{L}_2)_p) \\ &\qquad \otimes_{\mathbb{C}} \Lambda_{0,\mathrm{nov}} \otimes_{\mathbb{R}} \Lambda^{top}(\mathrm{Index}\,(3.8.i)). \end{aligned} \tag{3.16}$$

(3.16) is isomorphic to (2.30). The choice of isomorphism corresponds one to one to the choice of orientations of the index of (3.8.i). In case we fix a choice of the index virtual vector space of (3.8.i) for each p, we use (2.30) in place of (3.16).

Remark 3.17. In (3.16) we include the determinant of (3.8.i) in Floer's chain complex. The necessity of it becomes more apparent if we consider more general situation. Namely let us consider the case when L_1 may not be transversal to L_2 but is of clean intersection. (It means that the $L_1 \cap L_2$ is a submanifold and the $T_pL_1 \cap T_pL_2 \cong T_p(L_1 \cap L_2)$ for $p \in L_1 \cap L_2$ is of constant dimension.) In that case the right hand side of (3.16) will become

$$\begin{aligned}&\mathcal{LAG}((L_1, \tilde{s}_1, \mathfrak{L}_1), (L_2, \tilde{s}_2, \mathfrak{L}_2))\\ &\quad = \Gamma(L_1 \cap L_2; \mathrm{Hom}(\mathfrak{L}_1|_{L_1\cap L_2}, \mathfrak{L}_2|_{L_1\cap L_2})\\ &\qquad\qquad \otimes_{\mathbb{C}} \Lambda_{0,\mathrm{nov}} \otimes_{\mathbb{R}} \Lambda^{top}(\mathrm{Index}(3.8.i)).\end{aligned}$$

Here $\Lambda^{top}(\mathrm{Index}(3.8.i))$ is one dimensional real vector bundle on $L_1 \cap L_2$, which corresponds to a local system $\pi_1(L_1 \cap L_2) \to \{\pm 1\}$. See [FOOO] Chapter 6 §25.6 for detail.

Theorem 3.18. *A choice of (M, st)-relative spin structures on L_i and a choice of an orientation of the index virtual vector space of* (3.8.i) *induce an orientation of $\mathcal{CM}_{k+1}(L_0, \ldots, L_k; p_0, \ldots, p_k)$, in a canonical way.*

Here orientation of $\mathcal{CM}_{k+1}(L_0, \ldots, L_k; p_0, \ldots, p_k)$ means orientation in the sense of Kuranishi structure [FOn2].

Proof. The idea of the proof is a combination of ones in [FOOO] Chapter 6 and [FOn] §21, (the later is suggested already in Floer's paper [Fl3]). Let E_i be the index virtual vector space of (3.8.i). For each $\varphi : (D^2, \partial D^2) \to (M, L)$, let $E(\varphi)$ be the index virtual vector space of (3.6). φ induces paths $s_i : \partial_i\Sigma \to \mathrm{Lag}_n$ as explained during the proof of Theorem 3.2. Joining them with ℓ_i we obtain a loop $\ell : \Sigma \to \mathrm{Lag}_n$ as in the proof of Theorem 3.2. ℓ depends continuously on φ, so we write $\ell(\varphi)$. We thus find a family of elliptic operators

$$\overline{\partial} : W^{1,p}(D^2; \mathbb{C}^n; \ell(\varphi)) \to W^{0,p}(D^2; \mathbb{C}^n \otimes \Lambda^{0,1}) \tag{3.19}$$

parametrized by φ. The following result is proved in [FOOO] Chapter 6 Theorem 21.1 (see also D. Silva [Sil]).

Theorem 3.20. *The choice of (M, st)-relative spin structures on L_i induces an orientation of the index bundle of* (3.19).

Let $E'(\varphi)$ be the virtual vector space which is the index of (3.19). Let $E(\varphi)$ be the virtual vector space which is the index of (3.5). Let E_i

be the index of (3.8.i). Then by family of index gluing theorem (see for example [Fu8] §4) we have an isomorphism

$$E'(\varphi) \oplus \bigoplus_{i=0}^{k} E_i \cong E(\varphi). \tag{3.21}$$

Here $E'(\varphi)$, $E(\varphi)$ are virtual vector bundles and E_i are virtual vector spaces. Theorem 3.18 follows from Theorem 3.20 and (3.21). □

Remark 3.22. We may take an orientation of $\mathcal{CM}_{k+1}(L_0,\dots,L_k;p_0,\dots,p_k)$ such that (3.21) holds as an identify of oriented vector space. However in order to define an operator $\mathfrak{m}_k$ satisfying A_∞ formula, we need to change this orientation in a way depending on the $\dim E_i$ and n in a similar way to [FOOO] Chapter 6. Namely we put

$$E'(\varphi) \oplus (-1)^{\epsilon} \bigoplus_{i=0}^{k} E_i \cong E(\varphi),$$

$$\epsilon = \sum_{j=0}^{k} \sum_{\ell=1}^{j} \dim E_i.$$

(Compare the above formula to [FOOO] Remark 25.2 (1).)

The problem of degree and orientation being understood, we are ready to define $\mathfrak{m}_k$ in the case $L_i \ne L_{i+1}$.

We consider the case when the dimension

$$\begin{aligned}\dim &\mathcal{CM}_{k+1}(L_0,\dots,L_k;p_0,\dots,p_k)\\ &= n + (k+1) - \sum_{i=0}^{k} \eta_{(L_i,\tilde s_i),(L_{i+1},\tilde s_{i+1})}(p_i) - 3 = 0.\end{aligned}$$

This (together with (2.37)) implies

$$1 + \sum_{i=0}^{k-1} (\eta_{(L_i,\tilde s_i),(L_{i+1},\tilde s_{i+1})}(p_i) - 1) = \eta_{(L_0,\tilde s_0),(L_k,\tilde s_k)}(p_k) - 1. \tag{3.23}$$

Let us put $\deg p_i = \eta_{(L_i,\tilde s_i),(L_{i+1},\tilde s_{i+1})}(p_i)$ for $i = 0,\dots,k-1$, and $\deg p_k = \eta_{(L_0,\tilde s_0),(L_k,\tilde s_k)}(p_k)$. Then (3.23) implies that

$$1 + \sum (\deg p_i - 1) = \deg p_k - 1.$$

We are going to define the matrix elements

$$(3.24)\qquad \begin{aligned}&\langle \mathfrak{m}_k(p_0,\dots,p_{k-1}),p_k\rangle\\ &\in \mathrm{Hom}\left(\bigotimes_{i=0}^{k-1}\mathrm{Hom}((\mathfrak{L}_i)_{p_i},(\mathfrak{L}_{i+1})_{p_i}),\mathrm{Hom}((\mathfrak{L}_0)_{p_k},(\mathfrak{L}_k)_{p_k})\right)\otimes_{\mathbb{C}}\Lambda_{\mathrm{nov}}\end{aligned}$$

when (3.23) is satisfied.

Before defining (3.24), we explain how (3.24) defines $\mathfrak{m}_k$. We recall that an element of $\mathcal{LAG}((L_i,\tilde{s}_i,\mathfrak{L}_i),(L_{i+1},\tilde{s}_{i+1},\mathfrak{L}_{i+1}))$ is, by definition, a formal sum

$$\sum_{j=0}^{\infty} T^{\lambda_i^{(j)}}[v_i^{(j)}]$$

where

$$v_i^{(j)}\in \mathrm{Hom}((\mathfrak{L}_i)_{p_i^{(j)}},(\mathfrak{L}_{i+1})_{p_i^{(j)}}),\quad p_i^{(j)}\in L_i\cap L_{i+1}.$$

Then we define

$$\mathfrak{m}_k(v_1^{(j_1)},\dots,v_{k-1}^{(j_{k-1})}),$$

by

$$(3.25)\qquad \begin{aligned}&\mathfrak{m}_k(v_1^{(j_1)},\dots,v_{k-1}^{(j_{k-1})})\\ &=\sum_{j_1,\dots,j_k}\sum_{p_k\in L_0\cap L_k} T^{\lambda_1^{(j_1)}+\cdots+\lambda_{k-1}^{(j_{k-1})}}\\ &\qquad\qquad \langle \mathfrak{m}_k(p_0^{(j_1)},\dots,p_{k-1}^{(j_{k-1})}),p_k\rangle(v_1^{(j_1)}\otimes\cdots\otimes v_{k-1}^{(j_{k-1})}),\end{aligned}$$

using (3.24). Now we define (3.24).

Definition 3.26. We define (3.24) by

$$(3.27)\qquad \sum_{\varphi\in\mathcal{CM}_{k+1}(L_0,\dots,L_k;p_0,\dots,p_k)} T^{\mathrm{Re}\,E(\varphi)}e^{\sqrt{-1}\,\mathrm{Im}\,E(\varphi)}h_{\nabla}(\varphi(\partial D^2))\epsilon_\varphi$$

where E,h,ϵ are defined below. First we define E by

$$E(\varphi)=\int_{D^2}\varphi^*\Omega.$$

$\mathrm{Re}\,E$ and $\mathrm{Im}\,E$ are its real and imaginary parts, respectively.

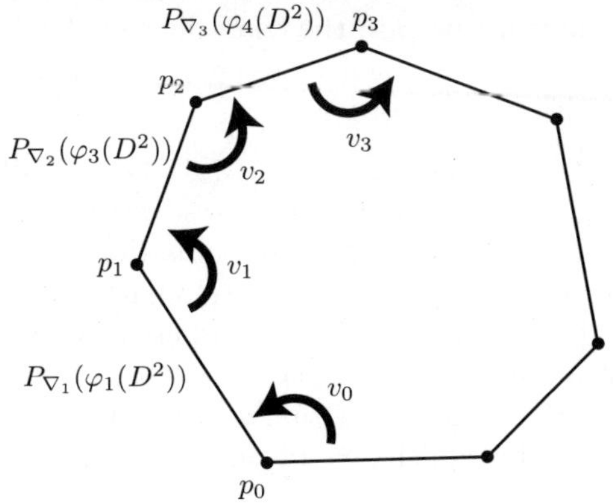

Figure 3.3.

Next we define h. Let $v_i \in \mathrm{Hom}((\mathfrak{L}_i)_{p_i}, (\mathfrak{L}_{i+1})_{p_i})$. Then

$$(3.28) \quad \begin{aligned} &h_\nabla(\varphi(\partial D^2))(v_0 \otimes \cdots \otimes v_{k-1}) \\ &= P_{\nabla_k}(\varphi(\partial_k D^2)) \circ v_{k-1} \circ \cdots \circ P_{\nabla_1}(\varphi(\partial_1 D^2)) \circ v_0. \end{aligned}$$

Here $P_{\nabla_i}(\varphi(\partial_i D^2)) : (\mathfrak{L}_i)_{p_i} \to (\mathfrak{L}_i)_{p_{i+1}}$ is the parallel transport along the path $\varphi(\partial_i D^2)$ of the bundle $\mathfrak{L}_i$ with respect to the connection ∇_i.

Finally $\epsilon_\varphi \in \{\pm 1\}$ is determined by the orientation of $\mathcal{CM}_{k+1}(L_0, \ldots, L_k; p_0, \ldots, p_k)$ defined by Theorem 3.20. (See also Remark 3.22.)

Lemma 3.29. (3.27) *is an element of* $\mathrm{Hom}((\mathfrak{L}_0)_{p_k}, (\mathfrak{L}_k)_{p_k}) \otimes \Lambda_{+,\mathrm{nov}}$.

Lemma 3.27 is a consequence of Gromov compactness theorem. (See [FOOO] Proposition 5.8 for the proof of a similar statement.) The following lemma is used in the proof of A_∞ formula.

Lemma 3.30. $T^{\mathrm{Re}\, E(\varphi)} e^{\sqrt{-1}\, \mathrm{Im}\, E(\varphi)} h_\nabla(\varphi(\partial D^2))$ *depends only on the homotopy class of* φ.

Proof. We can prove that $\mathrm{Re}\, E(\varphi)$ depends only on the homotopy class of φ by using Stokes' theorem and the fact that L_i is a Lagrangian submanifold. Homotopy independence of $e^{\sqrt{-1}\, \mathrm{Im}\, E(\varphi)} h_\nabla(\varphi(\partial D^2))$ follows from the condition that the curvature of ∇_i is $2\pi\sqrt{-1}B$ (Condition (2.1.2)). □

Remark 3.31. We remark that we are working under the hypothesis $L_i \neq L_{i+1}$ in this section. In this case we put $\mathfrak{m}_0 = 0$.

We also remark that in this case the coefficient of the operator belongs to the maximal ideal $\Lambda_{+,\mathrm{nov}}$, since the energy of nonconstant pseudo-holomorphic map is always positive.

Remark 3.32. Theorem 3.2 implies that $\mathcal{CM}_{k+1}(L_0,\dots,L_k;p_0,\dots,p_k)$ has a Kuranishi structure but it is in general not true that we can find a generic perturbation using J so that it is a manifold. So in order to make sense of (3.27) we need to take multivalued perturbation (multisection) as in [FOn2] and use it to define a fundamental chain over $\mathbb{Q}$ of our moduli space $\mathcal{CM}_{k+1}(L_0,\dots,L_k;p_0,\dots,p_k)$. In the situation of (3.27) where the moduli space is of dimension 0, it means that we need to put multiplicity $= \pm 1/\text{integer}$ in place of $\epsilon_\varphi \in \{\pm 1\}$.

However, in our situation, all of our Lagrangian submanifolds are semipositive in the sense defined in [FOOO] Chapter 3. Hence by using normally polynomial sections described in [FOOO] §A3, we can prove that $\mathcal{CM}_{k+1}(L_0,\dots,L_k;p_0,\dots,p_k)$ has a fundamental chain over $\mathbb{Z}$. In other words our operation $\mathfrak{m}_k$ is defined over $\Lambda_{\mathbb{Z},0,\mathrm{nov}}$.

The next task to be carried out might be the proof of A_∞ formula $\hat{d}\circ\hat{d}=0$. However the formula

$$\sum \pm \mathfrak{m}_*(x_0,\dots,x_{\ell-1},\mathfrak{m}_*(x_\ell,\dots,x_m),x_{m+1},\dots,x_k)=0 \tag{3.33}$$

does *not* holds if we consider only $\mathfrak{m}_*$ with $*>0$. In other words we can not prove A_∞ formula when we consider only L_i's with $L_i \ne L_{i+1}$. We need to include the case $L_i = L_{i+1}$ and $\mathfrak{m}_0$, which is nonzero in the general case. Namely, for example, we are *not* able to prove

$$\begin{aligned}0 = \;& \mathfrak{m}_1(\mathfrak{m}_3(x_0,x_1,x_2))\\ &+\mathfrak{m}_2(\mathfrak{m}_2(x_0,x_1),x_2)+(-1)^{\deg x_0+1}\mathfrak{m}_2(x_0,\mathfrak{m}_2(x_1,x_2))\\ &+\mathfrak{m}_3(\mathfrak{m}_1(x_0),x_1,x_2)+(-1)^{\deg x_0+1}\mathfrak{m}_3(x_0,\mathfrak{m}_1(x_1),x_2)\\ &+(-1)^{\deg x_0+1+\deg x_1+1}\mathfrak{m}_3(x_0,x_1,\mathfrak{m}_1(x_2)),\end{aligned} \tag{3.34}$$

which is (3.33) in case $k=2$, if we consider only $\mathfrak{m}_*$ with $*>0$. In place of (3.34), we will prove

$$\begin{aligned}0 = \;& \mathfrak{m}_1(\mathfrak{m}_3(x_0,x_1,x_2))\\ &+\mathfrak{m}_2(\mathfrak{m}_2(x_0,x_1),x_2)+(-1)^{\deg x_0+1}\mathfrak{m}_2(x_0,\mathfrak{m}_2(x_1,x_2))\\ &+\mathfrak{m}_3(\mathfrak{m}_1(x_0),x_1,x_2)+(-1)^{\deg x_0+1}\mathfrak{m}_3(x_0,\mathfrak{m}_1(x_1),x_2)\\ &+(-1)^{\deg x_0+1+\deg x_1+1}\mathfrak{m}_3(x_0,x_1,\mathfrak{m}_1(x_2))\\ &+\mathfrak{m}_4(\mathfrak{m}_0(1),x_0,x_1,x_2)+(-1)^{\deg x_0+1}\mathfrak{m}_4(x_0,\mathfrak{m}_0(1),x_1,x_2)\\ &+(-1)^{\deg x_0+1+\deg x_1+1}\mathfrak{m}_4(x_0,x_1,\mathfrak{m}_0(1),x_2)\\ &+(-1)^{\deg x_0+1+\deg x_1+1+\deg x_2+1}\mathfrak{m}_4(x_0,x_1,x_2,\mathfrak{m}_0(1)).\end{aligned} \tag{3.35}$$

Here, for example, the element $\mathfrak{m}_0(1)$ which appeared in $\mathfrak{m}_4(\mathfrak{m}_0(1), x_0, x_1, x_2)$ is an element of the $\Lambda_{0,nov}$ module $\mathcal{LAG}((L_0, \tilde{s}_0, \mathfrak{L}_0), (L_0, \tilde{s}_0, \mathfrak{L}_0))$. Thus $\mathfrak{m}_0$ and $\mathfrak{m}_4$ in (3.35) are not defined in this section. So we generalize the definition of this section and discuss the case when $L_i = L_{i+1}$ for some i, in the next section.

§4. Floer homology and A_∞ category III – the operator $\mathfrak{m}_k$ – the general case –

The discussion of this section is a combination of the argument of [FOOO] (especially its Chapter 4) and one of the last section. Namely, in the case when $c_0 = \cdots = c_{k+1} = c = (L, \tilde{s}, \mathfrak{L})$, the operator

$$\mathfrak{m}_k : \mathcal{LAG}[1](c, c) \otimes \cdots \otimes \mathcal{LAG}[1](c, c) \to \mathcal{LAG}[1](c, c)$$

$(k = 0, 1, \ldots)$ is the operator of the filtered A_∞ algebra structure constructed in [FOOO] Theorem 13.22.

We consider the following situation. Let $L_{(j)}$, $j = 1, \ldots, m$ be Lagrangian submanifolds such that $L_{(j)} \neq L_{(j+1)}$. Let $\ell_j \in \mathbb{Z}_{>0}$. We put $L_0 = \cdots = L_{\ell_1 - 1} = L_{(0)}$, $L_{\ell_1} = \cdots = L_{\ell_1 + \ell_2 - 1} = L_{(1)}$, $\ldots$, $L_{\sum_{j=0}^{m-1} \ell_j} = \cdots = L_{\sum_{j=0}^{m} \ell_j - 1} = L_{(m)}$. We take $\tilde{s}_i$, $\mathfrak{L}_i$ so that

$$c_i = (L_i, \tilde{s}_i, \mathfrak{L}_i) \in \mathfrak{Ob}_4(\mathcal{LAG}(M, \Omega, st, \widetilde{\mathrm{Lag}}(M))).$$

We write also

$$s_i^{(j)} = s_{\ell_0 + \cdots + \ell_{i-1} + j}, \quad \mathfrak{L}_i^{(j)} = \mathfrak{L}_{\ell_0 + \cdots + \ell_{i-1} + j},$$

sometimes. We are going to define:

$$\mathfrak{m}_k : \mathcal{LAG}[1](c_0, c_1) \otimes \cdots \otimes \mathcal{LAG}[1](c_{k-1}, c_k) \to \mathcal{LAG}[1](c_0, c_k).$$

(Here $k = 0, 1 \ldots$.) Note that we defined the graded $\Lambda_{0,nov}$ module $\mathcal{LAG}[1](c_i, c_{i+1})$ if $L_i \neq L_{i+1}$ in Definition 2.29. But in case $L_i = L_{i+1}$ we need to start with the definition of $\mathcal{LAG}[1](c_i, c_{i+1})$. Let us put $L = L_i = L_{i+1}$. We have $(\tilde{s}_i, \mathfrak{L}_i)$, $(\tilde{s}_{i+1}, \mathfrak{L}_{i+1})$ on L_i, L_{i+1}.

Roughly speaking the graded $\Lambda_{0,nov}$ module $\mathcal{LAG}((L, \mathfrak{L}_i), (L, \mathfrak{L}_{i+1}))$ we are going to define is a singular chain complex with local coefficient $\mathrm{Hom}(\mathfrak{L}_i, \mathfrak{L}_{i+1})$. (We remark that the curvature of $\mathfrak{L}_i$ coincides with the curvature of $\mathfrak{L}_{i+1}$ by (2.2.2). Hence $\mathrm{Hom}(\mathfrak{L}_i, \mathfrak{L}_{i+1})$ is a flat $U(1)$ bundle.) But we need to be a bit careful in defining it, so that the definition of $\mathfrak{m}_k$, which is based on the smooth correspondence, works. (See also [FOOO] §A1.)

Let $\sigma : \Delta^q \to L$ be a smooth (singular) simplex. Let $(1,0,\dots,0) \in \Delta^q$ be the base point. We take a trivialization $\sigma^*\mathfrak{L}_i \cong \Delta^q \times (\mathfrak{L}_i)_{\sigma(1,0,\dots,0)}$ by using a parallel transport along the σ image of the straight line joining $(1,0,\dots,0)$ with a given point on Δ^q. A *smooth singular simplex with local coefficient* $\mathrm{Hom}(\mathfrak{L}_i, \mathfrak{L}_{i+1})$ is a pair (σ, u) such that $\sigma : \Delta^q \to L$ and $u \in \mathrm{Hom}((\mathfrak{L}_i)_{\sigma(1,0,\dots,0)}, (\mathfrak{L}_{i+1})_{\sigma(1,0,\dots,0)})$. We now put:

$$S_q(L; \mathrm{Hom}(\mathfrak{L}_i, \mathfrak{L}_{i+1})) = \left\{ \sum a_i(\sigma_i, u_i) \Big| a_i \in \mathbb{C} \right\} \Big/ \sim_1 . \tag{4.1}$$

Here (σ_i, u_i) are smooth singular chain complexes with local coefficient, $\sum$ is a finite sum, and $\sim_1$ is defined by $a_i(\sigma_i, u_i) + a_i'(\sigma_i, u_i') \sim_1 (\sigma_i, a_i u_i + a_i' u_i')$. We can define a boundary operator ∂ on it using a trivialization $\sigma^*\mathfrak{L}_i \cong \Delta^q \times (\mathfrak{L}_i)_{\sigma(1,0,\dots,0)}$.

We need to divide it further by an appropriate equivalence relation and then take a countably generated subcomplex, in order to obtain the chain complex we use. Let us now describe this process.

Let $\mathrm{Hom}(\mathfrak{L}_i, \mathfrak{L}_{i+1})^*$ be the dual bundle of $\mathrm{Hom}(\mathfrak{L}_i, \mathfrak{L}_{i+1})$. We consider the vector bundle $\mathrm{Hom}(\mathfrak{L}_i, \mathfrak{L}_{i+1}) \otimes \Lambda^q(L)$ on L and let

$$W^{-\infty}(\mathrm{Hom}(\mathfrak{L}_i, \mathfrak{L}_{i+1}) \otimes \Lambda^q(L))$$

be the set of all distribution valued sections of it. Note an element of it can be identified with a linear map

$$C^\infty(\mathrm{Hom}(\mathfrak{L}_i, \mathfrak{L}_{i+1})^* \otimes \Lambda^{n-q}(L)) \to \mathbb{C}$$

by

$$T \mapsto \left(w \mapsto \int_L T \wedge w \right).$$

(We assume that L is oriented.) We use this identification to define a map

$$T : S_q(L; \mathrm{Hom}(\mathfrak{L}_i, \mathfrak{L}_{i+1})) \to W^{-\infty}(\mathrm{Hom}(\mathfrak{L}_i, \mathfrak{L}_{i+1}) \otimes \Lambda^{n-q}(L))$$

by putting

$$\int_L T(\sigma_i, u_i) \wedge w = \int_{\Delta^q} \langle u_i, \sigma^* w \rangle, \tag{4.2}$$

and extending it to a complex linear map in an obvious way. Note that we used the trivialization $\sigma^*\mathfrak{L}_i \cong \Delta^q \times (\mathfrak{L}_i)_{\sigma(1,0,\dots,0)}$ mentioned before to define the right hand side of (4.2).

We define an equivalence relation $S_q(L; \mathrm{Hom}(\mathfrak{L}_i, \mathfrak{L}_{i+1}))$ such that $x \sim_2 y$ if and only if $T(x) = T(y)$. We now put

$$\overline{S}_q(L; \mathrm{Hom}(\mathfrak{L}_i, \mathfrak{L}_{i+1})) = S_q(L; \mathrm{Hom}(\mathfrak{L}_i, \mathfrak{L}_{i+1}))/ \sim_2 .$$

Let P be a simplicial complex of dimension q and $f : P \to L$ be a piecewise smooth map. We assume that P has a base point. We assume that the pull back $f^* \mathrm{Hom}(\mathfrak{L}_i, \mathfrak{L}_{i+1})$ is trivial and we fix a trivialization. Let u be an element of the fiber of $f^* \mathrm{Hom}(\mathfrak{L}_i, \mathfrak{L}_{i+1})$ at the base point. Then, from f and u, we obtain an element of $\overline{S}_q(L; \mathrm{Hom}(\mathfrak{L}_i, \mathfrak{L}_{i+1}))$ in an obvious way. This element is independent of the subdivision of the simplicial complex P (as a chain). We denote it by $[P, f, u]$. Every element of $\overline{S}_q(L; \mathrm{Hom}(\mathfrak{L}_i, \mathfrak{L}_{i+1}))$ is realized in this way. We remark that, if we do not divide by the equivalence relation $\sim_2$, then the element we obtain will depend on the subdivision of the simplicial complex P.

We use cohomology rather than homology (since the product structure is the main issue here). We write

$$\overline{S}^q(L; \mathrm{Hom}(\mathfrak{L}_i, \mathfrak{L}_{i+1})) = \overline{S}_{n-q}(L; \mathrm{Hom}(\mathfrak{L}_i, \mathfrak{L}_{i+1})).$$

Next we need to take a countably generated (over $\mathbb{C}$) subcomplex of $\overline{S}^q(L; \mathrm{Hom}(\mathfrak{L}_i, \mathfrak{L}_{i+1}))$. We recall that we have fixed a countable set of Lagrangian submanifolds and we assumed that the Lagrangian submanifold part of the objects of our category is always in this set. We next choose a countably generated subcomplex of $\overline{S}^*(L)$ for each member L of the countable set of Lagrangian submanifolds we have chosen.

The condition that this subcomplex is assumed to satisfy, is rather delicate and is not mentioned here. (See [FOOO] §A1, §A5.) The reason we need to choose a countably generated subcomplex is that we need to use frequently Bair's category theorem to achieve transversality in various situations and in Bair's category theorem countability is an essential issue. The transversality here is not only a technical problem but also is related to many essential points of the story. It is related to the fact that, for example, the square of delta function is ill-defined, and hence is also related to the problem of infinity in quantum field theory.

We denote by $C^q(L; \mathrm{Hom}(\mathfrak{L}_i, \mathfrak{L}_{i+1}))$ the countably generated subcomplex we have chosen. We put

$$\mathcal{LAG}^q((L, \mathfrak{L}_i), (L, \mathfrak{L}_{i+1})) = C^q(L; \mathrm{Hom}(\mathfrak{L}_i, \mathfrak{L}_{i+1}))\hat{\otimes}_{\mathbb{C}}\Lambda_{0,\mathrm{nov}}.$$

Here $\hat{\otimes}_{\mathbb{C}}$ means that we take a completion by using topology induced by the filtration of $\Lambda_{0,\mathrm{nov}}$. (Note we assumed $L_i = L_{i+1} = L$.) In other words an element of $\mathcal{LAG}^q((L, \mathfrak{L}_i), (L, \mathfrak{L}_{i+1}))$ is realized by countable sum

$$\sum T^{\lambda_i}[P_i, f_i, u_i]$$

where $\lambda_i \to \infty$. Using the grading of our Lagrangian submanifold, we shift the degree as follows. Let $\tilde{s}_i$ and $\tilde{s}_{i+1}$ be two gradings of our Lagrangian submanifold L. Then, as mentioned in section 2, there exists a unique integer k such that $\tilde{s}_{i+1} = k \cdot \tilde{s}_i$. Here $\mathbb{Z}$ acts on $\widetilde{\mathrm{Lag}_n}$ as a deck transformation group. We put

$$k = \tilde{s}_{i+1} - \tilde{s}_i \in \mathbb{Z}. \tag{4.3}$$

Now we define:

Definition 4.4.

$$\begin{aligned} &\mathcal{LAG}^q((L, \tilde{s}_i, \mathfrak{L}_i), (L, \tilde{s}_{i+1}, \mathfrak{L}_{i+1})) \\ &= \quad C^{q+(\tilde{s}_{i+1}-\tilde{s}_i)}(L; \mathrm{Hom}(\mathfrak{L}_i, \mathfrak{L}_{i+1}))\hat{\otimes}_{\mathbb{C}}\Lambda_{0,\mathrm{nov}}. \end{aligned}$$

We thus defined $\mathcal{LAG}(c, c')$ in our general situation. We turn to the definition of our operators $\mathfrak{m}_k$.

Let $L_{(j)}$, ℓ_j, $(j = 1, \dots, m)$, $c_i = (L_i, \tilde{s}_i, \mathfrak{L}_i)$ $(i = 0, \dots, \sum_{j=0}^m \ell_j)$ be as in the beginning of this section. For i with $L_{(j)} = L_i \neq L_{i+1} = L_{(j+1)}$, we take $p_i = p_{(j)} \in L_{(j)} \cap L_{(j+1)}$ and

$$v_i \in \mathrm{Hom}((\mathfrak{L}_i)_{p_{(j)}}, (\mathfrak{L}_{i+1})_{p_{(j)}}).$$

For i with $L_{(j)} = L_i = L_{i+1}$, we take

$$x_i = [P_i, f_i, u_i] \in C^{g_i+(\tilde{s}_{i+1}-\tilde{s}_i)}(L_{(j)}; \mathrm{Hom}(\mathfrak{L}_i, \mathfrak{L}_{i+1})).$$

We are going to define

$$\begin{aligned} \mathfrak{m}_{\sum_{j=0}^{m-1}\ell_j}(x_0, \dots, \quad & x_{\ell_0-1}, v_{\ell_0}, x_{\ell_0+1}, \dots, x_{\ell_0+\ell_1-1}, v_{\ell_0+\ell_1}, \\ & x_{\ell_0+\ell_1-1}, \dots, x_{\ell_0+\dots+\ell_{m-1}-1}, \\ & v_{\ell_0+\dots+\ell_{m-1}}, x_{\ell_0+\dots+\ell_{m-1}+1}, \dots, x_{\ell_0+\dots+\ell_m-1}). \end{aligned} \tag{4.5}$$

Hereafter we write (4.5) as $\mathfrak{m}(x_0, \dots, x_{\ell_0+\dots+\ell_m-1})$, for simplicity. There are two cases:

(4.6.1) $L_{(m)} \neq L_{(0)}$.
(4.6.2) $L_{(m)} = L_{(0)}$.

Case (4.6.1): In this case we take $p_{(m)} \in L_{(m)} \cap L_{(0)}$ and are going to define the matrix element

$$\begin{aligned} &\langle \mathfrak{m}(x_0, \dots, x_{\ell_0+\dots+\ell_m-1}), p_{(m)} \rangle \\ &\qquad \in \mathrm{Hom}((\mathfrak{L}_{\ell_0+\dots+\ell_m-1})_{p_{(m)}}, (\mathfrak{L}_0)_{p_{(m)}}). \end{aligned} \tag{4.7}$$

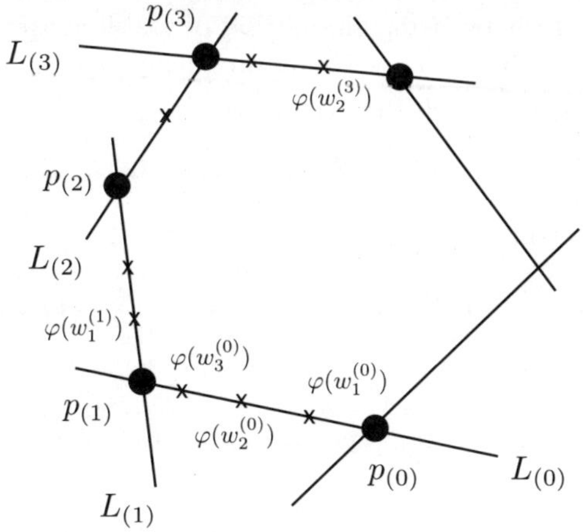

Figure 4.1.

We need to define (4.7) only in case the degree is correct. Namely in case

$$(4.8) \qquad \sum(g_i + 1) + \sum(\eta(p_{(j)}) + 1) = \eta(p_{(m)}) + 1.$$

To define (4.7) in case (4.8) is satisfied, we use a moduli space which is similar to but a bit more complicated than one we used in the last section. Let us define it now. Let us consider the system $((D^2, \vec{z}; \vec{w}^{(0)}, \ldots, \vec{w}^{(m)}), \varphi)$ such that

(4.9.1) $((D^2, \vec{z}), \varphi) \in \tilde{\mathcal{M}}_{m+1}(L_{(0)}, \ldots, L_{(m)}; p_{(1)}, \ldots, p_{(m)})$, where the right hand side is as in (3.1).

(4.9.2) $\vec{w}^{(j)} = (w_1^{(j)}, \ldots, w_{\ell_j - 1}^{(j)})$. $w_i^{(j)} \in \partial_j D^2$. Here $\partial_j D^2$ is as in the last section.

(4.9.3) If $i \neq i'$ then $w_i^{(j)} \neq w_{i'}^{(j)}$. $w_1^{(j)}, \ldots, w_{\ell_j - 1}^{(j)}$ respects the order of $\partial_j D^2$.

The totality of such $((D^2, \vec{z}; \vec{w}^{(0)}, \ldots, \vec{w}^{(m)}), \varphi)$ is denoted by

$$\tilde{\mathcal{M}}_{m+1}(L_{(0)}, \ldots, L_{(m)}; p_{(1)}, \ldots, p_{(m)}; \ell_0, \ldots, \ell_m).$$

We divide it by an obvious action of $PSL(2; \mathbb{R})$ and denote the quotient space by

$$\mathcal{M}_{m+1}(L_{(0)}, \ldots, L_{(m)}; p_{(1)}, \ldots, p_{(m)}; \ell_0, \ldots, \ell_m).$$

We can compactify it by using the notion of stable maps. (See [FOOO] §3 for its definition in the case Riemann surface has a boundary.) We denote the compactification by

$$\mathcal{CM}_{m+1}(L_{(0)},\dots,L_{(m)};p_{(1)},\dots,p_{(m)};\ell_0,\dots,\ell_m).$$

We define evaluation maps

$$\vec{ev}=(\vec{ev}^{(0)},\dots,\vec{ev}^{(m)}),\quad \vec{ev}^{(j)}=(ev^{(j)}_1,\dots,ev^{(j)}_{\ell_j-1}),$$

such that

$$\vec{ev}^{(j)}:\mathcal{CM}_{m+1}(L_{(0)},\dots,L_{(m)};p_{(1)},\dots,p_{(m)};\ell_0,\dots,\ell_m)\to L_{(j)}^{\ell_j-1}$$

is

$$\vec{ev}^{(j)}_i[(D^2,\vec{z};\vec{w}^{(0)},\dots,\vec{w}^{(m)}),\varphi]=\varphi(w^{(j)}_i).$$

We consider the fiber product

$$\mathcal{CM}_{m+1}(L_{(0)},\dots,L_{(m)};p_{(1)},\dots,p_{(m)};\ell_0,\dots,\ell_m) \tag{4.10}$$
$$\qquad\qquad {}_{\vec{ev}}\times_{f_*}\prod_j\prod_{i=1}^{\ell_j-1}P_{\ell_0+\dots+\ell_{j-1}+i}.$$

Now we have:

Lemma 4.11. *If* (4.8) *is satisfied then* (4.10) *has a Kuranishi structure of dimension* 0. *The choice of relative spin structure on* $L_{(j)}$ *and the choices of the orientations of the index virtual bundle of* (3.8.i), *determine an orientation of* (4.10) *in a canonical way.*

Proof. The proof of Lemma 4.11 is a straight forward combination of the argument of [FOOO] Chapters 5, 6 and one in the last section. We remark that to prove Lemma 4.11 we need to restrict the chains P_i so that only countably many of them are studied, since we need to use Bair's category theorem to prove Lemma 4.11.

We need to choose the orientation by modifying the fiber product orientation in a way combining [FOOO] Chapter 6 and (3.21) as follows:

We remark that each P_i is regarded as a cochain rather than a chain. This means that we coorient it. Namely for each $x\in P_i$ such that $f_i:P_i\to L$ (f_i is the map defining P_i as a differential form valued distribution) is an immersion at x, we have an orientation of the normal bundle $N_{f_i(x)}f_i(P_i)$.

For each marked point z_i we define E_i as follows. If $L_i\neq L_{i+1}$ then we define E_i as in (3.21) namely it is the index of (3.8.i). In case $L_i\neq$

L_{i+1}, then z_i corresponds to some $P_{i'}$. Then we put $E_i = N_{f_j(x)} f_j(P_j)$ here $f_j(x) = \varphi(z_i)$. Then (3.21) holds. So we define an orientation by (3.22). □

Now to define the matrix element (4.7), we need to define a weight. We put

$$E((D^2, \vec{z}; \vec{w}^{(0)}, \dots, \vec{w}^{(m)}), \varphi) = \int_{D^2} \varphi^* \omega.$$

Then the "absolute value" of the weight we put is $T^{E((D^2,\vec{z};\vec{w}^{(0)},\dots,\vec{w}^{(m)}),\varphi)}$. We next are going to define the "phase factor"

$$\begin{aligned} &H(((D^2, \vec{z}; \vec{w}^{(0)}, \dots, \vec{w}^{(m)}), \varphi); \vec{u}, \vec{v}) \\ &\in \ \mathrm{Hom}((\mathfrak{L}_{\ell_0+\dots+\ell_m-1})_{p_{(m)}}, (\mathfrak{L}_0)_{p_{(m)}}). \end{aligned}$$

Here $\vec{u}$ denotes the totality of $u_i \in \mathrm{Hom}((\mathfrak{L}_i)_{q_i}, (\mathfrak{L}_{i+1})_{q_i})$ here q_i is the image by f_i of the base point of P_i, and $\vec{v}$ denotes the totality of $v_i \in \mathrm{Hom}((\mathfrak{L}_i)_{p_{(j)}}, (\mathfrak{L}_{i+1})_{p_{(j)}})$. We put

$$\alpha_i = \begin{cases} u_i & \text{if } L_i = L_{i+1}, \\ v_i & \text{if } L_i \neq L_{i+1}. \end{cases}$$

Now we put

$$\begin{aligned} &P_\nabla(\varphi(\partial D^2)) \\ &= \ P_{\nabla_{\sum_{j=0}^m \ell_j}}(\varphi(\partial_{\sum_{j=0}^m \ell_j} D^2)) \circ \alpha_{\sum_{j=0}^m \ell_j - 1} \circ \dots \circ \alpha_1 \circ P_{\nabla_1}(\varphi(\partial_1 D^2)). \end{aligned}$$

We remark here that we use the trivialization of $f_i^* \mathrm{Hom}(\mathfrak{L}_i, \mathfrak{L}_{i+1})$ to regard u_i as an element of $\mathrm{Hom}((\mathfrak{L}_i)_{w_{i'}^{(j)}}, (\mathfrak{L}_{i+1})_{w_{i'}^{(j)}})$ where $\ell_0 + \dots + \ell_{j-1} + i' = i$. (A priori, u_i is a homomorphism between the fibers at $q_i = f_i(1, 0, \dots, 0)$.)

Now we define

$$\begin{aligned} &H(((D^2, \vec{z}; \vec{w}^{(0)}, \dots, \vec{w}^{(m)}), \varphi); \vec{u}, \vec{v}) \\ &= \ \exp\left(2\pi\sqrt{-1} \int_{D^2} \varphi^* B\right) P_\nabla(\varphi(\partial D^2)). \end{aligned}$$

Definition 4.12. We assume that (4.6.1) and (4.8) are satisfied. Then we define the matrix element (4.7) by

$$\sum_{[(D^2,\vec{z};\vec{w}^{(0)},\dots,\vec{w}^{(m)}),\varphi] \in (4.10)} \epsilon_\varphi T^{E((D^2,\vec{z};\vec{w}^{(0)},\dots,\vec{w}^{(m)}),\varphi)} H(((D^2, \vec{z}; \vec{w}^{(0)}, \dots, \vec{w}^{(m)}), \varphi); \vec{u}, \vec{v}). \tag{4.13}$$

Here $\epsilon_\varphi = \pm 1$ is determined by the orientation of the moduli space (4.10).

It is a consequence of Gromov compactness that (4.13) is an element of

$$\mathrm{Hom}((\mathfrak{L}_{\ell_0+\cdots+\ell_m-1})_{p_{(m)}}, (\mathfrak{L}_0)_{p_{(m)}})\hat{\otimes}\Lambda_{\mathrm{nov}}.$$

We thus defined the operator $\mathfrak{m}_k$ in Case (4.6.1).

Case (4.6.2): This case is similar to the case of (4.6.1) and we proceed as follows.

Let us define a moduli space

$$\mathcal{M}_{m+1}(L_{(0)},\dots,L_{(m)};p_{(1)},\dots,p_{(m-1)};\ell_0,\dots,\ell_m)$$

as follows. We have $p_{(j)} \in L_{(j)} \cap L_{(j+1)}$ for $j = 0,\dots,m-1$. We consider the system $((D^2,\vec{z};\vec{w}^{(0)},\dots,\vec{w}^{(m)},u),\varphi)$ such that

(4.14.1) Put $\vec{z} = (z_0,\dots,z_m)$, $\vec{z}_- = (z_0,\dots,z_{m-1})$. Then $((D^2,(z_-)),\varphi) \in \tilde{\mathcal{M}}_m(L_{(0)},\dots,L_{(m)};p_{(1)},\dots,p_{(m-1)})$. Where the right hand side is as in (3.1).

(4.14.2) $\vec{w}^{(j)} = (w_1^{(j)},\dots,w_{\ell_j-1}^{(j)})$. $w_i^{(j)} \in \partial_j D^2$. Here $\partial_j D^2$ is as in the last section.

(4.14.3) If $i \neq i'$ then $w_i^{(j)} \neq w_{i'}^{(j)}$. $w_1^{(j)},\dots,w_{\ell_j-1}^{(j)}$ respects the order of $\partial_j D^2$.

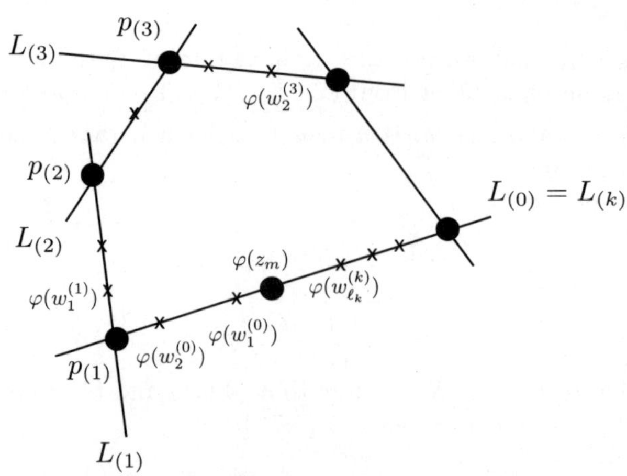

Figure 4.2.

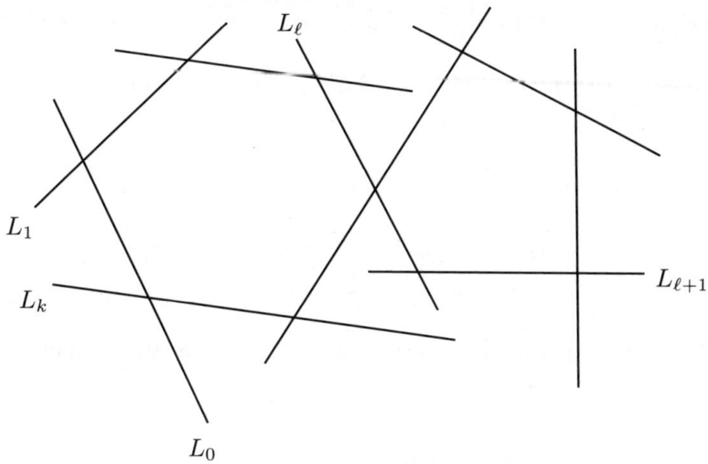

Figure 4.3.

We split this moduli space according to the homotopy type of the map φ. We denote a homotopy class by β and let

$$\mathcal{M}_{m+1}(\beta; L_{(0)}, \dots, L_{(m)}; p_{(1)}, \dots, p_{(m-1)}; \ell_0, \dots, \ell_m)$$

be the corresponding component of the moduli space. We define an evaluation map

$$\vec{ev}^{(j)} : \mathcal{CM}_{m+1}(\beta; L_{(0)}, \dots, L_{(m)}; p_{(1)}, \dots, p_{(m-1)}; \ell_0, \dots, \ell_m) \to L_{(j)}^{\ell_j - 1},$$

in a way similar to the case (4.6.1). Using it we define the fiber product.

$$\begin{aligned}
&\mathfrak{M}_\beta(p_{(1)}, \dots, p_{(m-1)}; P_1, \dots, P_{\sum \ell_j}) \\
&= \mathcal{CM}_{m+1}(\beta; L_{(0)}, \dots, L_{(m)}; p_{(1)}, \dots, p_{(m-1)}; \ell_0, \dots, \ell_m) \\
&\qquad {}_{\vec{ev}} \times_{f_*} \prod_j \prod_{i=1}^{\ell_j - 1} P_{\ell_0 + \cdots + \ell_{j-1} + i}.
\end{aligned} \tag{4.15}$$

We also define another evaluation map at the remaining mark point z_m. Namely we define

$$\begin{aligned}
ev : \mathcal{CM}_{m+1}(\beta; L_{(0)}, \dots, L_{(m)}; p_{(1)}, \dots, p_{(m-1)}; \ell_0, \dots, \ell_m) \to L_{(m)} \\
= L_{(0)}
\end{aligned}$$

by

$$ev((D^2, \vec{z}; \vec{w}^{(0)}, \dots, \vec{w}^{(m)}, u), \varphi) = \varphi(z_m).$$

It induces

$$ev : \mathfrak{M}_\beta(p_{(1)}, \dots, p_{(m-1)}; P_1, \dots, P_{\sum \ell_j}) \to L_{(0)}.$$

Now we put

$$\begin{aligned}\sum_\beta T^{E((D^2,\vec z;\vec w^{(0)},\dots,\vec w^{(m)}),\varphi_\beta)} \\ H(((D^2, \vec z; \vec w^{(0)}, \dots, \vec w^{(m)}), \varphi_\beta); \vec u, \vec v) \\ ev_*[\mathfrak{M}_\beta(p_{(1)}, \dots, p_{(m-1)}; P_1, \dots, P_{\sum \ell_j})],\end{aligned} \tag{4.16}$$

where φ_β is a map with homotopy class β. (4.16) is an element of

$$C^{g+(\tilde s_k - \tilde s_0)}(L; \mathrm{Hom}(\mathfrak{L}_{(0)}, \mathfrak{L}_{(m)})).$$

$(g = \ell_0 + \cdots + \ell_m - 1.)$ We define

$$\mathfrak{m}(x_0, \dots, x_{\ell_0+\cdots+\ell_m-1}) = (4.16).$$

Now we have:

Theorem 4.17. *The operation $\mathfrak{m}_k$ defined above satisfies A_∞ relations.*

Since the detail of the definition of module of morphisms and operations are already discussed, the proof of Theorem 4.17 is in fact a straight forward generalization of the argument of [FOOO] and one of [Fu1], [Fu4]. So we do not repeat it. The main idea is that the degeneration of elements of $\mathcal{M}_{k+1}$ can be described as in the Figure 4.3 and they correspond to the terms in the A_∞ formula.

§5. Unit and Homotopy unit

We have thus constructed operations $\mathfrak{m}_k$ which satisfy the A_∞ relations. To complete the construction of our filtered A_∞ category $\mathcal{LAG}(M, \omega)$ we need to construct a unit. There is a delicate problem related to transversality to construct a unit of our A_∞ category $\mathcal{LAG}(M, \omega)$. This problem is solved in [FOOO] §20. We discuss an outline of it here (together with its slight generalization. Namely we generalize from the case A_∞ algebra discussed in [FOOO] to A_∞ category).

Let $(L, \tilde s, \mathfrak{L})$ be an object of $\mathcal{LAG}(M, \omega)$. Then, by definition, the module of morphisms, $\mathcal{LAG}((L, \tilde s, \mathfrak{L}), (L, \tilde s, \mathfrak{L}))$ is a subcomplex of the de-Rham complex $W^{-\infty}(L; \mathbb{C} \otimes \Lambda^*(L))$ of distribution valued forms, since $\mathrm{Hom}(\mathfrak{L}, \mathfrak{L})$ together with its connection is trivial. Then $1 \in W^{-\infty}(L; \mathbb{C} \otimes$

$\Lambda^0(L)$) is an element of $\mathcal{LAG}^0((L,\tilde{s},\mathfrak{L}),(L,\tilde{s},\mathfrak{L}))$. Actually 1 is the element corresponding to the fundamental chain $[L]$ by our identification (4.2), the Poincaré duality. So we write $[L]$ rather than 1.

"Theorem 5.1"[3]. *$[L]$ is a unit of our filtered A_∞ category $\mathcal{LAG}(M,\omega)$.*

"*Proof*"[4]. We need to show

$$\mathfrak{m}_k(x_1,\dots,x_{i-1},[L],x_{i+1},\dots,x_k)=0 \tag{5.2}$$

for $k>2$. We also need to show

$$\mathfrak{m}_1([L])=0,\quad (-1)^{\deg x}\mathfrak{m}_2([L],x)=\mathfrak{m}_2(x,[L])=x. \tag{5.3}$$

(5.3) can be "proved" in a way similar to (5.2). So we explain the idea of the "proof" of (5.2). There are two cases to study. Namely the transversal case (which corresponds §3) and the nontransversal case (which corresponds to §4). Actually the general case is a mixture. In the case when $L_0=\cdots=L_k$ the argument is explained in detail in [FOOO] Chapter 5 (and is outlined in [FOOO] Chapter 2 section 7). So we restrict ourselves to the following special case. Suppose we have mutually transversal three Lagrangian submanifolds L_0, L_1, L_2. We put $L_3=L_2$. We also assume $\mathfrak{L}_3=\mathfrak{L}_2$. Let $p_i\in L_i\cap L_{i+1}$, $i=0,1$ and let $p_2\in L_1\cap L_2$. Let $u_i\in \mathrm{Hom}(\mathfrak{L}_i,\mathfrak{L}_{i+1})_{p_i}$, $i=0$, 1 and $u_2\in\mathrm{Hom}(\mathfrak{L}_0,\mathfrak{L}_2)_{p_2}$. We want to show that the matrix element

$$\langle\mathfrak{m}_3((p_0,u_0),(p_1,u_1),[L_2]),(p_2,u_2)\rangle \tag{5.4}$$

is zero. (This is a part of Formula (5.2) to be shown.) We recall that the matrix element (5.4) is defined as follows. We consider the set of $(\varphi,(z_0,z_1,z_2),w)$ such that the following holds:

(5.5.1) $\varphi: D^2\to M$ is a pseudoholomorphic map.

(5.5.2) z_0, z_1, w, $z_2\in\partial D^2$. And they are in this order, with respect to the usual cyclic orientation on ∂D^2.

(5.5.3) $\varphi(\overline{z_0,z_1})\in L_1$, $\varphi(\overline{z_1,z_2})\in L_2$, $\varphi(\overline{z_2,z_0})\in L_0$.

We divide the set of all such $(\varphi,(z_0,z_1,z_2),w)$ by an obvious $PSL(2;\mathbb{R})$ action and let

$$\mathcal{M}_3(L_0,L_1,L_2;p_0,p_1,p_2;0,0,1)$$

[3]We put this statement in the quote since it is not correct as it is stated. We will state a precise theorem later (Theorem 5.13).

[4]We put proof in the quote since it contains a gap. The correct proof of the theorem will be explained later. We need to go into the detail of the framework of Kuranishi structure to make it precise.

be the quotient space. (This definition is a special case of the definition in the last section.) We compactify it to $\mathcal{CM}_3(L_0, L_1, L_2; p_0, p_1, p_2; 0, 0, 1)$. We define an evaluation map

$$ev : \mathcal{CM}_3(L_0, L_1, L_2; p_0, p_1, p_2; 0, 0, 1) \to L_2$$

by

$$ev(\varphi, (z_0, z_1, z_2), w) = \varphi(w).$$

Now, by definition, the matrix element (5.4) is given by the order counted with sign of the (finite) set

$$\mathcal{CM}_3(L_0, L_1, L_2; p_0, p_1, p_2; 0, 0, 1) \times_{L_2} L_2, \tag{5.6}$$

in the case when the moduli space and the fiber product is transversal and the virtual dimension of (5.6) is zero. We need to show that this number is zero.

To "prove" it we consider the moduli space

$$\mathcal{CM}_3(L_0, L_1, L_2; p_0, p_1, p_2) = \mathcal{CM}_3(L_0, L_1, L_2; p_0, p_1, p_2; 0, 0, 0)$$

defined in section 3. This moduli space consists of isomorphism classes of $(\varphi, (z_0, z_1, z_2))$ satisfying the same conditions as (5.5). So we can define a map

$$\pi : \mathcal{CM}_3(L_0, L_1, L_2; p_0, p_1, p_2; 0, 0, 1) \to \mathcal{CM}_3(L_0, L_1, L_2; p_0, p_1, p_2) \tag{5.7}$$

by

$$\pi(\varphi, (z_0, z_1, z_2), w) = (\varphi, (z_0, z_1, z_2)).$$

It is easy to see that the fiber of the map (5.7) is one dimensional and is parametrized by the position of $w \in \overline{p_2 p_0}$.

It follows that

$$\begin{aligned} &\operatorname{Virdim} \mathcal{CM}_3(L_0, L_1, L_2; p_0, p_1, p_2; 0, 0, 1) \\ &= \quad \operatorname{Virdim} \mathcal{CM}_3(L_0, L_1, L_2; p_0, p_1, p_2) + 1. \end{aligned}$$

One the other hand, by the definition of fiber product and virtual dimension, we have, in general

$$\operatorname{Virdim} A \times_B C = \operatorname{Virdim} A + \operatorname{Virdim} C - \operatorname{Virdim} B.$$

Hence

$$\begin{aligned} &\operatorname{Virdim} \mathcal{CM}_3(L_0, L_1, L_2; p_0, p_1, p_2; 0, 0, 1) \times_{L_2} L_2 \\ &= \operatorname{Virdim} \mathcal{CM}_3(L_0, L_1, L_2; p_0, p_1, p_2; 0, 0, 1). \end{aligned}$$

We are studying the case when the virtual dimension of (5.6) is zero. Therefore we have

$$\text{(5.8)} \qquad \operatorname{Virdim} \mathcal{CM}_3(L_0, L_1, L_2; p_0, p_1, p_2) = -1.$$

(5.8) implies that (if everything is transversal then)

$$\mathcal{CM}_3(L_0, L_1, L_2; p_0, p_1, p_2)$$

is empty. Hence $\mathcal{CM}_3(L_0, L_1, L_2; p_0, p_1, p_2; 0, 0, 1)$ is also empty.

Thus we are done ? □

Actually there is a gap in the above argument. (5.8) implies that the space $\mathcal{CM}_3(L_0, L_1, L_2; p_0, p_1, p_2)$ is empty, only in case it is transversal. Using the theory of Kuranishi structure and multivalued perturbation, we can always make it transversal. However the trouble is whether we can make this perturbation to be compatible with the map π in (5.7). Namely if we take a perturbation compatible with (5.7), then it is not consistent with other maps we mention below.

The boundary of moduli spaces

$$\mathcal{CM}_m(L_0, \ldots, L_m; p_{(0)}, \ldots, p_{(m)}; \ell_1, \ldots, \ell_m)$$

can be described by products of similar moduli spaces (but with smaller m or ℓ_i). This consistency is essential to show A_∞ formula, Theorem 4.17. It is possible to make the perturbations consistent with this identification of the boundary to other moduli spaces. However, we cannot find, in general, a perturbation which is compatible to both. Namely in general there is no perturbation which is compatible with π and the identification of the boundary of the moduli space with the product of the other moduli spaces.

This problem looks rather technical. But it is quite delicate and essential point. It is explained in detail in [FOOO] Chapter 5 and is solved there. In [FOOO], the case we have only one Lagrangian submanifold is discussed. But the general case is completely parallel. So we only give a statement here.

To state the precise version of "Theorem 5.1", we need to introduce a notion of homotopy unit. To explain it, let us start with a geometric model of this notion. We first review the following well know notion in topology.

Definition 5.9. An *H-space* is a space X with a map $p : X \times X \to X$ and a base point $e \in X$ such that $p(x, e) = p(e, x) = x$.

Homotopy H space is (X, p, e) such that the restrictions of p to $\{e\} \times X \cong X$ and to $X \times \{e\} \cong X$ are homotopic to identity.

Lemma 5.10. *Homotopy H space is homotopy equivalent to an H space.*

Proof. Let us consider $X \cup [0,1]$ and identify 0 with e. We get a space X_+ which is obviously homotopy equivalent to X. By assumption there exists a homotopies

$$H_l : X \times [0,1] \to X, \quad H_r : X \times [0,1] \to X$$

such that

$$H_l(x,0) = p(e,x), \quad H_r(x,0) = p(x,e), \quad H_l(x,1) = H_r(x,1) = x.$$

We extend p to $X_+ \times X_+$ by putting

$$p(t,x) = H_l(x,t), \quad p(x,t) = H_r(x,t).$$

Then $(X_+, p, 1)$ is an H space. □

Now we translate the definition of X_+ in the proof above into the algebraic language and define homotopy unit as follows. Let $\mathcal{C}$ be an A_∞ category without unit. Let

$$\mathbf{e}_c \in \mathcal{C}^0(c,c)$$

be elements. We define

$$\mathcal{C}(c,c)_+ = \mathcal{C}(c,c) \oplus R \cdot \mathbf{e}_{c+} \oplus R \cdot \mathbf{f}_c$$

such that $\deg \mathbf{e}_{c+} = 0$, $\deg \mathbf{f}_c = -1$. We extend $\mathfrak{m}_1$ to $\mathcal{C}(c,c)_+$ by

$$\begin{aligned} \mathfrak{m}_1(\mathbf{f}_c) &= \mathbf{e}_{c+} - \mathbf{e}_c, \\ \mathfrak{m}_1(\mathbf{e}_{c+}) &= 0. \end{aligned}$$

We put $\mathcal{C}(c,c')_+ = \mathcal{C}(c,c')$. We then define $B\mathcal{C}_+$ as in §1.

Definition 5.11. We say that $\mathbf{e}_c$ is a *homotopy unit* if we can extend $\mathfrak{m}_k$ to $B\mathcal{C}_+$ so that it will become an A_∞ category with unit $\mathbf{e}_{c+}$.

We remark that, by the definition of unit, the extension of $\mathfrak{m}_k$ to $B\mathcal{C}_+$ is automatically determined in the case $\mathbf{e}_{c+}$ is included in the formula. So to extend $\mathfrak{m}_k$ we only need to define

$$\begin{aligned} \mathfrak{m}_k(&x_{1,1},\dots,x_{1,\ell_1},\mathbf{f}_c,x_{2,1},\dots,x_{2,\ell_2}, \\ &,\mathbf{f}_c,x_{3,1},\dots,x_{m-1,\ell_{m-1}},\mathbf{f}_c,x_{m,1},\dots,x_{m,\ell_m}), \end{aligned} \tag{5.12}$$

where $k = \sum_{i=1}^m (\ell_i + 1) - 1$. We can write down the equation they are supposed to satisfy by rewriting A_∞ formula. See [FOOO] §20 for such a formula.

We can define a notion of homotopy unit of filtered A_∞ category in the same way. Now we have:

Theorem 5.13. *$[L]$ is a homotopy unit of $\mathcal{LAG}(M,\omega)$.*

The proof is a straight forward generalization of the argument in [FOOO] §20. So we discuss it only very briefly. As we mentioned above, the reason that $[L]$ fails to be a unit (or in other words the reason (5.2) can be nonzero) is that the perturbation is not compatible with maps (5.7) (and its analogues). Let us restrict ourselves to the case of (5.4).

On the other hand, we can find another perturbation so that it is compatible with π but is not compatible with other operations. We now choose these two perturbations and take a homotopy between them. We now have a moduli space using this one parameter family of perturbations. Taking its fundamental chain (or counting its order with sign) we obtain a matrix element

$$\left\langle \mathfrak{m}_3((p_0,u_0),(p_1,u_1),\mathbf{f}_{[L_2]}),(p_2,u_2)\right\rangle .$$

The other operations in (5.12) can be defined in a similar way.

Using Theorem 5.13 we can modify our filtered A_∞ category $\mathcal{LAG}(M,\omega)$, (which has only a homotopy unit) so that it has an (exact) unit. From now one, we write $\mathcal{LAG}(M,\omega)$ for this modified one.

Chapter 2: Homological algebra of A_∞ category

§6. Twisted complex and derived A_∞ category

We have constructed our main example in §2 $\sim$ §5. So we go back to the continuation of Section 1 and further study algebraic formalism. In this section we follow Bondal-Kapranov [BoK] and Kontsevich [Ko1] (see also [Fu7] §16), to define twisted complex and derived A_∞ category.

In the case of abelian category, its derived category $\mathcal{C}$ is constructed in the following way. First, we consider the category $\mathcal{CC}$ of chain complex of objects of $\mathcal{C}$. We then consider the weak equivalence between objects of $\mathcal{CC}$. (Weak equivalence is a chain map which induces an isomorphism in cohomology.) We divide our category $\mathcal{CC}$ by weak equivalence and obtain $\mathbb{D}\mathcal{C}$. The category $\mathbb{D}\mathcal{C}$ is not an abelian category, but has an operation which is an algebraic version of the construction of the mapping cone (in topology). This construction of mapping cone gives a notion of distinguished triple on $\mathbb{D}\mathcal{C}$. Then $\mathbb{D}\mathcal{C}$ will become a triangulated category. (See [GM], [KaS], [Ha] etc. for detail.)

To generalize this construction to A_∞ category, we need to generalize the notion of chain complex (of elements of $\mathcal{C}$) to twisted complex.

Remark 6.1. The twisted complex is defined by Bondal-Kapranov [BoK] in the case when $\mathfrak{m}_k = 0$ for $k \geq 3$. Kontsevitch [Ko1] mentioned its generalization to A_∞ category and suggested its application to mirror symmetry. The twisted complex was also applied in [Fu7] to mirror symmetry. (The author learned some part of the contents of this section from P. Seidel's talks and papers [Se2], [Se3].)

Let $\mathcal{C}$ be an A_∞ category. We first increase the objects of it a bit in the following way. Let c be an object of $\mathcal{C}$ and k be an integer. We add an object $c[k]$ and put

$$\mathcal{C}(c'[\ell], c[k]) = \mathcal{C}(c', c)[\ell - k]. \tag{6.2}$$

We add all of $c[k]$ and define $\mathfrak{m}_k$ as follows. Let

$$x_i \in \mathcal{C}^{\deg x_i - k_{i-1} + k_i}(c_{i-1}, c_i). \cong \mathcal{C}^{\deg x_i}(c_{i-1}[k_{i-1}], c_i[k_i]).$$

We write $s^* x_i$ when we regard it as an element of $\mathcal{C}^{\deg x_i}(c_{i-1}[k_{i-1}], c_i[k_i])$, and write x_i when we regard it as an element of $\mathcal{C}^{\deg x_i + k_i - k_{i-1}}(c_{i-1}, c_i)$. Now we put

$$\mathfrak{m}_k(s^* x_1, \dots, s^* x_k) = (-1)^{k_0} s^* \mathfrak{m}_k(x_1, \dots, x_k). \tag{6.3}$$

Then we have

$$\begin{aligned}
&\sum_{1 \leq \ell < m \leq k} (-1)^{\deg s^* x_1 + \dots + \deg s^* x_{\ell-1} + \ell - 1} \\
&\qquad \mathfrak{m}_{k-m+\ell+1}(s^* x_1, \dots, \mathfrak{m}_{m-\ell+1}(s^* x_\ell, \dots, s^* x_m), \dots, s^* x_k) \\
= &\sum_{1 \leq \ell < m \leq k} (-1)^{\deg x_1 + \dots + \deg x_{\ell-1} + \ell - k_0 + k_{\ell-1} - 1} \\
&\qquad \mathfrak{m}_{k-m+\ell+1}(s^* x_1, \dots, \mathfrak{m}_{m-\ell+1}(s^* x_\ell, \dots, s^* x_m), \dots, s^* x_k) \\
= &\sum_{1 \leq \ell < m \leq k} (-1)^{\deg x_1 + \dots + \deg x_{\ell-1} + \ell - 1} \\
&\qquad s^*(\mathfrak{m}_{k-m+\ell+1}(x_1, \dots, \mathfrak{m}_{m-\ell+1}(x_\ell, \dots, x_m), \dots, x_k)) \\
= &\ 0.
\end{aligned}$$

Thus A_∞ formula holds.

We add more objects to $\mathcal{C}$ as follows. Let us consider formally the direct sum $c_1 \oplus \dots \oplus c_k$ and regard it as an object of our A_∞ category. We then define

$$\mathcal{C}(c_1 \oplus \dots \oplus c_k, c'_1 \oplus \dots \oplus c'_m) = \bigoplus_{i=1}^{k} \bigoplus_{j=1}^{m} \mathcal{C}(c_i, c'_j).$$

We define operations $\mathfrak{m}_k$ in an obvious way.

From now on, we always extend the set of objects of A_∞ category in this way. Then, for any objects c, c', we have $c[k]$ and $c \oplus c'$. The object $c \oplus c'$ is the direct sum in the sense of additive category.

Actually in our main example, the object $c[k]$ exists already since it corresponds to the same Lagrangian submanifold, $U(1)$ bundle, etc. as c but with different grading $\tilde{s}$.

On the other hand, the direct sum $c \oplus c'$ (already) exists if c and c' correspond to Lagrangian submanifolds L, L' such that $L \cap L' = \emptyset$. In this case, $c \oplus c'$ corresponds $L \cup L'$ together with line bundle, framing, relative spin structure induced ones from c and c'.

However, in case $L \cap L' \neq \emptyset$, the disjoint union $L \cup L'$ is not an embedded Lagrangian submanifold and hence is not an objects of our A_∞ category. (If we include immersed Lagrangian submanifold as an objects, then again $c \oplus c'$ will be included (geometrically).) Here we simply add $c \oplus c'$ formally.

Now we start the construction of twisted complex and derived A_∞ category. Let $k_1 < k_2$ be integers. We consider a finitely many objects $c_i \in \mathfrak{Ob}(\mathcal{C})$, $(i = k_1, \dots, k_2)$ and elements

$$x_{i,j} \in \mathcal{C}[1]^0(c_i[i], c_j[j]) \cong \mathcal{C}^{1+i-j}(c_i, c_j)$$

for each $i < j$.

Definition 6.4. We say $(c_{k_1}, \dots, c_{k_2}; (x_{i,j}))$ is a *twisted complex* if for each $i < j$, the equation:

$$\sum_{m \geq 1} \sum_{i=\ell_0 < \cdots < \ell_m = j} \mathfrak{m}_m(x_{\ell_0,\ell_1}, \dots, x_{\ell_{m-1},\ell_m}) = 0 \tag{6.5}$$

is satisfied. The set of all twisted complex is denoted by $\mathfrak{Ob}(\mathcal{DC})$.

Example 6.6. If $x_{i,j} = 0$ for $j \neq i+1$ then (6.5) is

$$\mathfrak{m}_2(x_{i,i+1}, x_{i+1,i+2}) = 0, \tag{6.7.1}$$

$$\mathfrak{m}_1(x_{i,i+1}) = 0. \tag{6.7.2}$$

(6.7.2) implies that $x_{i,i+1}$ is a cocycle. Since $\mathfrak{m}_2$ is the composition of morphisms (upto sign), (6.7.1) implies that

$$0 \to c_{k_1} \xrightarrow{x_{k_1,k_1+1}} \cdots \xrightarrow{x_{k_2-1,k_2}} c_{k_2} \to 0$$

is a chain complex in our additive category $\mathcal{C}$. (Note $\deg x_{i,i+1} = 0$ if we regard it as an element of $\mathcal{C}(c_i, c_{i+1})$ (and not of $\mathcal{C}(c_i[i], c_{i+1}[i+1])$).

We are going to show that twisted complex consist a triangulated category. (See [GM], [Ha], [KaS] for the definition of triangulated category.)

We first define an additive category whose object is a twisted complex. Let $\mathbf{c}^{(1)} = (c^{(1)}_{k_1}, \ldots, c^{(1)}_{k_2}; (x^{(1)}_{i,j}))$, $\mathbf{c}^{(2)} = (c^{(2)}_{k_1}, \ldots, c^{(2)}_{k_2}; (x^{(2)}_{i,j}))$ be twisted complexes. We first define a morphism between them. We put

$$C^k(\mathbf{c}^{(1)}, \mathbf{c}^{(2)}) = \bigoplus_{i,j} \mathcal{C}[1]^k(c^{(1)}_i[i], c^{(2)}_j[j]). \tag{6.8}$$

We define a boundary operator $\hat{\mathfrak{m}}_1$ on it by

$$\hat{\mathfrak{m}}_1(y_{i,j}) = (z_{i,j})$$

where

$$z_{i,j} = \sum_{i\le a<b\le j} \sum_{m_1\ge 1, m_2\ge 1} \sum_{i=\ell^{(1)}_0<\cdots<\ell^{(1)}_{m_1}=a<b=\ell^{(2)}_0<\cdots<\ell^{(2)}_{m_2}=j} \mathfrak{m}_m\left(x^{(1)}_{\ell^{(1)}_0,\ell^{(1)}_1}, \ldots, x^{(1)}_{\ell^{(1)}_{m_1-1},\ell^{(1)}_{m_1}}, y_{a,b}, x^{(2)}_{\ell^{(2)}_0,\ell^{(2)}_1}, \ldots, x^{(2)}_{\ell^{(2)}_{m_2-1},\ell^{(2)}_{m_2}}\right).$$

Lemma 6.9. $\hat{\mathfrak{m}}_1 \circ \hat{\mathfrak{m}}_1 = 0$.

Since the formulas which we need to prove Lemma 6.9 are rather long, we develop some notations before starting calculation. We define

$$\mathbf{x}^{(1)}_{i,j} \in B\mathcal{C}[1]^0(c^{(1)}_i[i], c^{(1)}_j[j])$$

by

$$\mathbf{x}^{(1)}_{i,j} = \sum_{m\ge 1} \sum_{i=\ell_0<\cdots<\ell_m=j} x^{(1)}_{\ell_0,\ell_1} \otimes \cdots \otimes x^{(1)}_{\ell_{m-1},\ell_m}.$$

Then (6.5) is

$$\mathfrak{m}(\mathbf{x}^{(1)}_{i,j}) = \mathfrak{m}(\mathbf{x}^{(2)}_{i,j}) = 0. \tag{6.10}$$

Here and hereafter $\mathfrak{m} : B\mathcal{C}[1] \to B_1\mathcal{C}[1]$ is the operation which is $\mathfrak{m}_k$ on $B_k\mathcal{C}[1]$. The definition of $\hat{\mathfrak{m}}_1$ is rewritten as

$$z_{i,j} = \sum_{i\le a<b\le j} \mathfrak{m}(\mathbf{x}^{(1)}_{i,a} \otimes y_{a,b} \otimes \mathbf{x}^{(2)}_{b,j}). \tag{6.11}$$

We may further simplify the notation as follows. We consider $B\mathcal{C}[1]$ and define a product $\bullet$ on it as follows. Let $\mathbf{x} \in B\mathcal{C}[1](a,b)$, $\mathbf{y} \in B\mathcal{C}[1](c,d)$,

Then:

$$\mathbf{x} \bullet \mathbf{y} = \begin{cases} \mathbf{x} \otimes \mathbf{y} & \text{if } b = c, \\ 0 & \text{if not.} \end{cases}$$

We put

$$\mathbf{x}_{i,*} = \sum_j \mathbf{x}_{i,j}, \quad \mathbf{x}_{*,j} = \sum_i \mathbf{x}_{i,j}, \quad \mathbf{x}_{*,*} = \sum_{i,j} \mathbf{x}_{i,j}.$$

We define $y_{i,*}, y_{*,j}, y_{*,*}, z_{i,*}, z_{*,j}, z_{*,*}$ in a similar way. Then (6.10) is equivalent to

$$\hat{d}(\mathbf{x}^{(0)}_{*,*}) = \hat{d}(\mathbf{x}^{(1)}_{*,*}) = 0. \tag{6.12}$$

(6.11) can be written as

$$z_{*,*} = \mathfrak{m}(\mathbf{x}^{(1)}_{*,*} \bullet y_{*,*} \bullet \mathbf{x}^{(2)}_{*,*}). \tag{6.13}$$

Proof of Lemma 6.9. We have

$$\hat{\mathfrak{m}}_1(\hat{\mathfrak{m}}_1(y_{i,j})) = (w_{i,j})$$

where

$$w_{*,*} = \mathfrak{m}(\mathbf{x}^{(1)}_{*,*} \bullet \mathfrak{m}(\mathbf{x}^{(1)}_{*,*} \bullet y_{*,*} \bullet \mathbf{x}^{(2)}_{*,*}) \bullet \mathbf{x}^{(2)}_{*,*}).$$

(Here we assumed $c_i \neq c_j$ for $i \neq j$ for simplicity.) Note the degree of $x_{i,j}$ are all zero (after shifted). Hence A_∞ formula implies

$$\begin{aligned} &\mathfrak{m}(\mathbf{x}^{(1)}_{*,*} \bullet \mathfrak{m}(\mathbf{x}^{(1)}_{*,*} \bullet y_{*,*} \bullet \mathbf{x}^{(2)}_{*,*}) \bullet \mathbf{x}^{(2)}_{*,*}) \\ &\quad + \mathfrak{m}(\hat{d}(\mathbf{x}^{(1)}_{*,*}) \bullet y_{*,*} \bullet \mathbf{x}^{(2)}_{*,*}) + (-1)^{\deg y_{*,*}+1} \mathfrak{m}(\mathbf{x}^{(1)}_{*,*} \bullet y_{*,*} \bullet \hat{d}(\mathbf{x}^{(2)}_{*,*})) = 0. \end{aligned}$$

The second and the third terms vanish by (6.10). The lemma follows. □

Definition 6.14. We say an element of $\mathcal{C}(\mathbf{c}^{(1)}, \mathbf{c}^{(2)})$ a *morphism in* $\mathcal{DC}$. It is said to be a *closed morphism* if it is $\hat{\mathfrak{m}}_1$ closed. We also put:

$$\mathbb{D}\mathcal{C}(\mathbf{c}^{(1)}, \mathbf{c}^{(2)}) = H^*(\mathcal{C}(\mathbf{c}^{(1)}, \mathbf{c}^{(2)}); \hat{\mathfrak{m}}_1).$$

We next define (higher) compositions:

$$\hat{\mathfrak{m}}_k : C(\mathbf{c}^{(0)}, \mathbf{c}^{(1)}) \otimes \cdots \otimes C(\mathbf{c}^{(k-1)}, \mathbf{c}^{(k)}) \to C(\mathbf{c}^{(0)}, \mathbf{c}^{(k)}) \tag{6.15}$$

by

$$\hat{\mathfrak{m}}_k(y^{(1)}_{*,*}, \dots, y^{(k)}_{*,*}) = \mathfrak{m}(\mathbf{x}^{(0)}_{*,*} \bullet y^{(1)}_{*,*} \bullet \mathbf{x}^{(1)}_{*,*} \bullet \cdots y^{(k)}_{*,*} \bullet \mathbf{x}^{(k)}_{*,*}). \tag{6.16}$$

Theorem 6.17. *We have an A_∞ category $\mathcal{DC}$ such that the set of objects is $\mathfrak{Ob}(\mathcal{DC})$, the set of morphisms is given by* (6.8), *and the operations are given by* (6.16).

Proof. We are only to check the A_∞ formula. The proof of it is quite similar to the proof of Lemma 6.9 and is left to the reader. □

We next are going to define the mapping cone in our category $\mathcal{DC}$. For this purpose, we introduce a systematic way to associate an object of $\mathcal{DC}$ to each object of $\mathcal{DDC}$. (More precisely we can define an A_∞ functor $\mathcal{DDC} \to \mathcal{DC}$. See the next section for the definition of A_∞ functor.)

Let $\mathbf{c}^{(k)} = ((c_i^{(k)}; i \in I^k), (x_{i,j}^{(k)}))$ be a twisted complex, where $k \in I$. (I, I^k be subsets of $\mathbb{Z}$ of the form $\{a, a+1, a+2, \dots, b\}$.) Let

$$\mathbf{y}^{(k,n)} \in (\mathcal{DC})[1]^0(\mathbf{c}^{(k)}[k], \mathbf{c}^{(n)}[n]) = \bigoplus_{\ell,m} \mathcal{C}[1]^0(c_\ell^{(k)}[k+\ell], c_m^{(n)}[n+m]).$$

We write its $\mathcal{C}[1]^0(c_\ell^{(k)}[k+\ell], c_m^{(n)}[n+m])$ component by $y_{\ell,m}^{(k,n)}$.

We assume $\mathfrak{c} = ((\mathbf{c}^{(k)}; k \in I), (\mathbf{y}^{(k,n)}))$ is a twisted complex of the A_∞ category $\mathcal{DC}$. We are going to construct an object $|\mathfrak{c}|$ of $\mathcal{DC}$. The construction is an analogy of the construction of double complex.

We put

$$|\mathfrak{c}|_i = \bigoplus_{k+\ell=i} c_\ell^{(k)}. \tag{6.18}$$

Let $i < j$. We are going to define $z_{i,j} \in \mathcal{C}^0[1](|\mathfrak{c}|_i[i], |\mathfrak{c}|_j[j])$. Let k, ℓ, m, n be integers such that $k + \ell = i$, $m + n = j$. We define $z_{i,j}^{(k,\ell,m,n)}$, the $\mathcal{C}(c_\ell^{(k)}, c_m^{(n)})$ component of $z_{i,j}$ by:

$$z_{i,j}^{(k,\ell,m,n)} = \begin{cases} y_{\ell,m}^{(k,n)} & \text{if } k < n \\ x_{\ell,m} & \text{if } k = n, \ell < m \\ 0 & \text{otherwise.} \end{cases} \tag{6.19}$$

Lemma 6.20. $|\mathfrak{c}| = (|\mathfrak{c}|_i, z_{i,j})$ *is a twisted complex.*

Proof. The condition (6.5) for $|\mathfrak{c}|$ reduces to the condition (6.5) for $\mathfrak{c}$ and ones for $\mathbf{c}_i$. □

Suppose we have two twisted complexes $\mathbf{c}$, $\mathbf{c}'$ and $\mathbf{y} \in \mathcal{C}^0(\mathbf{c}, \mathbf{c}') = \mathcal{C}[1]^0(\mathbf{c}[-1], \mathbf{c}')$, be a closed morphism. We put $\mathbf{c}_{-1} = \mathbf{c}$ and $\mathbf{c}_0 = \mathbf{c}'$. Then $(\mathbf{c}_{-1}, \mathbf{c}_0; \mathbf{y})$ is an object of $\mathcal{DDC}$. Hence $|\mathbf{c}_{-1}, \mathbf{c}_0; \mathbf{y}|$ is an object of $\mathcal{DC}$. We call it the *mapping cone* of $\mathbf{y} : \mathbf{c} \to \mathbf{c}'$ and write it as $\mathrm{Cone}(\mathbf{c}, \mathbf{c}'; \mathbf{y})$.

Lemma 6.21. *There exists* $\mathcal{I} : \mathbf{c}' \to \mathrm{Cone}(\mathbf{c}, \mathbf{c}'; \mathbf{y})$, $\mathcal{J} : \mathrm{Cone}(\mathbf{c}, \mathbf{c}'; \mathbf{y}) \to \mathbf{c}[1]$ *such that, for any twisted complex* $\mathbf{b}$, *we have a long exact sequence*

$$\begin{aligned} H^*(\mathcal{DC}(\mathbf{b}, \mathbf{c}), \mathfrak{m}_1) &\overset{R_{\mathbf{y}}}{\to} H^*(\mathcal{DC}(\mathbf{b}, \mathbf{c}'), \mathfrak{m}_1) \\ &\overset{\mathcal{I}}{\to} H^*(\mathcal{DC}(\mathbf{b}, \mathrm{Cone}(\mathbf{c}, \mathbf{c}'; \mathbf{y})), \mathfrak{m}_1) \\ &\overset{\mathcal{J}}{\to} H^*(\mathcal{DC}(\mathbf{b}, \mathbf{c}[1]), \mathfrak{m}_1) \\ &\overset{R_{\mathbf{y}}}{\to} H^*(\mathcal{DC}(\mathbf{b}, \mathbf{c}'[1]), \mathfrak{m}_1) \to \cdots \end{aligned}$$

Here $R_{\mathbf{y}}$ *is induced by the right multplication by* $\mathbf{y}$.

Proof. We remark that

$$\mathrm{Cone}(\mathbf{c}, \mathbf{c}'; \mathbf{y})_i = c_{i+1} \oplus c'_i.$$

Hence we can define $\mathcal{I}$ as an inclusion $c'_i \to c_{i+1} \oplus c'_i \cong \mathrm{Cone}(\mathbf{c}, \mathbf{c}'; \mathbf{y})_i$. And we define $\mathcal{J}$ as a projection $\mathrm{Cone}(\mathbf{c}, \mathbf{c}'; \mathbf{y})_i \cong c_{i+1} \oplus c'_i \to c_{i+1}$. It is easy to see that $\mathcal{I}$ and $\mathcal{J}$ are morphisms in $\mathcal{DC}$. By definition we have an exact sequence of chain complex

$$\begin{aligned} 0 \to (\mathcal{DC}(\mathbf{b}, \mathbf{c}'), \mathfrak{m}_1) &\overset{\mathcal{I}}{\to} (\mathcal{DC}(\mathbf{b}, \mathrm{Cone}(\mathbf{c}, \mathbf{c}'; \mathbf{y})), \mathfrak{m}_1) \\ &\overset{\mathcal{J}}{\to} (\mathcal{DC}(\mathbf{b}, \mathbf{c}[1]), \mathfrak{m}_1) \to 0. \end{aligned}$$

By definition, the operator $H^*(\mathcal{DC}(\mathbf{b}, \mathbf{c}), \mathfrak{m}_1) \to H^*(\mathcal{DC}(\mathbf{b}, \mathbf{c}'), \mathfrak{m}_1)$ of the associated long exact sequence is $R_{\mathbf{y}}$. □

So far we defined a notion corresponding to chain complex and to mapping cone, in the case of A_∞ category. Usually to construct the derived category from the category of chain complex, we need to divide it by weak equivalence. We define a similar notion, homotopy equivalence between two objects of A_∞ category.

Definition 6.22. Let $\mathcal{C}$ be an A_∞ category and $c, c' \in \mathfrak{Ob}(\mathcal{C})$. Let $x \in \mathcal{C}^0(c, c')$. We say that x is a *homotopy equivalence* if there exists $y \in \mathcal{C}^0(c', c)$ such that

(6.23.1) $\mathfrak{m}_1(x) = \mathfrak{m}_1(y) = 0$.

(6.23.2) $\mathfrak{m}_2(y, x) - \mathbf{e}_c \in \mathrm{Im}\, \mathfrak{m}_1$, $\mathfrak{m}_2(x, y) - \mathbf{e}_{c'} \in \mathrm{Im}\, \mathfrak{m}_1$.

Two objects $c, c' \in \mathfrak{Ob}(\mathcal{C})$ are said to be *homotopy equivalent* to each other if there exists a homotopy equivalence between them.

Lemma 6.24. *Let* $c, c' \in \mathfrak{Ob}(\mathcal{C})$, $x \in \mathcal{C}^0(c, c')$ *with* $\mathfrak{m}_1(x) = 0$. *Then the following five conditions are equivalent to each other.*

(6.25.1) *x is a homotopy equivalence.*

(6.25.2) *For each* $b \in \mathfrak{Ob}(\mathcal{C})$ *the map* $R_x : \mathcal{C}(b,c) \to \mathcal{C}(b,c')$, $z \mapsto \mathfrak{m}_2(z,x)$ *induces an isomorphism on homology.*
(6.25.3) *The map* $R_x : \mathcal{C}(c',c) \to \mathcal{C}(c',c')$ *induces a surjection on homology and* $R_x : \mathcal{C}(c,c) \to \mathcal{C}(c,c')$ *induces an injection on homology.*
(6.25.4) *For each* $b \in \mathfrak{Ob}(\mathcal{C})$ *the map* $L_x : \mathcal{C}(c',b) \to \mathcal{C}(c,b)$, $z \mapsto \mathfrak{m}_2(x,z)$ *induces an isomorphism on homology.*
(6.25.5) *The map* $L_x : \mathcal{C}(c',c) \to \mathcal{C}(c,c)$ *induces a surjection on homology and* $L_x : \mathcal{C}(c',c') \to \mathcal{C}(c,c')$ *induces an injection on homology.*

Proof. (6.25.1)⇒(6.25.2): $\mathfrak{m}_1(x) = 0$ together with A_∞ formulae implies that R_x is a chain map. Let y be as in (6.23). Then it is easy to see that $R_x \circ R_y$, $R_y \circ R_x$ induces identity in homology. (6.25.2) follows.
(6.25.2)⇒(6.25.3): Obvious.
(6.25.3)⇒(6.25.1): By (6.25.3) there exists y such that $\mathfrak{m}_2(y,x) - \mathbf{e}_{c'} \in \operatorname{Im}\mathfrak{m}_1$. Hence

$$\begin{aligned}
\mathfrak{m}_2(\mathfrak{m}_2(x,y),x) &\equiv -(-1)^{\deg' x}\mathfrak{m}_2(x,\mathfrak{m}_2(y,x)) \mod \operatorname{Im}\mathfrak{m}_1 \\
&\equiv (-1)^{\deg x}\mathfrak{m}_2(x,\mathbf{e}_{c'}) \mod \operatorname{Im}\mathfrak{m}_1 \\
&\equiv x \mod \operatorname{Im}\mathfrak{m}_1 \\
&\equiv \mathfrak{m}_2(\mathbf{e}_c,x) \mod \operatorname{Im}\mathfrak{m}_1.
\end{aligned}$$

Hence $\mathfrak{m}_2(x,y) \equiv \mathbf{e}_c \mod \operatorname{Im}\mathfrak{m}_1$.

The proof of equivalence between (6.25.1), (6.25.4), (6.25.5) is similar. □

Lemma 6.24 implies that the composition of homotopy equivalences is a homotopy equivalence.

We define a category $\mathbb{D}\mathcal{C}$ as follows. (It is a category in the usual sense and is an additive category.) Its object is a homotopy equivalence class of the objects of A_∞ category $\mathcal{DC}$. Morphism between them is defined by Definition 6.14. By (6.25.2), (6.25.4), a homotopy equivalence induces an isomorphism on $\operatorname{Hom}(\mathbf{c}^{(1)},\mathbf{c}^{(2)})$ in Definition 6.14 hence the set of morphisms is well defined. The composition of the morphisms is induced by $\mathfrak{m}_2$. The A_∞ formula implies that the composition in the homology level is (exactly) associative.

(Actually to perform this construction in a rigorous way we need to define and use a notion of quotient category. Since it is standard we do not discuss it. See for example [GM], [KaS].)

We next define distinguished triangles in $\mathbb{D}\mathcal{C}$ and prove that $\mathbb{D}\mathcal{C}$ will be a triangulated category. We use the notion of mapping cone for this purpose. To do so, we need to show that the homotopy equivalence classes of $\operatorname{Cone}(\mathbf{c}^{(1)},\mathbf{c}^{(2)};\mathbf{y})$ depends only of the homotopy class of $(\mathbf{c}^{(1)},\mathbf{c}^{(2)};\mathbf{y})$. More precisely we prove the following.

Lemma 6.26. *Let* $\mathbf{u}_1 \in \mathcal{DC}(\mathbf{c}^{(1)}, \mathbf{c}'^{(1)})$, $\mathbf{u}_2 \in \mathcal{DC}(\mathbf{c}^{(2)}, \mathbf{c}'^{(2)})$ *be homotopy equivalences. We assume* $\mathfrak{m}_2(\mathbf{u}_1, \mathbf{y}') - \mathfrak{m}_2(\mathbf{y}, \mathbf{u}_2) \in \operatorname{Im}\mathfrak{m}_1$. *Then* $\operatorname{Cone}(\mathbf{c}^{(1)}, \mathbf{c}^{(2)}; \mathbf{y})$ *is homotopy equivalent to* $\operatorname{Cone}(\mathbf{c}'^{(1)}, \mathbf{c}'^{(2)}; \mathbf{y}')$ *in the* A_∞ *category* $\mathcal{DC}$.

Proof. We put

$$\mathfrak{m}_2(\mathbf{u}_1, \mathbf{y}') - \mathfrak{m}_2(\mathbf{y}, \mathbf{u}_2) = \mathfrak{m}_1(\mathbf{z}).$$

Then $(\mathbf{u}_1, \mathbf{u}_2, \mathbf{z})$ defines a closed morphism

$$\mathcal{DC}((\mathbf{c}^{(1)}, \mathbf{c}^{(2)}; \mathbf{y}), (\mathbf{c}'^{(1)}, \mathbf{c}'^{(2)}; \mathbf{y}')).$$

Hence we have a closed morphism

$$\operatorname{Cone}(\mathbf{c}^{(1)}, \mathbf{c}^{(2)}; \mathbf{y}) \to \operatorname{Cone}(\mathbf{c}'^{(1)}, \mathbf{c}'^{(2)}; \mathbf{y}'). \tag{6.27}$$

Then we obtain a commutative diagram:

$$\begin{array}{ccccc}
\mathbf{c}^{(1)} & \xrightarrow{\ \mathbf{y}\ } & \mathbf{c}^{(2)} & \longrightarrow & Cone(\mathbf{c}^{(1)}, \mathbf{c}^{(2)}; \mathbf{y}) \\
\downarrow & & \downarrow & & \downarrow \\
\mathbf{c}'^{(1)} & \xrightarrow{\ \mathbf{y}'\ } & \mathbf{c}'^{(2)} & \longrightarrow & Cone(\mathbf{c}'^{(1)}, \mathbf{c}'^{(2)}; \mathbf{y}')
\end{array}$$

Diagram 6.1.

comparing exact sequences in Lemma 6.21. Then Lemma 6.21 and five lemma implies that (6.27) induces an isomorphism

$$\mathcal{DC}(\mathbf{b}, \operatorname{Cone}(\mathbf{c}^{(1)}, \mathbf{c}^{(2)}; \mathbf{y})) \cong \mathcal{DC}(\mathbf{b}, \operatorname{Cone}(\mathbf{c}'^{(1)}, \mathbf{c}'^{(2)}; \mathbf{y}'))$$

for any $\mathbf{b}$. The lemma now follows from Lemma 6.24. □

Using Lemma 6.26 we can define a notion of distinguished triangle in $\mathbb{D}\mathcal{C}$ as follows.

Definition 6.28.

$$[\mathbf{c}^{(1)}] \to [\mathbf{c}^{(2)}] \to [\operatorname{Cone}(\mathbf{c}^{(1)}, \mathbf{c}^{(2)}; \mathbf{y})] \to [\mathbf{c}^{(1)}[1]]$$

is said to be a *distinguished triangle.*

Theorem 6.29. $\mathbb{D}\mathcal{C}$ *is a triangulated category.*

Figure 6.1.

Figure 6.2.

We omit the proof, which is rather a straightforward check of the axiom.

Definition 6.30. $\mathbb{D}\mathcal{C}$ is called the *derived A_∞ category* of our A_∞ category $\mathcal{C}$.

In case we start with filtered A_∞ category $\mathcal{C}$ we first construct an A_∞ category as in Definition 1.14 and then construct derived A_∞ category. We call it also the derived A_∞ category of $\mathcal{C}$ and write it also as $\mathbb{D}\mathcal{C}$.

Note that the objects of derived A_∞ category of a filtered A_∞ category $\mathcal{C}$ can be regarded as $\mathbf{c} = ((c_i; i \in \{k_1, k_1+1, \dots, k_2\}), (x_{i,j}^{(k)}))$. But, in this case, $x_{i,i}$ may be nonzero, ($x_{i,i}$ should be in $\mathcal{C}[1]^0(c_i, c_i)$ and moreover $x_{i,i} \in \cup_{\lambda>0} F^\lambda \mathcal{C}[1]^0(c_i, c_i)$.) $x_{i,j}$ is supposed to satisfy an equation

$$\sum_m \sum_{\ell_0,\dots,\ell_m} \mathfrak{m}_m(x_{\ell_0,\ell_1}, \dots, x_{\ell_{m-1},\ell_m}) = 0, \tag{6.31}$$

which is similar to (6.5). However, in (6.31), $m = 0, 1, \dots$, $\ell_0 \le \ell_1 \le \dots \le \ell_m$. (Namely the case $\ell_i = \ell_{i+1}$ is included.) Actually, in the case $\ell_0 = \dots = \ell_m = \ell$, (6.31) is

$$\mathfrak{m}(e^{x_{\ell,\ell}}) = 0,$$

that is the definition $x_{\ell,\ell}$ to be a bounding chain.

An element $\mathbf{c} = ((c_i; i \in \{k_1, k_1+1, \dots, k_2\}), (x_{i,j}^{(k)}; i \le j))$ can be described by Figure 6.1 above. On the other hand, an element $((c_i; i \in \{k_1, k_1+1, \dots, k_2\}), (x_{i,j}^{(k)}; i < j))$ can be described by Figure 6.2 above.

§7. A_∞ functor and natural transformation

In sections 6, we consider more general objects than the original objects (that is the twisted complex) of our A_∞ category $\mathcal{C}$. In this section, we further generalize it and define a notion, A_∞ functor $\mathcal{C} \to \mathcal{CH}$, which was defined in [Fu4] over $\mathbb{Z}_2$ coefficient. We simplify the description of [Fu4] using Bar complex and also we put here the precise sign, which was not discussed in [Fu4].

Definition 7.1. Let $\mathcal{C}_1$, $\mathcal{C}_2$ be A_∞ categories. An A_∞ *functor* $\mathcal{F}$ from $\mathcal{C}_1$ to $\mathcal{C}_2$ is a collection of $\mathcal{F}_k$, $k = 1, 2, \dots$ such that

(7.2.1) $\mathcal{F}_0 : \mathfrak{Ob}(\mathcal{C}_1) \to \mathfrak{Ob}(\mathcal{C}_2)$ is a map.
(7.2.2) For $c_1, c_2 \in \mathfrak{Ob}(\mathcal{C}_1)$,

$$\mathcal{F}_k(c_1, c_2) : B_k\mathcal{C}_1[1](c_1, c_2) \to \mathcal{C}_2[1](\mathcal{F}_0(c_1), \mathcal{F}_0(c_2))$$

is a homomorphism of degree 0.
(7.2.3) We extend $\mathcal{F}_k(c_1, c_2)$ to a coalgebra homomorphism

$$\hat{\mathcal{F}}(c_1, c_2) : B\mathcal{C}_1[1](c_1, c_2) \to B\mathcal{C}_2[1](\mathcal{F}_0(c_1), \mathcal{F}_0(c_2)). \tag{7.3.1}$$

Then $\hat{\mathcal{F}}(c_1, c_2)$ is a chain map with respect to the boundary operator d in Definition 1.1.

Note

$$\begin{aligned}
&\hat{\mathcal{F}}(x_1 \otimes \cdots \otimes x_k) \\
&\quad = \sum_m \sum_{0=\ell_1<\ell_2<\cdots<\ell_m=k} \mathcal{F}_{\ell_2-\ell_1-1}(x_{\ell_1+1} \otimes \cdots \otimes x_{\ell_2}) \otimes \cdots \\
&\quad\quad \cdots \otimes \mathcal{F}_{\ell_m-\ell_{m-1}-1}(x_{\ell_{m-1}+1} \otimes \cdots \otimes x_{\ell_m}).
\end{aligned}$$

Our homomorphism $\hat{\mathcal{F}}$ on $B_0\mathcal{C}_1[1]$ is defined as follows. We remark

$$B_0\mathcal{C}_1[1](c_1, c_2) \cong \begin{cases} R & c_1 = c_2, \\ 0 & c_1 \neq c_2. \end{cases}$$

We put

$$\hat{\mathcal{F}}(x) = \begin{cases} x & \text{if } x \in B_0\mathcal{C}_1[1](c, c), \\ 0 & \text{if } x \in B_0\mathcal{C}_1[1](c_1, c_2), c_1 \neq c_2. \end{cases} \tag{7.3.2}$$

We next give an example of A_∞ functor, that is a representable functor. We first define an A_∞ category $\mathcal{CH}$ for this purpose.

Definition 7.4. $\mathfrak{Ob}(\mathcal{CH})$ is the set of (all) chain complexes of free R nodules[5]. Let $(C,d),(C',d) \in \mathfrak{Ob}(\mathcal{CH})$. Then

$$\mathcal{CH}^k((C,d),(C',d)) = \bigoplus_{\ell} \mathrm{Hom}_R(C^\ell, C'^{\ell+k}).$$

We define

$$\begin{aligned} &(7.5.1) & \mathfrak{m}_1(x) &= d\circ x + (-1)^{\deg x+1} x\circ d, \\ &(7.5.2) & \mathfrak{m}_2(x,y) &= (-1)^{\deg x(\deg y+1)} y\circ x. \end{aligned}$$

We put $\mathfrak{m}_k = 0$ for $k \geq 3$.

Remark 7.6. The sign in (7.5.2) is the same as one we need to regard differential graded algebra as an A_∞ algebra. (See Example-Lemma 1.7.)

Proposition 7.7. *$\mathcal{CH}$ is an A_∞ category.*

Proof. It is easy to check $\mathfrak{m}_1\circ\mathfrak{m}_1 = 0$. We calculate

$$\begin{aligned} (\mathfrak{m}_1\circ\mathfrak{m}_2)(x,y) &= (-1)^{\deg x(\deg y+1)} d\circ y\circ x \\ &\quad +(-1)^{\deg x+\deg y+1}(-1)^{\deg x(\deg y+1)} y\circ x\circ d \\ &= (-1)^{\deg x\deg y+\deg x} d\circ y\circ x \\ &\quad +(-1)^{\deg x(\deg y+1)+\deg y+1} y\circ d\circ x \\ &\quad -(-1)^{(\deg x+1)(\deg y+1)} y\circ d\circ x \\ &\quad -(-1)^{\deg x+\deg y}(-1)^{\deg x(\deg y+1)} y\circ x\circ d \\ &= -(-1)^{\deg x+1}\mathfrak{m}_2(x,\mathfrak{m}_1(y)) - \mathfrak{m}_2(\mathfrak{m}_1(x),y). \end{aligned}$$

This implies $\mathfrak{m}\circ\hat{d} = 0$ on B_2. We next calculate

$$\begin{aligned} &\mathfrak{m}_2(\mathfrak{m}_2(x,y),z) + (-1)^{\deg x+1}\mathfrak{m}_2(x,\mathfrak{m}_2(y,z)) \\ &\quad= (-1)^{\deg x(\deg y+1)+(\deg x+\deg y)(\deg z+1)} z\circ(y\circ x) \\ &\qquad +(-1)^{\deg x+1+\deg y(\deg z+1)+\deg x(\deg y+\deg z+1)}(z\circ y)\circ x \\ &\quad= 0, \end{aligned}$$

which is the third and the last part of the A_∞ formulae to be checked. □

We next define:

[5]To avoid Russell paradox in set theory, we fix a sufficiently large set (a universe) and consider only free R modules contained in this set.

Definition 7.8. Let $\mathcal{C}$ be an A_∞ category. We define its *opposite* A_∞ *category* $\mathcal{C}^o$ as follows.

(7.9.1) $\mathfrak{Ob}(\mathcal{C}^o) = \mathfrak{Ob}(\mathcal{C})$.

(7.9.2) Let $c, c' \in \mathfrak{Ob}(\mathcal{C}^o) = \mathfrak{Ob}(\mathcal{C})$. We put $\mathcal{C}^o(c,c') = \mathcal{C}(c',c)$.

(7.9.3) We define (higher) composition operators $\mathfrak{m}_k^o$ of $\mathcal{C}^o$ by:

$$\mathfrak{m}_k^o(x_1,\dots,x_k) = (-1)^\epsilon \mathfrak{m}_k(x_k,\dots,x_1),$$

where

$$\epsilon = \sum_{1\le i<j\le k} (\deg x_i + 1)(\deg x_j + 1) + 1.$$

Lemma 7.10. *$\mathcal{C}^o$ is an A_∞ category.*

Proof. First we introduce some notations to simplify the formula. We put $\mathbf{x} = x_1 \otimes \cdots \otimes x_k \in B_k\mathcal{C}$ and

$$\Delta^{m-1}\mathbf{x} = \sum_a \mathbf{x}_a^{(1)} \otimes \cdots \otimes \mathbf{x}_a^{(m)}. \tag{7.11}$$

Here

$$\Delta^{m-1} = \cdots \circ (\Delta \otimes 1 \otimes 1) \circ (\Delta \otimes 1) \circ \Delta.$$

Let

$$\deg \mathbf{x} = \deg x_1 + \cdots + \deg x_k \tag{7.12.1}$$

be the degree of $\mathbf{x}$ and

$$\deg' \mathbf{x} = \deg x_1 + \cdots + \deg x_k + k \tag{7.12.2}$$

be its degree after shifted. We use notations (7.11), (7.12.1), (7.12.2) frequently for the rest of this article. We put

$$\mathbf{x}^{op} = x_k \otimes \cdots \otimes x_1, \tag{7.13.1}$$

and

$$\epsilon(\mathbf{x}) = \sum_{1\le i<j\le k} (\deg x_i + 1)(\deg x_j + 1). \tag{7.13.2}$$

The A_∞ formula for $\mathfrak{m}$ can be written as

$$\sum_a (-1)^{\deg' \mathbf{x}_a^{(1)}} \mathfrak{m}(\mathbf{x}_a^{(1)}, \mathfrak{m}(\mathbf{x}_a^{(2)}), \mathbf{x}_a^{(3)}) = 0. \tag{7.14}$$

We have

$$\mathfrak{m}(\mathbf{x}_a^{(1)}, \mathfrak{m}(\mathbf{x}_a^{(2)}), \mathbf{x}_a^{(3)}) = (-1)^{\epsilon(\mathbf{x}_a^{(1)})+\epsilon(\mathbf{x}_a^{(2)})+\epsilon(\mathbf{x}_a^{(3)})+\epsilon_1(a)} \mathfrak{m}^o(\mathbf{x}_a^{(3)op}, \mathfrak{m}^o(\mathbf{x}_a^{(2)op}), \mathbf{x}_a^{(1)op}), \tag{7.15}$$

where $\epsilon(\mathbf{x}_a^{(j)})$ are as in (7.13.2) and

$$\epsilon_1(a) = (\deg' \mathbf{x}_a^{(1)} + \deg' \mathbf{x}_a^{(3)})(\deg' \mathbf{x}_a^{(2)} + 1) + \deg' \mathbf{x}_a^{(1)} \deg' \mathbf{x}_a^{(3)}.$$

We remark that

$$\epsilon(\mathbf{x}_a^{(1)}) + \epsilon(\mathbf{x}_a^{(2)}) + \epsilon(\mathbf{x}_a^{(3)}) + (\deg' \mathbf{x}_a^{(1)} + \deg' \mathbf{x}_a^{(3)}) \deg' \mathbf{x}_a^{(2)} + \deg' \mathbf{x}_a^{(1)} \deg' \mathbf{x}_a^{(3)} = \epsilon(\mathbf{x}),$$

and is independent of a. Hence (7.14) and (7.15) imply

$$\sum_a (-1)^{\deg' \mathbf{x}_a^{(3)op}} \mathfrak{m}^o(\mathbf{x}_a^{(3)op}, \mathfrak{m}^o(\mathbf{x}_a^{(2)op}), \mathbf{x}_a^{(1)op}) = 0.$$

This is the A_∞ formula of $\mathfrak{m}^o$ to be checked. □

Definition 7.16. Let $\mathcal{C}$ be an A_∞ category and $c \in \mathfrak{Ob}(\mathcal{C})$. We construct an A_∞ functor $\mathcal{F}^c = \mathrm{Hom}(\cdot, c) : \mathcal{C} \to \mathcal{CH}^o$ as follows.

(7.17.1) $\mathcal{F}_0^c(b) = \mathcal{C}(b,c)$. (We take $\mathfrak{m}_1$ as the boundary operator.)

(7.17.2) $\mathcal{F}_k^c(x_1, \dots, x_k)(y) = \mathfrak{m}_{k+1}(x_1, \dots, x_k, y)$. Here

$$y \in \mathcal{C}(b,c), b_1, \dots, b_{k+1} = b \in \mathfrak{Ob}(\mathcal{C}), x_i \in \mathcal{C}(b_i, b_{i+1}).$$

Proposition 7.18. *$\mathcal{F}^c$ is an A_∞ functor.*

Proof. We calculate

$$\begin{aligned} & \mathfrak{m}_2^o(\mathcal{F}(\mathbf{x}_a^{(1)}), \mathcal{F}(\mathbf{x}_a^{(2)})) \\ = & -(-1)^{(\deg \mathcal{F}(\mathbf{x}_a^{(1)})+1)(\deg \mathcal{F}(\mathbf{x}_a^{(2)})+1)} \mathfrak{m}_2(\mathcal{F}(\mathbf{x}_a^{(2)}), \mathcal{F}(\mathbf{x}_a^{(1)})) \\ = & -(-1)^{(\deg \mathcal{F}(\mathbf{x}_a^{(1)})+1)(\deg \mathcal{F}(\mathbf{x}_a^{(2)})+1)+(\deg \mathcal{F}(\mathbf{x}_a^{(1)})+1)\deg \mathcal{F}(\mathbf{x}_a^{(2)})} \\ & \mathcal{F}(\mathbf{x}_a^{(1)}) \circ \mathcal{F}(\mathbf{x}_a^{(2)}) \\ = & -(-1)^{\deg' \mathbf{x}_a^{(1)}} \mathcal{F}(\mathbf{x}_a^{(1)}) \circ \mathcal{F}(\mathbf{x}_a^{(2)}). \end{aligned}$$

We recall $\mathfrak{m}_k = 0$ for $k \ge 3$ in $\mathcal{CH}$. Hence the condition that $\hat{\mathcal{F}}$ is a chain map is

$$\begin{aligned} 0 = & \sum_a (-1)^{\deg' \mathbf{x}_a^{(1)}} \mathcal{F}(\mathbf{x}_a^{(1)}) \circ \mathcal{F}(\mathbf{x}_a^{(2)}) \\ & +(-1)^{\deg' \mathbf{x}} \mathcal{F}(\mathbf{x}) \circ \mathfrak{m}_1 + \mathfrak{m}_1 \circ \mathcal{F}(\mathbf{x}) \\ & + \sum_a (-1)^{\deg' \mathbf{x}_a^{(1)}} \mathcal{F}(\mathbf{x}_a^{(1)}, \mathfrak{m}(\mathbf{x}_a^{(2)}), \mathbf{x}_a^{(3)}). \end{aligned} \tag{7.19}$$

(We use also $\mathfrak{m}_1^o = -\mathfrak{m}_1$ to deduce (7.19).) We plug in y to the right hand side of (7.19). Then the first term of the right hand side will be

$$\sum_a (-1)^{\deg' \mathbf{x}_a^{(1)}} \mathfrak{m}(\mathbf{x}_a^{(1)}, \mathfrak{m}(\mathbf{x}_a^{(2)}, y)). \tag{7.20}$$

The second and the third terms of the right hand side will be

$$(-1)^{\deg' \mathbf{x}} \mathfrak{m}(\mathbf{x}, \mathfrak{m}_1(y)) + \mathfrak{m}_1(\mathfrak{m}(\mathbf{x}, y)). \tag{7.21}$$

The fourth term will be

$$\sum_a (-1)^{\deg' \mathbf{x}_a^{(1)}} \mathfrak{m}(\mathbf{x}_a^{(1)}, \mathfrak{m}(\mathbf{x}_a^{(2)}), \mathbf{x}_a^{(3)}, y). \tag{7.22}$$

(7.20) + (7.21) + (7.22) = 0 is the A_∞ formula for $\mathfrak{m}$. We thus proved (7.19). □

We next define a similar but a bit different A_∞ functor $\mathfrak{Rep}_0(c)$: $\mathcal{C}^o \to \mathcal{CH}$. (At this stage $\mathfrak{Rep}_0(c)$ is just a symbol. We will define $\mathfrak{Rep}_k$ in section 9.) For this purpose, we prove the following.

Definition-Lemma 7.23. *For each A_∞ functor $\mathcal{F} : \mathcal{C}_1 \to \mathcal{C}_2$, we can construct its opposite A_∞ functor $\mathcal{F}^o : \mathcal{C}_1^o \to \mathcal{C}_2^o$ as follows.*
(7.24.1) $\mathcal{F}_0^o = \mathcal{F}_0$.
(7.24.2) $\mathcal{F}_k^o(\mathbf{x}) = (-1)^{\epsilon(\mathbf{x})} \mathcal{F}_k(\mathbf{x}^{op})$. *Here we use notation* (7.13).

Proof. We need to check

$$\begin{aligned} &\textstyle\sum_\ell \sum_a \mathfrak{m}_\ell^o(\mathcal{F}^o(\mathbf{x}_a^{(1)}), \dots, \mathcal{F}^o(\mathbf{x}_a^{(\ell)})) \\ &= \textstyle\sum_a (-1)^{\deg' \mathbf{x}_a^{(1)}} \mathcal{F}^o(\mathbf{x}_a^{(1)}, \mathfrak{m}^o(\mathbf{x}_a^{(2)}), \mathbf{x}_a^{(3)}). \end{aligned} \tag{7.25}$$

The left hand side of (7.25) is

$$\sum_\ell \sum_a (-1)^{\epsilon(\mathbf{x})+1} \mathfrak{m}_\ell(\mathcal{F}(\mathbf{x}_a^{(\ell)op}), \dots, \mathcal{F}(\mathbf{x}_a^{(1)op})). \tag{7.26}$$

In a way similar to the proof of Lemma 6.8, we can check that the right hand side of (7.25) is

$$\sum_a (-1)^{\deg' \mathbf{x}_a^{(3)}} \mathcal{F}(\mathbf{x}_a^{(3)op}, \mathfrak{m}(\mathbf{x}_a^{(2)op}), \mathbf{x}_a^{(1)op}). \tag{7.27}$$

(7.26) = (7.27) is the condition that $\mathcal{F}$ is an A_∞ functor. □

In view of (7.24) and (7.17), we can define an A_∞ functor $\mathfrak{Rep}_0(c)$: $\mathcal{C}^o \to \mathcal{CH}$ as follows.

Definition 7.28.

$$\mathfrak{Rep}_0(c)_0(b_0) = \mathcal{C}(b_0, c). \tag{7.29.1}$$

Let $\mathbf{x} \in B_k\mathcal{C}^o(b_0, b_k) = B_k\mathcal{C}(b_k, b_0)$, $y \in \mathfrak{Rep}_0(c)_0(b_0) = \mathcal{C}(c, b_0)$. Then

$$\mathfrak{Rep}_0(c)_k(\mathbf{x})(y) = (-1)^{\epsilon(\mathbf{x})}\mathcal{F}^c(\mathbf{x}^{op})(y) = (-1)^{\epsilon(\mathbf{x})}\mathfrak{m}(\mathbf{x}^{op}, y). \tag{7.29.2}$$

We next apply the construction of $\mathcal{F}^c$ and $\mathfrak{Rep}_0(c)$ to the opposite A_∞ category $\mathcal{C}^o$ and define ${}^c\mathcal{F}$, $\mathfrak{OpRep}_0(c)$ as follows.

Definition 7.30. ${}^c\mathcal{F} : \mathcal{C} \to \mathcal{CH}$ is defined by

(7.31.1) ${}^c\mathcal{F}_0(b_0) = \mathcal{C}(c, b_0)$.

(7.31.2) ${}^c\mathcal{F}_k(\mathbf{x})(y) = \mathfrak{m}^o_{k+1}(\mathbf{x}, y) = -(-1)^{\epsilon(\mathbf{x})+\deg' y \deg' \mathbf{x}}\mathfrak{m}_{k+1}(y, \mathbf{x}^{op})$,

where $\mathbf{x} \in B_k\mathcal{C}^o(b_0, b_k) = B_k\mathcal{C}(b_k, b_0)$, $y \in {}^c\mathcal{F}_0(b_0) \in \mathcal{C}(c, b_0)$.

$\mathfrak{OpRep}_0(c) : \mathcal{C}^o \to \mathcal{CH}^o$ is defined by

(7.32.1) $\mathfrak{OpRep}_0(c)(b_0) = \mathcal{C}(c, b_0)$.

(7.32.2)
$$\begin{aligned}\mathfrak{OpRep}_k(\mathbf{x})(y) &= (-1)^{\epsilon(\mathbf{x})}\, {}^c\mathcal{F}_k(\mathbf{x})(y)\\ &= -(-1)^{\deg' y \deg' \mathbf{x}}\mathfrak{m}_{k+1}(y, \mathbf{x}^{op}).\end{aligned}$$

It follows from construction that ${}^c\mathcal{F}$ and $\mathfrak{Rep}_0(c)$ are A_∞ functors.

Definition 7.33. We say an A_∞ functor : $\mathcal{C} \to \mathcal{CH}^o$, $\mathcal{C}^o \to \mathcal{CH}$, $\mathcal{C} \to \mathcal{CH}$, $\mathcal{C}^o \to \mathcal{CH}^o$ to be *representable* if it is homotopic to $\mathcal{F}^c$, $\mathfrak{Rep}_0(c)$, ${}^c\mathcal{F}$ and $\mathfrak{OpRep}_0(c)$, respectively. (Homotpy between A_∞ functors will be defined in the next section.)

We next generalize the constructions above to the case when c is a twisted complex. For this purpose we first define a composition of two A_∞ functors. Let $\mathcal{C}_1$, $\mathcal{C}_2$, $\mathcal{C}_3$ be A_∞ categories and $\mathcal{F} : \mathcal{C}_1 \to \mathcal{C}_2$, $\mathcal{G} : \mathcal{C}_2 \to \mathcal{C}_3$ be A_∞ functors.

Definition 7.34. The composition $\mathcal{G} \circ \mathcal{F} : \mathcal{C}_1 \to \mathcal{C}_3$ is defined as follows.

$$\begin{aligned}
&(7.35.1)\ (\mathcal{G}\circ\mathcal{F})_0 &&= \mathcal{G}_0 \circ \mathcal{F}_0\\
&\widehat{\mathcal{G}\circ\mathcal{F}}(c_1, c_2) &&= \widehat{\mathcal{G}}(\mathcal{F}_0(c_1), \mathcal{F}_0(c_2)) \circ \widehat{\mathcal{F}}(c_1, c_2)\\
&(7.35.2) &&: B\mathcal{C}_1(c_1, c_2) \to B\mathcal{C}_3(\mathcal{G}_0(\mathcal{F}_0(c_1)), \mathcal{G}_0(\mathcal{F}_0(c_2))).
\end{aligned}$$

It is easy to see that the composition is an A_∞ functor.

We remark that there is an obvious A_∞ functor $\mathcal{C} \to \mathcal{DC}$.

Definition 7.36. Let $\mathbf{c}$ be an twisted complex of A_∞ category $\mathcal{C}$. We then consider the composition

$$\mathcal{C} \to \mathcal{DC} \xrightarrow{\mathcal{F}^{\mathbf{c}}} \mathcal{CH}^o \tag{7.37}$$

where $\mathcal{F}^{\mathbf{c}}$ is defined by applying Definition 6.15 to A_∞ category $\mathcal{DC}$. We write also $\mathcal{F}^{\mathbf{c}}$ the composition (7.37). Similarly we define ${}^{\mathbf{c}}\mathcal{F}$ as the composition

$$\mathcal{C}^o \to (\mathcal{DC})^o \overset{{}^{\mathbf{c}}\mathcal{F}}{\to} \mathcal{CH}^o,$$

where $\mathbf{c}$ is a twisted complex of $\mathcal{C}^o$. (Actually the twisted complex of $\mathcal{C}^o$ can be constructed from one on $\mathcal{C}$ by changing the sign of the maps $x_{i,j}$ appropriately. We leave the reader the problem to find the correct sign.) We define $\mathfrak{Rep}_0(\mathbf{c})$, $\mathfrak{OpRep}_0(\mathbf{c})$ in a similar way.

Definition 7.38. An A_∞ functor : $\mathcal{C} \to \mathcal{CH}^o$, $\mathcal{C}^o \to \mathcal{CH}$, $\mathcal{C} \to \mathcal{CH}$, $\mathcal{C}^o \to \mathcal{CH}^o$ is said to be *derived representable* if it is homotopic to $\mathcal{F}^{\mathbf{c}}$, $\mathfrak{Rep}_0(\mathbf{c})$, ${}^{\mathbf{c}}\mathcal{F}$, $\mathfrak{OpRep}_0(\mathbf{c})$, respectively.

We next explain that A_∞ functors are natural generalization of the notion of differential graded modules. Let $(\mathcal{A}, d, \wedge)$ be a differential graded algebra. It defines an A_∞ algebra as in Example-Lemma 1.7. We write it $(\mathcal{A}, \mathfrak{m}_1, \mathfrak{m}_2)$. ($\mathfrak{m}_3 = \cdots = 0$.) We may regard it as an A_∞ category $\mathcal{A}$ with one object.

Lemma 7.39. *Homotopy classes of A_∞ functors $\mathcal{F} : \mathcal{A} \to \mathcal{CH}^0$ such that $\mathcal{F}_k = 0$ for $k \geq 2$, correspond one to one to the homotopy equivalence classes of graded differential left module over $(\mathcal{A}, d, \wedge)$.*

Proof. Let $(D, d, \cdot)$ be a graded differential module over $(\mathcal{A}, d, \wedge)$. We define

$$\mathcal{F}_0^d(c_0) = (D, d).$$

Here c_0 is the unique object of $\mathcal{A}$. we define

$$\mathcal{F}_1^d : \mathcal{A} \to \mathrm{Hom}((D, d), (D, d))$$

by

$$\mathcal{F}_1^d(x)(v) = (-1)^{\deg x(\deg v + 1)} x \cdot v.$$

As in Example-Lemma 1.7, we can easily check that $\mathcal{F}_1^d$ is a chain map and

$$\mathcal{F}_1^d(\mathfrak{m}_2(x, x')) = \mathcal{F}_1^d(x') \circ \mathcal{F}_1^d(x).$$

Hence by putting $\mathcal{F}_k^d = 0$, $k > 1$ we find an A^∞ functor $\mathcal{F} : \mathcal{A} \to \mathcal{CH}^0$.

The converse can be proved in a similar way. □

We remark that a representable functor $\mathcal{F} : \mathcal{A} \to \mathcal{CH}^0$ corresponds to $\mathcal{A}$ itself (regarded as an $\mathcal{A}$ module) by Lemma 7.39.

Definition 7.40. A *left A_∞ module* of an A_∞ algebra $\mathcal{C}$ is an A_∞ functor $\mathcal{C} \to \mathcal{CH}^0$. A *right A_∞ module* of an A_∞ algebra $\mathcal{C}$ is an A_∞ functor $\mathcal{C} \to \mathcal{CH}$.

In a way similar to the proof of Lemma 7.39, we can check that this definition coincides with one given in [FOOO] §14.

Let $\mathcal{C}$ be an A_∞ category and c, c' be objects of it. There is an obvious A_∞ functor $\mathcal{C}(c) \to \mathcal{C}$. (Here we regard $\mathcal{C}(c)$ an A_∞ algebra, that is an A_∞ category with single object c.) We compose it with $\mathcal{F}^{c'}$: $\mathcal{C} \to \mathcal{CH}$ and we obtain a left A_∞ $\mathcal{C}(c)$ module, which is $\mathcal{C}(c, c')$ as an R module. In a similar way, $\mathcal{C}(c, c')$ has a structure of right $\mathcal{C}(c')$ module. In other words, $\mathcal{C}(c, c')$ is a left $\mathcal{C}(c)$ and right $\mathcal{C}(c')$ bimodule in the sense defined in [FOOO] §14.

In the case of $\mathcal{LAG}(M, \omega)$, this implies that we have a $\mathcal{LAG}(L, b)$, $\mathcal{LAG}(L', b')$ bimodule $\mathcal{LAG}((L, b), (L', b'))$. (Here b and b' are bounding chains.) This is what is constructed in [FOOO] §14. The homology of $\mathcal{LAG}((L, b), (L', b'))$ is the Floer homology between two Lagrangian submanifolds.

Remark 7.41. Let X be a scheme over R. We can associate the following category $\mathcal{C}(X)$, which is an A_∞ category. (Note that it satisfies $\mathfrak{m}_k = 0$ for $k \neq 2$, and that $\mathfrak{m}_2$ is commutative.)

The object of $\mathcal{C}(X)$ is an affine open subsets $U_A = \mathrm{Spec}(A) \subset X$. The set of morphisms from U_A to U_B is $\{0\}$ unless $U_B \subseteq U_A$. In the case $U_B \subseteq U_A$, the set of the morphisms $\mathcal{C}(X)(U_A, U_B)$ is the ring A. If $U_C \subseteq U_B \subseteq U_A$ then $A \subseteq B \subseteq C$. $\mathfrak{m}_2$ is defined as $A \otimes B \to B$ (the product of ring B).

A functor from $\mathcal{C}(X)$ to the category of R modules can be identified with a presheaf on X.

We further study relations of A_∞ functor to twisted complex.

Lemma 7.42. *Let $\mathcal{F} : \mathcal{C}_1 \to \mathcal{C}_2$ be an A^∞ functor. Then there exists an A^∞ functor $\mathcal{DF} : \mathcal{DC}_1 \to \mathcal{DC}_2$ such that the following diagram commutes.*

Sketch of the proof. Let $(c_{k_1}, \ldots, c_{k_2}; (x_{i,j}))$ to be a twisted complex of $\mathcal{C}_1$. We put

$$y_{a,b} = \sum_k \sum_{a=i_1<\cdots<i_k=b} \mathcal{F}_k(x_{i_1}, \ldots, x_{i_k}).$$

It is easy to see that $(\mathcal{F}_0(c_{k_1}), \ldots, \mathcal{F}_0(c_{k_2}); (y_{i,j}))$ is a twisted complex of $\mathcal{C}_2$. We put

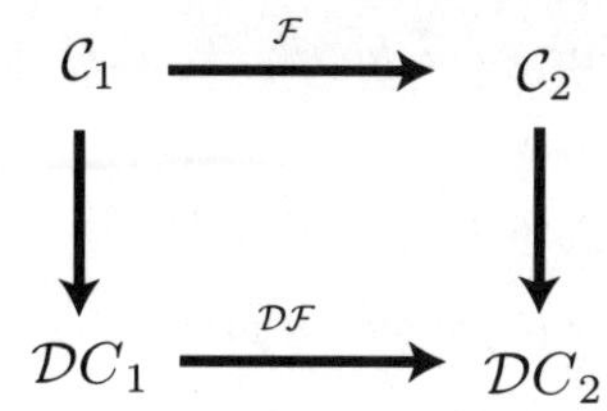

Diagram 7.1.

$$(\mathcal{D}\mathcal{F})_0(c_{k_1},\dots,c_{k_2};(x_{i,j})) = (\mathcal{F}_0(c_{k_1}),\dots,\mathcal{F}_0(c_{k_2});(y_{i,j})).$$

We omit the definition of $(\mathcal{D}\mathcal{F})_k$, $k \ge 1$. □

Lemma 7.43. *There exists an A^∞ functor $\mathcal{P} : \mathcal{D}\mathcal{D}\mathcal{F} \to \mathcal{D}\mathcal{F}$ such that $\mathcal{P}_0(\mathbf{c}) = |\mathbf{c}|$. Here $|\mathbf{c}|$ is defined in Lemma* 6.20.

The proof is straightforward and is left to the reader.

We now proceed to the definition of an A_∞ category $\mathfrak{Funk}(\mathcal{C}_1,\mathcal{C}_2)$ whose objects are A_∞ functors. (Here $\mathcal{C}_1$, $\mathcal{C}_2$ are A_∞ categories.) This section is almost the same as [Fu4] §10. However we put signs to every formula and check the formula with signs. (In [Fu4] we worked over $\mathbb{Z}_2$.) The presentation of the proof is improved also.

Let $\mathcal{F}^1$, $\mathcal{F}^2$ be A_∞ functors from $\mathcal{C}_1$ to $\mathcal{C}_2$. Let $a, b \in \mathfrak{Ob}(\mathcal{C}_1)$. Let ℓ be an integer. We consider a family of operators

$$T_k(a,b) : B_k\mathcal{C}_1[1](a,b) \to \mathcal{C}_2[1](\mathcal{F}^1_0(a),\mathcal{F}^2_0(b)) \tag{7.44}$$

of degree t. (Here $k = 1,2,\dots$ for $a \ne b$ and $k = 0,1,2,\dots$ for $a = b$.) We write $t - 1 = \mathfrak{deg}T$, $t = \mathfrak{deg}'T$. We use $\mathfrak{deg}$ in place of deg here to avoid confusion. $\mathfrak{deg}T$ will be a degree of T as a pre natural transformation as we will define below (Definition 7.49). $\deg T_k(a,b)$ is a degree as a homomorphism between graded modules. We remark that $\mathfrak{deg}'T = \deg T_k(a,b)$. We also remark that, for $\mathbf{x} \in B_k\mathcal{C}[1](a,b)$, we have

$$\deg' T_k(a,b)(\mathbf{x}) = \mathfrak{deg}'T + \deg'\mathbf{x} = \mathfrak{deg}T + \deg'\mathbf{x} + 1.$$

For $a', b' \in \mathfrak{Ob}(\mathcal{C}_2)$, let

$$\pi_{a',b'} : B\mathcal{C}_2[1](a',b') \to \mathcal{C}_2[1](a',b')$$

be the projection.

Lemma 7.45. *For each family $T_k(a,b)$ there exists uniquely a family*

$$\hat T(a,b) : B\mathcal{C}_1[1](a,b) \to B\mathcal{C}_2[1](\mathcal{F}_0^1(a), \mathcal{F}_0^2(b)),$$

of homomorphisms with the following properties.

$$\pi_{\mathcal{F}_0^1(a),\mathcal{F}_0^2(b)} \circ \hat T(a,b) = T_k(a,b) \quad \text{on } B_k\mathcal{C}_1[1](a,b). \tag{7.46.1}$$

$$\Delta \circ \hat T(a,b) = \sum_c (\hat{\mathcal{F}}^1 \hat\otimes \hat T(c,b) + \hat T(a,c) \hat\otimes \hat{\mathcal{F}}^2) \circ \Delta. \tag{7.46.2}$$

Here $\hat\otimes$ is defined by $(A\hat\otimes B)(\mathbf{x},\mathbf{y}) = (-1)^{\deg B \deg' \mathbf{x}} A(\mathbf{x}) \otimes B(\mathbf{y})$. (Note $\deg B = \mathfrak{deg}' B$ *and* $\deg' \mathbf{x}$ *is the degrees after shifted.)*

Proof. Let $\mathbf{x} \in B\mathcal{C}_1[1](a,b)$. We use notation (7.11). We define

$$\hat T(\mathbf{x}) = \sum_a (-1)^{\deg T \deg' \mathbf{x}_a^{(1)}} \hat{\mathcal{F}}^1(\mathbf{x}_a^{(1)}) \otimes T(\mathbf{x}_a^{(2)}) \otimes \hat{\mathcal{F}}^3(\mathbf{x}_a^{(3)}). \tag{7.47}$$

It is easy to check (7.46). Uniqueness is also easy to show. □

Hereafter we write T etc. in place of $T_k(a,b)$ etc. in case no confusion can occur.

Lemma 7.48. *For each family $T_k(a,b)$ with $t = \mathfrak{deg}' T_k(a,b) =$* $\deg T$, *there exists a family $(\delta T)_k(a,b)$ such that*

$$\widehat{\delta T} = \hat d \circ \hat T + (-1)^{t+1} \hat T \circ \hat d.$$

Proof. We calculate

$$\begin{aligned}
&\Delta \circ (\hat d \circ \hat T + (-1)^{t+1} \hat T \circ \hat d) \\
&= (\hat d \hat\otimes 1 + 1 \hat\otimes \hat d) \circ \Delta \circ \hat T + (-1)^t \Delta \circ \hat T \circ \hat d \\
&= (\hat d \hat\otimes 1 + 1 \hat\otimes \hat d) \circ (\hat{\mathcal{F}}^1 \hat\otimes \hat T + \hat T \hat\otimes \hat{\mathcal{F}}^2) \circ \Delta \\
&\qquad + (-1)^{t+1} (\hat{\mathcal{F}}^1 \hat\otimes \hat T + \hat T \hat\otimes \hat{\mathcal{F}}^2) \circ (\hat d \hat\otimes 1 + 1 \hat\otimes \hat d) \circ \Delta \\
&= (\hat{\mathcal{F}}^1 \hat\otimes (\hat d \circ \hat T + (-1)^{t+1} \hat T \circ \hat d) + (\hat d \circ \hat T + (-1)^{t+1} \hat T \circ \hat d) \hat\otimes \hat{\mathcal{F}}^2) \circ \Delta.
\end{aligned}$$

Lemma 7.48 then follows from Lemma 7.45. □

Definition 7.49. We say a family T as in (7.44), a *pre natural transformation* from $\mathcal{F}^1$ to $\mathcal{F}^2$. We let $\mathfrak{Funk}(\mathcal{F}^1, \mathcal{F}^2)$ be the set of all pre natural transformations. (It is a graded module over the commutative ring R we work on.)

We define a boundary operator $\mathfrak{M}_1 = -\delta$ on it by Lemma 7.48.

We say that T is a *natural transformation* or A_∞ *transformation* if it is δ closed.

We remark that $\mathfrak{deg}(T) = \deg(T_0(1))$, where $T_0(1) \in \mathcal{C}[1](c,c)$.

We put minus sign in $\mathfrak{M}_1 = -\delta$ since we need it to show Theorem 9.1 later.

Corollary 7.50. $\delta \circ \delta = 0$. *In other words,* $(\mathfrak{Funk}(\mathcal{F}^1, \mathcal{F}^2), \mathfrak{M}_1)$ *is a chain complex.*

Corollary 7.50 is immediate from Lemma 7.45. We use the symbol $\mathfrak{M}$ to denote the operations on $\mathfrak{Funk}(\mathcal{F}^1, \mathcal{F}^2)$ in order to distinguish it from operations on $\mathcal{C}_1$, $\mathcal{C}_2$.

Remark 7.51. If T is a natural transformation, then $T_0(a) = T_0(a)(1)$ is a closed morphism in $\mathcal{C}_2(\mathcal{F}_0^1(a), \mathcal{F}_0^2(a))$. (Here $1 \in B_0\mathcal{C}_1[1] =$ the coefficient ring R.) Moreover we have

$$\mathfrak{m}_2(T_1(a,b)(x), T_0(b)) \equiv \pm \mathfrak{m}_2(T_0(a), T_1(a,b)(x)) \mod \operatorname{Im} \mathfrak{m}_1.$$

for each $x \in \mathcal{C}_1(a,b)$. Thus natural transformation in our sense defines a natural transformation in the usual sense in homology level.

We next define $\mathfrak{m}_k$, $k \geq 2$. Let $\mathcal{F}^i$, $i = 0, \dots, k$ be A_∞ functors from $\mathcal{C}_1$ to $\mathcal{C}_2$ and $T^i \in \mathfrak{Funk}(\mathcal{F}^{i-1}, \mathcal{F}^i)$ be pre natural transformations. We put $t_i = \mathfrak{deg}' T^i$. Let $\mathbf{x} \in B(\mathcal{C}_1[1])$. We consider

$$\Delta^{2k+1}\mathbf{x} = \sum_a \mathbf{x}_a^{(1)} \otimes \cdots \otimes \mathbf{x}_a^{(2k+1)}.$$

and put

$$\begin{aligned} &\hat{T}(\mathbf{x}) = \\ (7.52)\quad &-\sum (-1)^{\epsilon_a} \mathfrak{m}\left(\hat{\mathcal{F}}^0(\mathbf{x}_a^{(1)}), T^1(\mathbf{x}_a^{(2)}), \dots, T^k(\mathbf{x}_a^{(2k)}), \hat{\mathcal{F}}^k(\mathbf{x}_a^{(2k+1)})\right), \end{aligned}$$

where

$$(7.53)\qquad \epsilon_a = \sum_{j=1}^{k} \sum_{i=1}^{2j-1} t_j \deg' \mathbf{x}_a^{(i)}.$$

It is easy to see that $\hat{T}$ satisfies (7.46.2). Hence we can use it to define a pre natural transformation T. We also remark that $\mathfrak{deg} T = t_1 + \cdots t_k$ and is $\mathfrak{deg}' T = t_1 + \cdots t_k + 1$.

Definition 7.54. We put $\mathfrak{M}_k(T^1, \dots, T^k) = T$.

We remark that $\mathfrak{M}_k$ is of degree one in the sense of $\mathfrak{deg}'$. Namely we have

$$\mathfrak{deg}' T = \mathfrak{deg}' T^1 + \cdots + \mathfrak{deg}' T^k + 1.$$

We remark that the overall minus sign in (7.52) will be necessary for Theorem 9.1 to hold.

Theorem-Definition 7.55. *There exists an A_∞ category $\mathfrak{Funk}(\mathcal{C}_1,\mathcal{C}_2)$ such that its object is an A_∞ functor : $\mathcal{C}_1 \to \mathcal{C}_2$, the set of morphisms are the set of pre natural transformations, and that (higher) compositions are defined by Definition* 7.54.

Proof. The A_∞ formula we are going to check is:

$$0 = \sum_{1\le m\le \ell\le k} (-1)^{t_1+\dots+t_{m-1}}\mathfrak{M}_{k-\ell+m}(T^1,\dots,T^{m-1}, \mathfrak{M}_{\ell-m+1}(T^m,\dots,T^\ell),T^{\ell+1},\dots,T^k)(\mathbf{x}), \tag{7.56}$$

where $t_i = \mathfrak{deg}'T^i$. To prove (7.56), we compare it with

$$0 = (-1)^{\epsilon_1(a)}(\mathfrak{m}\circ\hat{d})(\hat{\mathcal{F}}^0(\mathbf{x}_a^{(1)}),T^1(\mathbf{x}_a^{(2)}),\dots,T^k(\mathbf{x}_a^{(2k)}),\hat{\mathcal{F}}^k(\mathbf{x}_a^{(2k+1)})), \tag{7.57}$$

where

$$\epsilon_1(a) = \sum_{j=1}^{k}\sum_{i=1}^{2j-1} t_j \deg' \mathbf{x}_a^{(i)}. \tag{7.58}$$

The formula (7.57) follows from the A_∞ formula of $\mathcal{C}_2$. We study the terms appearing in (7.56) and (7.57).

We first study the terms appearing in (7.57). There are two types of them. One is:

Type 1: Let $1 \le m < \ell \le k$. Then we have

$$\begin{aligned}\mathfrak{m}\Big(&\hat{\mathcal{F}}^0(\mathbf{x}_m^{(1)}),T^1(\mathbf{x}_a^{(2)}),\dots,\hat{\mathcal{F}}^{m-1}(\mathbf{x}_{a,b}^{(2m-1),(1)}),\\ &\mathfrak{m}\Big(\hat{\mathcal{F}}^{m-1}(\mathbf{x}_{a,b}^{(2m-1),(2)}),T^m(\mathbf{x}_a^{(2m)}),\dots\\ &\dots,T^{\ell-1}(\mathbf{x}_a^{(2\ell-2)}),\hat{\mathcal{F}}^\ell(\mathbf{x}_{a,b}^{(2\ell-1),(1)})\Big)\\ &\hat{\mathcal{F}}^\ell(\mathbf{x}_{a,b}^{(2\ell-1),(2)}),\dots,T^k(\mathbf{x}_a^{(2k)}),\hat{\mathcal{F}}^k(\mathbf{x}_a^{(2k+1)})\Big).\end{aligned} \tag{7.59}$$

Here we put

$$\Delta(\mathbf{x}_a^{(i)}) = \sum_b \mathbf{x}_{a,b}^{(i),(1)} \otimes \mathbf{x}_{a,b}^{(i),(2)}.$$

For a moment we do not put sign in the formula. We will check the sign carefully later.

Type 2: This is an analogue of (7.59) in the case when $1 \le m = \ell \le k$. That is:

$$\begin{aligned}(7.60)\qquad &\mathfrak{m}(\hat{\mathcal{F}}^0(\mathbf{x}_a^{(1)}), T^1(\mathbf{x}_a^{(2)}), \dots, T^m(\mathbf{x}_a^{(2m)}),\\ &\qquad \hat{d}(\hat{\mathcal{F}}^m(\mathbf{x}_a^{(2m+1)})), T^{m+1}(\mathbf{x}_a^{(2m+2)}),\\ &\qquad\qquad \dots, T^k(\mathbf{x}_a^{(2k)}), \hat{\mathcal{F}}^k(\mathbf{x}_a^{(2k+1)})).\end{aligned}$$

Now we turn to the terms appearing in (7.56).

Type 3: This is exactly the same as (7.59).

The other types of terms in (7.56) are the cases when $\mathfrak{M}_1$ appears. We remark that $\mathfrak{M}_1$ appears in (7.56) in case $\ell = m + k - 1$ or $\ell = m$. The terms of Types 4, 5 below correspond to the case either $\ell = m+k-1$ and the terms of Type 6 correspond to the case $\ell = m$.

Let us first consider the case $\ell = m + k - 1$. We recall

$$\begin{aligned}(7.61)\qquad (\mathfrak{M}_1(\mathfrak{M}_k(T^1, \dots, T^k)))(\mathbf{x}) &= (\mathfrak{m} \circ \mathfrak{M}_k(T^1, \dots, T^k)^\wedge)(\mathbf{x})\\ &\quad +(-1)^{t_1+\cdots+t_k}(\mathfrak{M}_k(T^1, \dots, T^k) \circ \hat{d})(\mathbf{x}).\end{aligned}$$

We remark that $\mathfrak{deg}'\mathfrak{M}_k(T^1, \dots, T^k) = t_1 + \cdots + t_k + 1$. Note that the first term of (7.61) is already included in Type 3. So we only need to consider the second term. They are one of the following two types.

Type 4:

$$\begin{aligned}(7.62)\qquad &\mathfrak{m}(\hat{\mathcal{F}}^0(\mathbf{x}_a^{(1)}), T^1(\mathbf{x}_a^{(2)}), \dots, T^m(\mathbf{x}_a^{(2m)}),\\ &\quad \hat{\mathcal{F}}^m(\hat{d}(\mathbf{x}_a^{(2m+1)})), T^{m+1}(\mathbf{x}_a^{(2m+2)}), \dots, T^k(\mathbf{x}_a^{(2k)}), \hat{\mathcal{F}}^k(\mathbf{x}_a^{(2k+1)})).\end{aligned}$$

Type 5:

$$\begin{aligned}(7.63)\qquad &\mathfrak{m}(\hat{\mathcal{F}}^0(\mathbf{x}_a^{(1)}), T^1(\mathbf{x}_a^{(2)}), \dots, \hat{\mathcal{F}}^{m-1}(\mathbf{x}_a^{(2m-1)}),\\ &\quad T^m(\hat{d}(\mathbf{x}_a^{(2m)})), \dots, T^k(\mathbf{x}_a^{(2k)}), \hat{\mathcal{F}}^k(\mathbf{x}_a^{(2k+1)})).\end{aligned}$$

We next consider the case $\ell = m$. we recall

$$\begin{aligned}(7.64)\qquad &\mathfrak{M}_k(T^1, \dots, \mathfrak{M}_1(T^m), \dots, T^k)(\mathbf{x})\\ &\quad = \mathfrak{M}_k(T^1, \dots, \mathfrak{m} \circ \hat{T}^m, \dots, T^k)(\mathbf{x})\\ &\quad +(-1)^{t_m}\mathfrak{M}_k(T^1, \dots, T^m \circ \hat{d}, \dots, T^k)(\mathbf{x}).\end{aligned}$$

The first term again is included in Type 3. The second term then gives:

Type 6: This type is actually the same as Type 5.

We have finished describing all the types of the terms in (7.56). We now can prove the case when the coefficient ring is $\mathbb{Z}_2$ immediately. Namely terms of Types 1 and 3 cancel each other and terms of Types 2 and 4 cancel each other, since $\hat{\mathcal{F}}$ is a chain map, and terms of Types 5 and 6 cancel each other.

Let us now check the sign.

The sign in Type 3 is given by $(-1)^{\epsilon_2(a)+t_1+\cdots+t_{m-1}}$ where

$$
\begin{aligned}
\epsilon_2(a) \;&=\; \sum_{j=m}^{\ell} t_j \left(\deg' \mathbf{x}_{a,b}^{(2m-1),(2)} + \sum_{i=2m}^{2j-1} \deg' \mathbf{x}_a^i \right) \\
&\quad + \sum_{j=1}^{m-1} \sum_{i=1}^{2j-1} t_j \deg' \mathbf{x}_a^i \\
&\quad + (t_m + \cdots + t_\ell + 1) \times \\
&\quad \times \left(\sum_{i=0}^{2m-2} \deg' \mathbf{x}_a^i + \deg' \mathbf{x}_{a,b}^{(2m-1),(1)} \right) \\
&\quad + \sum_{j=\ell+1}^{k} \sum_{i=1}^{2j-1} t_j \deg' \mathbf{x}_a^i .
\end{aligned}
\tag{7.65}
$$

Note that the first line in (7.65) comes from $\mathfrak{M}_{\ell-m+1}(T^m,\dots,T^\ell)$ in (7.56) and the rest comes from $\mathfrak{M}_{k-\ell+m}$.

We calculate (7.65) and obtain

$$
\epsilon_2(a) - \epsilon_1(a) = \sum_{i=0}^{2m-2} \deg' \mathbf{x}_a^i + \deg' \mathbf{x}_{a,b}^{(2m-1),(1)}.
$$

On the other hand, the sign in the case of Type 1 in (7.57) is $(-1)^{\epsilon_1(a)+\epsilon_3(a)}$ where

$$
\begin{aligned}
\epsilon_3(a) \;&=\; \deg' \mathbf{x}_m^{(1)} + \deg' T^1(\mathbf{x}_a^{(2)}) + \cdots + \deg' \mathbf{x}_{a,b}^{(2m-1),(1)} \\
&\equiv\; \sum_{i=0}^{2m-2} \deg' \mathbf{x}_a^i + \deg' \mathbf{x}_{a,b}^{(2m-1),(1)} + \sum_{i=1}^{m-1} t_i \mod 2.
\end{aligned}
$$

Thus Types 1 and 3 coincides together with sign.

We next consider Types 2 and 4. We find using (7.61) that the sign of both of them is

$$
\epsilon_1(a) + \sum_{i=1}^{m} t_i + \sum_{j=1}^{2m} \deg' \mathbf{x}_a^j
$$

and coincides to each other.

Next the sign of the terms of Type 5 is $(-1)^{\epsilon_4(a)}$, where

$$\epsilon_4(a) = \epsilon_1(a) + \sum_{i=1}^{m} t_i + \sum_{j=1}^{2m+1} \deg' \mathbf{x}_a^j.$$

We study Type 6. Note that we have sign $(-1)^{t_1+\cdots+t_{m-1}}$ in (7.56). So using (7.64) we find that the sign of the terms of Type 6 is $(-1)^{\epsilon_4(a)+1}$. The proof of A_∞ formula (7.56) is now complete.

Finally we define unit, that is the identity transform $\mathrm{Id}^{\mathcal{F}}$ for each $\mathcal{F} \in \mathfrak{Ob}(\mathcal{C}_1, \mathcal{C}_2)$. Let $\mathbf{e}_c \in \mathcal{C}_2^0(c,c)$ be the unit in $\mathcal{C}_2$. Namely we assume that (1.5) is satisfied for it. We put

$$\mathrm{Id}_0^{\mathcal{F}}(a) = -\mathbf{e}_{\mathcal{F}_0(a)} \in \mathcal{C}_2^0(\mathcal{F}_0(a), \mathcal{F}_0(a)), \tag{7.66.1}$$

$$\mathrm{Id}_k^{\mathcal{F}} = 0, \qquad \text{for } k \geq 1. \tag{7.66.2}$$

(Note that $\mathrm{Id}_1^{\mathcal{F}}(x) \neq x$.) It is easy to see from definition that $\mathrm{Id}^{\mathcal{F}}$ satisfies (1.5) for $\mathfrak{M}$. (We remark that we need minus sign in (7.66) since we put overall minus sign in the definition of $\mathfrak{M}_k$.) □

§8. Homotopy equivalence and A_∞ analogue of J. H. C. Whitehead Theorem

We now define homotopy equivalence between A_∞ categories.

Definition 8.1. Two A_∞ functors $\mathcal{F}^1, \mathcal{F}^2 : \mathcal{C}_1 \to \mathcal{C}_2$ are said to be *homotopic* to each other if they are homotopy equivalent as objects of $\mathfrak{Funt}(\mathcal{C}_1, \mathcal{C}_2)$ in the sense of Definition 6.22.

Definition 8.2. The identity functor $\mathrm{Id}^{\mathcal{C}} : \mathcal{C} \to \mathcal{C}$ is defined as follows.

(8.3.1) $\mathrm{Id}_0^{\mathcal{C}}(c) = c$.
(8.3.2) $\mathrm{Id}_1^{\mathcal{C}}(x) = x$.
(8.3.3) $\mathrm{Id}_k^{\mathcal{C}} = 0$ for $k \geq 2$.

Remark 8.4. The identity functor is similar to but is different from the identity transformation defined at the end of the last section.

Definition 8.5. An A_∞ functor $\mathcal{F} : \mathcal{C}_1 \to \mathcal{C}_2$ is said to be a *homotopy equivalence* if there exists an A_∞ functor $\mathcal{F}' : \mathcal{C}_2 \to \mathcal{C}_1$ such that the composition $\mathcal{F} \circ \mathcal{F}'$ is homotopic to $\mathrm{Id}^{\mathcal{C}_2}$ and that $\mathcal{F}' \circ \mathcal{F}$ is homotopic to $\mathrm{Id}^{\mathcal{C}_1}$.

Two A_∞ categories are said to be *homotopy equivalent* to each other if there exists a homotopy equivalence between them.

Now the main result of this section is:

Theorem 8.6. *Let $\mathcal{F} : \mathcal{C}_1 \to \mathcal{C}_2$ be an A_∞ functor such that:*
(8.7.1) $\mathcal{F}_1 : \mathcal{C}_1(c_1, c_1') \to \mathcal{C}_2(\mathcal{F}_0(c_1), \mathcal{F}_0(c_1'))$ *induces an isomorphism on* $\mathfrak{m}_1$ *homology:*
(8.7.2) *For any* $c_2 \in \mathfrak{Ob}(\mathcal{C}_2)$ *there exists* $c_1 \in \mathfrak{Ob}(\mathcal{C}_1)$ *such that* $\mathcal{F}_0(c_1)$ *is homotpy equivalent to* c_2.

Then $\mathcal{F}$ *is a homotopy equivalence.*

Remark 8.8. In the case of A_∞ algebras, Theorem 8.9 was proved in [FOOO] §A5. (8.7) is used as a *definition* of homotopy equivalence in [KoS1]. It seems that Kontsevich-Soibelman announced in [KoS1] that they will prove in [KoS2] a result similar to Theorem 8.6.

To prove Theorem 8.6, we start with the following special case.

Proposition 8.9. *We assume* (8.7.1) *and that* $\mathcal{F}_0 : \mathfrak{Ob}(\mathcal{C}_1) \to \mathfrak{Ob}(\mathcal{C}_2)$ *is a bijection. Then* $\mathcal{F}$ *is a homotopy equivalence.*

Proof. The proof is similar to the argument of [FOOO] §A5. We need to construct an A_∞ functor $\mathcal{F}' : \mathcal{C}_2 \to \mathcal{C}_1$ and a natural transformation $T : \mathcal{F} \circ \mathcal{F}' \to \mathrm{Id}^{\mathcal{C}_2}$. For this purpose, we construct $\mathcal{F}'_k : B_k\mathcal{C}_2[1] \to B_1\mathcal{C}_1[1]$ and $T_k : B_k\mathcal{C}_2[1] \to B_1\mathcal{C}_2[1]$ inductively on k. To describe the induction hypothesis, we define the notions, A_k functor and A_k transformation.

Let $\mathcal{C}$, $\mathcal{C}'$ be A_∞ categories, $\mathcal{G}_0 : B_0\mathcal{C} \to B_0\mathcal{C}'$ be a map and $\mathcal{G}_\ell : B_\ell\mathcal{C}[1](c_1, c_2) \to B_1\mathcal{C}'[1](c_1, c_2)$, $1 \le i \le k$ be R module homomorphisms of degree 0.

We put:

$$B_{a,\dots,b}\mathcal{C}[1] = \frac{\bigoplus_{i=0}^{a} B_i\mathcal{C}[1]}{\bigoplus_{i=0}^{b-1} B_i\mathcal{C}[1]}.$$

The boundary operator of the chain complex $B\mathcal{C}[1]$ induces a boundary operator on $B_{a,\dots,b}\mathcal{C}[1]$. Hence $B_{a,\dots,b}\mathcal{C}[1]$ is a chain complex.

Lemma 8.10. *There exists uniquely a coalgebra homomorphism*

$$\hat{\mathcal{G}} : B\mathcal{C}[1] \to B\mathcal{C}'[1]$$

such that
(8.11.1) $\hat{\mathcal{G}} = \hat{\mathcal{G}}_{0,\dots,k}$ *on* $\bigoplus_{i=0}^{k} B_i\mathcal{C}[1]$.
(8.11.2) $B_1\mathcal{C}'[1]$ *component of* $\hat{\mathcal{G}}(B_i\mathcal{C}[1])$ *is* 0 *for* $i > k$.

Proof. We define $\mathcal{G} : B\mathcal{C}[1] \to B_1\mathcal{C}'[1]$ by $\mathcal{G} = \mathcal{G}_\ell$ on $B_\ell\mathcal{C}[1]$ $1 \le \ell \le k$, and $\mathcal{G} = 0$ on $B_\ell\mathcal{C}[1]$ $\ell > k$. We now define $\hat{\mathcal{G}}$ in the same way as (7.3). Namely

$$\hat{\mathcal{G}}(\mathbf{x}) = \sum_{i,a} \mathcal{G}(\mathbf{x}_{i,a}^{(1)}) \otimes \cdots \otimes \mathcal{G}(\mathbf{x}_{i,a}^{(i)}).$$

Here

$$\Delta^i(\mathbf{x}) = \sum_a \mathbf{x}_{i,a}^{(1)} \otimes \cdots \otimes \mathbf{x}_{i,a}^{(i)}.$$

(The definition of $\hat{\mathcal{G}}$ on $B_0\mathcal{C}[1]$ is the same as (7.3.2).) It is easy to check (8.11). □

$\hat{\mathcal{G}}$ induces a homomorphism $:B_{a,\dots,b}\mathcal{C}[1] \to B_{a,\dots,b}\mathcal{C}[1]$ for each $a < b$. We denote it by $\mathcal{G}_{a,\dots,b}$.

Definition 8.12. We say $\mathcal{G}_i$ $i \le k$ is an A_k *functor* if $\mathcal{G}_{0,\dots,k}$ is a chain map.

Lemma 8.13. *If $\mathcal{G}_i$, $i \le k$ is an A_k functor then $\mathcal{G}_{a,\dots,b}$ is a chain map for $b - a \le k - 1$.*

The proof is easy and is left to the reader.

Let us next denfine an A_k transformation. Let $\mathcal{G}$ and $\mathcal{G}'$ be A_k functors from $\mathcal{C}$ to $\mathcal{C}'$. Let $S_\ell : B_\ell\mathcal{C}[1](c_1, c_2) \to B_1\mathcal{C}'[1](\mathcal{G}_0(c_1), \mathcal{G}'_0(c_2))$, $1 \le \ell \le k$ be R module homomorphisms of degree $s + 1$. It induces

$$S_{0,\dots,k} : \bigoplus_{i=0}^{k} B_i\mathcal{C}[1] \to \bigoplus_i B_i\mathcal{C}'[1] \tag{8.14}$$

as follows. We define $S : B\mathcal{C}[1] \to B_1\mathcal{C}'[1]$ by $S = S_\ell$ on $B_\ell\mathcal{C}[1]$, $0 \le \ell \le k$ and $S = 0$ on $B_\ell\mathcal{C}$, $\ell > k$. We then put:

$$\begin{aligned} &S_{0,\dots,k}(\mathbf{x}) \\ &= \textstyle\sum_a (-1)^{s \deg' \mathbf{x}_a^{(1)}} \mathcal{G}_{0,\dots,k}(\mathbf{x}_a^{(1)}) \otimes S(\mathbf{x}_a^{(2)}) \otimes \mathcal{G}'_{0,\dots,k}(\mathbf{x}_a^{(3)}). \end{aligned} \tag{8.15}$$

We remark here in (8.15) the case $\mathbf{x}_a^{(2)} = 1 \in B_0\mathcal{C}[1]$ is included. As a consequence, the image of $S_{0,\dots,k}$ is not contained in $\bigoplus_{i=0}^k B_i\mathcal{C}[1]$.

Definition 8.16. We say S_ℓ, $0 \le \ell \le k$, is an A_k *transformation* if

$$\hat{d} \circ S_{0,\dots,k} + (-1)^{s+1} S_{0,\dots,k} \circ \hat{d} = 0,$$

on $\bigoplus_{i=0}^k B_i\mathcal{C}[1]$. We put $\mathfrak{deg}'S = s$.

Now we go back to the proof of Proposition 8.9. We are going to construct an A_k functor $\mathcal{F}'_\ell$ $\ell \le k$, and an A_k transformation T_ℓ $\ell \le k$, from $\mathcal{F} \circ \mathcal{F}'$ to $\mathrm{Id}^{\mathcal{C}_2}$ by induction on k. (We remark that the composition of two A_k functors is well defined and is an A_k functor.)

We start with the case $k = 0$. We put

$$\mathcal{F}'_0 = \mathcal{F}_0^{-1} : \mathfrak{Ob}(\mathcal{C}_2) \to \mathfrak{Ob}(\mathcal{C}_1). \tag{8.17}$$

We remark that

$$B_0\mathcal{C}[1] = \bigoplus_{c \in \mathfrak{Ob}(\mathcal{C})} R$$

hence a map $\mathfrak{Ob}(\mathcal{C}_2) \to \mathfrak{Ob}(\mathcal{C}_1)$ induces a homomorphism $B_0\mathcal{C}_2[1] \to B_0\mathcal{C}_1[1]$.

We next put

$$T_0(c) = \mathbf{e}_c \in \mathcal{C}_1(\mathcal{F}_0(\mathcal{F}'_0(c)), c) = \mathcal{C}_1(c, c). \tag{8.18}$$

Now we assume that we have an A_{k-1} transformation $\mathcal{F}'_\ell$, $0 \le \ell \le k-1$ and an A_{k-1} transformation T_ℓ, $0 \le \ell \le k-1$ and will consider the case of k. It follows from Lemma 8.13 that $\mathcal{F}'_\ell$, $0 \le \ell \le k-1$ determine a chain map

$$\mathcal{F}'_{2,\dots,k} : B_{2,\dots,k}\mathcal{C}_2[1] \to B_{2,\dots,k}\mathcal{C}_1[1]. \tag{8.19}$$

Using the obvious R module isomorphism

$$B_{2,\dots,k}\mathcal{C}[1] = \bigoplus_{i=2}^{k} B_i\mathcal{C}[1]$$

(8.19) induces an R module homomorphism

$$\tilde{\mathcal{F}}'_{2,\dots,k} : \bigoplus_{i=2}^{k} B_i\mathcal{C}[1] \to \bigoplus_{i=1}^{k} B_i\mathcal{C}[1]. \tag{8.20}$$

Note however that (8.20) is *not* a chain map in general.

It is easy to see that (8.20) coincides with $\tilde{\mathcal{F}}'_{1,\dots,k-1}$ on $\bigoplus_{i=2}^{k-1} B_i\mathcal{C}$. Hence we extend (8.20) to

$$\tilde{\mathcal{F}}'_{1,\dots,k} : \bigoplus_{i=1}^{k} B_i\mathcal{C} \to \bigoplus_{i=1}^{k} B_i\mathcal{C}. \tag{8.21}$$

We remark however that (8.21) is *not* a chain map in general. Nevertheless, using the fact that $\mathcal{F}'_{2,\dots,k}$ and $\mathcal{F}'_{1,\dots,k-1}$ are chain maps (Lemma 8.13), we can easily prove the following:

Lemma 8.22. *The image of* $\tilde{\mathcal{F}}'_{1,\dots,k} \circ \hat{d} - \hat{d} \circ \tilde{\mathcal{F}}'_{1,\dots,k}$ *is contained in* $B_1\mathcal{C}_1[1]$. *Moreover* $\tilde{\mathcal{F}}'_{1,\dots,k} \circ \hat{d} - \hat{d} \circ \tilde{\mathcal{F}}'_{1,\dots,k}$ *vanishes on* $\bigoplus_{i=1}^{k-1} B_i\mathcal{C}_2[1]$.

We put

$$\Xi = \tilde{\mathcal{F}}'_{1,\dots,k} \circ \hat{d} - \hat{d} \circ \tilde{\mathcal{F}}'_{1,\dots,k} : B_k\mathcal{C}_2[1] \to B_1\mathcal{C}_1[1].$$

Note that $\mathfrak{m}_1$ defines an R module homomorphism

$$d_1 : B_k\mathcal{C}_2[1] \to B_k\mathcal{C}_2[1],$$

by

$$\begin{aligned} & d_1(x_1 \otimes \cdots \otimes x_k) \\ &= \sum_i (-1)^{\deg' x_1 + \cdots \deg' x_{i+1}} x_1 \otimes \cdots \otimes \mathfrak{m}_1(x_i) \otimes \cdots \otimes x_k. \end{aligned}$$

We can prove $d_1 d_1 = 0$ easily.

We define a boundary operator on $\mathrm{Hom}(B_k\mathcal{C}_2[1], B_1\mathcal{C}_1[1])$ by

$$d\phi = \mathfrak{m}_1 \circ \phi + (-1)^{\deg' \phi} \phi \circ d_1. \tag{8.23}$$

Lemma 8.22 then implies

$$d\Xi = 0.$$

We moreover have the following:

Lemma 8.24. Ξ *is contained in the image of* d.

Proof. We define a boundary operator d on $\mathrm{Hom}(B_k\mathcal{C}_2[1], B_1\mathcal{C}_2[1])$ in a way similar to (8.23). Then $\mathcal{F}_1$ induces a chain homomorphism $\mathcal{F}_{1,*}$: $\mathrm{Hom}(B_k\mathcal{C}_2, B_1\mathcal{C}_1) \to \mathrm{Hom}(B_k\mathcal{C}_2, B_1\mathcal{C}_2)$. By assumption $\mathcal{F}_{1,*}$ induces an isomorphism on homology. So it suffices to show that $\mathcal{F}_1 \circ \Xi$ is a d boundary.

We next rewrite the Definition 8.16 (Sublemma 8.26). We need some notations for it.

We define $T : B\mathcal{C}_1[1] \to B_1\mathcal{C}_1[1]$ by $T = T_\ell$ on $B_\ell\mathcal{C}_1[1]$, $0 \le \ell \le k-1$ and $T = 0$ on $B_\ell\mathcal{C}_1[1]$, $\ell > k$. We define $\overline{T} : B\mathcal{C}_1[1] \to B_1\mathcal{C}_1[1]$ by $\overline{T} = T_\ell$ on $B_\ell\mathcal{C}_1[1]$, $1 \le \ell \le k-1$ and $T = 0$ on $B_\ell\mathcal{C}_1[1]$, $\ell > k$ or $\ell = 0$.

We then define $\hat{T}$ and $\hat{\overline{T}}$ in the same way as (8.17). Namely we put

$$\begin{aligned} \hat{T}(\mathbf{x}) &= \textstyle\sum_a (-1)^{\deg' \mathbf{x}_a^{(1)}} \hat{\mathcal{F}}'(\hat{\mathcal{F}}(\mathbf{x}_a^{(1)})) \otimes T(\mathbf{x}_a^{(2)}) \otimes \mathbf{x}_a^{(3)}, \\ \hat{\overline{T}}(\mathbf{x}) &= \textstyle\sum_a (-1)^{\deg' \mathbf{x}_a^{(1)}} \hat{\mathcal{F}}'(\hat{\mathcal{F}}(\mathbf{x}_a^{(1)})) \otimes \overline{T}(\mathbf{x}_a^{(2)}) \otimes \mathbf{x}_a^{(3)}. \end{aligned} \tag{8.25}$$

We consider a filtration $\mathfrak{F}$ (the number filtration) of $BC_1[1]$ such that $\mathfrak{F}^k BC_1[1] = \bigoplus_{i=0}^k B_i C_1[1]$. Note that $\hat{T}$ does not preserve this filtration since $T_0(B_0 C_1[1]) \subset B_1 C_1[1]$ is nonzero. But $\hat{\overline{T}}$ does preserve this filtration. Therefore $\hat{\overline{T}}$ induces an R module homomorphism $B_{a,\dots,b} C_1[1] \to B_{a,\dots,b} C_1[1]$. We write it as $\overline{T}_{a,\dots,b}$.

We next recall that $\mathcal{F}' : BC_2[1] \to B_1 C_2[1]$ is equal to $\mathcal{F}'_\ell$ on $B_\ell C_2[1]$ $1 \le \ell \le k-1$, and is equal to 0 on $B_\ell C_2[1]$ $\ell \ge k$.

Sublemma 8.26. *T_ℓ $0 \le \ell \le k-1$, is an A_{k-1} transformation, if and only if*

$$\mathcal{F} \circ \tilde{\mathcal{F}}'_{1,\dots,k-1} - \mathrm{id} - \mathfrak{m} \circ \overline{T}_{1,\dots,k-1} - \overline{T} \circ \hat{d} = 0. \tag{8.27}$$

Here (8.27) is a formula for an element in

$$\mathrm{Hom}(B_{1,\dots,k-1} C_2[1], B_1 C_2[1]),$$

and

$$\mathrm{id} \in \mathrm{Hom}(B_1 C_2[1], B_1 C_2[1]) \subset \mathrm{Hom}(B_{1,\dots,k-1} C_2[1], B_1 C_2[1]).$$

Proof. We consider

$$\tilde{\Psi} = -\hat{d} \circ T_{0,\dots,k-1} + T_{0,\dots,k-1} \circ \hat{d} \in \mathrm{Hom}(B_{0,\dots,k-1} C_2[1], BC_2[1]). \tag{8.28}$$

Let Ψ be its $\mathrm{Hom}(B_{0,\dots,k-1} C_2[1], B_1 C_2[1])$ component. It is easy to see

$$\tilde{\Psi}(\mathbf{x}) = \sum_a \mathbf{x}_a^{(1)} \otimes \Psi(\mathbf{x}_a^{(2)}) \otimes \mathbf{x}_a^{(3)}.$$

Hence $\tilde{\Psi}$ is zero if and only if Ψ is zero. On the other hand $\tilde{\Psi}$ is zero if and only if T_ℓ, $0 \le \ell \le k-1$ is an A_{k-1} transformation, by definition. Hence, to prove Sublemma 8.26, it suffices to show that Ψ is the left hand side of (8.27). This can be easily seen by using (8.18) and the fact that $\mathbf{e}_c$ is the unit as follows.

Let us consider the $\mathrm{Hom}(B_{0,\dots,k-1} C_2[1], B_1 C_2[1])$ component of (8.28). We find that the sum of the terms of it which contains T_0 is

$$\mathbf{x} \mapsto \sum_a \mathfrak{m}\left((-1)^{\deg' \mathbf{x}_a^{(1)}} \hat{\mathcal{F}}(\hat{\mathcal{F}'}(\mathbf{x}_a^{(1)})) \otimes \mathbf{e} \otimes \mathbf{x}_a^{(2)}\right). \tag{8.29}$$

The terms of the right hand side of (8.29) are zero unless $\mathbf{x}_a^{(2)} = 1$ or $\mathbf{x}_a^{(1)} = 1$. In the first case it is

$$(-1)^{\deg' \mathbf{x}} \mathfrak{m}_2(\hat{\mathcal{F}}(\hat{\mathcal{F}'}(\mathbf{x})), \mathbf{e}) = -\hat{\mathcal{F}}(\hat{\mathcal{F}'}(\mathbf{x})).$$

In the second case, it is

$$\mathfrak{m}(\mathbf{e} \otimes \mathbf{x}) = \begin{cases} \mathbf{x} & \text{if } \mathbf{x} \in \mathcal{C}_1[1], \\ 0 & \text{otherwise.} \end{cases}$$

Thus they correspond to the first and second terms of (8.27). The terms of the $\mathrm{Hom}(B_{0,\dots,k-1}\mathcal{C}_2[1], B_1\mathcal{C}_1[1])$ component of (8.28) which do not contain T_0 correspond to the third and fourth terms of (8.27). (Note $\mathfrak{deg} T = 0$.)

The proof of Sublemma 8.26 is complete. □

We go back to the proof of Lemma 8.24. (8.27) implies

$$\mathcal{F}_{a,\dots,b} \circ \mathcal{F}'_{a,\dots,b} - \mathrm{id} - \hat{d} \circ \overline{T}_{a,\dots,b} - \overline{T}_{a,\dots,b} \circ \hat{d} = 0,$$

if $b - a < k - 1$. (Here id is the indentity on $B_{a,\dots,b}\mathcal{C}_2[1]$.) We put

$$\Phi = \mathcal{F}_{1,\dots,k} \circ \tilde{\mathcal{F}}'_{1,\dots,k} - \mathrm{id} - \hat{d} \circ \overline{T}_{,\dots,k} - \overline{T}_{1,\dots,k} \circ \hat{d}.$$

Since Φ induces zero on $B_{2,\dots,k}\mathcal{C}_2[1]$ and on $B_{1,\dots,k-1}\mathcal{C}_2[1]$ it follows that

$$\Phi \in \mathrm{Hom}(B_k\mathcal{C}_2[1], B_1\mathcal{C}_2[1]).$$

In other words, $\Phi = 0$ on $B_{1,\dots,k-1}\mathcal{C}_2[1]$ and the image of Φ is contained in $B_1\mathcal{C}_2[1]$.

Now we calculate:

$$\begin{aligned} \mathcal{F}_1 \circ \Xi &= \mathcal{F}_{1,\dots,k} \circ \Xi \\ &= \mathcal{F}_{1,\dots,k} \circ \tilde{\mathcal{F}}'_{1,\dots,k} \circ \hat{d} - \mathcal{F}_{1,\dots,k} \circ \hat{d} \circ \tilde{\mathcal{F}}'_{1,\dots,k} \\ &= \mathcal{F}_{1,\dots,k} \circ \tilde{\mathcal{F}}'_{1,\dots,k} \circ \hat{d} - \mathfrak{m}_1 \circ \mathcal{F}_{1,\dots,k} \circ \tilde{\mathcal{F}}'_{1,\dots,k} \\ &= (\mathrm{id} + \hat{d} \circ \overline{T}_{1,\dots,k-1} + \overline{T}_{1,\dots,k-1} \circ \hat{d} + \Phi) \circ \hat{d} \\ &\quad - \hat{d} \circ (\mathrm{id} + \hat{d} \circ \overline{T}_{1,\dots,k-1} + \overline{T}_{1,\dots,k-1} \circ \hat{d} + \Phi) \\ &= -d\Phi. \end{aligned}$$

The proof of Lemma 8.24 is complete. □

We define $\mathcal{F}'_k \in \mathrm{Hom}(B_k\mathcal{C}_2[1], B_1\mathcal{C}_1[1])$ such that

$$d\mathcal{F}'_k = -\Xi. \tag{8.30}$$

It is now easy to see that $\mathcal{F}'_\ell$, $0 \le \ell \le k$ is an A_k functor.

We remark that $\mathcal{F}'_k$ satisfying (8.30) is not unique. Namely we can take any element Ψ in the kernel of $d : \mathrm{Hom}(B_k\mathcal{C}_2[1], B_1\mathcal{C}_1[1]) \to$

$\mathrm{Hom}(B_k\mathcal{C}_2[1], B_1\mathcal{C}_1[1])$ and can replace $\mathcal{F}'_k$ by $\mathcal{F}'_k + \Psi$. We will use this freedom in the next step.

Now we are going to construct T_k. Using T_ℓ, $\ell = 1, \dots, k-1$ (and $\mathcal{F}'_\ell$) we define

$$\overline{T}_{a,\dots,b} : B_{a,\dots,b}\mathcal{C}_2[1] \to B_{a,\dots,b}\mathcal{C}_2[1].$$

Using T_ℓ, $\ell = 0, 1, \dots, k-1$ (and $\mathcal{F}'_\ell$) we define

$$\hat{T}_{1,\dots,k-1} : \bigoplus_{\ell=0}^{k-1} B_\ell\mathcal{C}_2[1] \to \bigoplus_{\ell=0}^{\infty} B_\ell\mathcal{C}_2[1].$$

$\hat{T}_{1,\dots,k-1}$ is a chain map by induction hypothesis. We consider

$$\overline{T}_{1,\dots,k} : \bigoplus_{i=1}^{k} B_i\mathcal{C}_2[1] \to \bigoplus_{i=1}^{k} B_i\mathcal{C}_2[1].$$

We put

$$\mathfrak{T} = \mathcal{F}_{1,\dots,k} \circ \mathcal{F}'_{1,\dots,k} - \mathrm{id} - \hat{d} \circ \overline{T}_{1,\dots,k} - \overline{T}_{1,\dots,k} \circ \hat{d}. \tag{8.31}$$

In the same way as the proof of Sublemma 8.26, we can use the fact that $\hat{T}_{1,\dots,k-1}$ is a chain map, to show the following:

Lemma 8.32. *The image of $\mathfrak{T}$ is contained in $B_1\mathcal{C}_2[1]$. Moreover $\mathfrak{T}$ vanishes on $\bigoplus_{i=1}^{k-1} B_i\mathcal{C}_2[1]$.*

We may regard $\mathfrak{T} \in \mathrm{Hom}(B_k\mathcal{C}_2[1], B_1\mathcal{C}_2[1])$ by Lemma 8.32. Then, by definition and Lemma 8.32 we have $d\mathfrak{T} = 0$.

Lemma 8.33. *We can choose $\mathcal{F}'_k$ so that $\mathfrak{T} \in \mathrm{Im}\, d$.*

Proof. If we replace $\mathcal{F}'_k$ by $\mathcal{F}'_k + \Psi$, then, by (8.31), $\mathfrak{T}$ is replaced by $\mathfrak{T} + \mathcal{F}_1 \circ \Psi$. The lemma now follows from the fact that $\mathcal{F}_1$ induces an isomorphism

$$H(\mathrm{Hom}(B_k\mathcal{C}_2[1], B_1\mathcal{C}_1[1]), d) \cong H(\mathrm{Hom}(B_k\mathcal{C}_2[1], B_1\mathcal{C}_2[1]), d).$$

□

Lemma 8.33 and Sublemma 8.26 immediately imply that we can choose T_k such that T_ℓ, $\ell = 0, \dots, k$ is an A_k transformation.

Thus we have constructed $\mathcal{F}'$ and $T : \mathcal{F} \circ \mathcal{F}' \to \mathrm{Id}^{\mathcal{C}_2}$. We next show that T is a homotopy equivalence. We prove it by using the following general result:

Proposition 8.34. *Let $\mathcal{G}$, $\mathcal{G}'$ be A_∞ functors $\mathcal{C} \to \mathcal{C}'$. Let $T : \mathcal{G} \to \mathcal{G}'$ be a natural transformation. We assume that $T_0(c) : \mathcal{G}_0(c) \to \mathcal{G}'_0(c)$ is a homotopy equivalence for any c.*

Then there exits a natural transformation $T' : \mathcal{G}' \to \mathcal{G}$, and pre natural transformations $S : \mathcal{G} \to \mathcal{G}$, $S' : \mathcal{G}' \to \mathcal{G}'$ such that

$$\mathrm{Id}^{\mathcal{G}'} - \mathfrak{M}_2(T', T) = \mathfrak{M}_1(S), \quad \mathrm{Id}^{\mathcal{G}} - \mathfrak{M}_2(T, T') = \mathfrak{M}_1(S').$$

Proof. The proof is similar to the construction of $\mathcal{F}'$ and T above. Namely we construct

$$T'_k : B_k\mathcal{C}[1] \to B_1\mathcal{C}'[1], \quad S'_k : B_k\mathcal{C}[1] \to B_1\mathcal{C}'[1]$$

inductively. More precisely we prove the following lemma by induction on k.

Lemma 8.35.*k*. *Suppose T'_ℓ, for $\ell = 0, \dots, k$ is an A_k transformation and $S'_\ell : B_\ell\mathcal{C}[1] \to B_1\mathcal{C}[1]$ for $\ell = 0, \dots, k$ are R module homomorphisms. We define $T'_{(k)}$, $S'_{(k)}$ by putting $T'_{(k),i} = S'_{(k),i} = 0$ for $i > k$, and $T'_{(k),i} = T'_i$, $S'_{(k),i} = S'_i$ for $i \le k$. Then we have:*

$$\mathrm{Id}^{\mathcal{G}'} - \mathfrak{M}_2(T'_{(k)}, T) = \mathfrak{M}_1(S'_{(k)}), \quad \text{on } B_k\mathcal{C}_2[1]. \tag{8.36}$$

Proof. For $k = 0$, we let $T'_0(c) \in \mathcal{C}_2(\mathcal{G}'(c), \mathcal{G}(c))$ be a homotopy inverse to $T_0(c)$. Then there exists $S'_0(c) \in \mathcal{C}_2(\mathcal{G}'(c), \mathcal{G}'(c))$ such that

$$\mathfrak{m}_1(T'_0(c)) = 0, \quad \mathfrak{m}_1(S'_0(c)) = \mathfrak{m}_2(T'_0(c), T_0(c)). \tag{8.37}$$

Thus, we have proved Lemma 8.35.0.

We assume Lemma 8.35.$k-1$ and will prove Lemma 8.35.k.

Sublemma 8.38. *$\mathfrak{M}_1(S'_{(k-1)})$ is zero on $B_{k-1}\mathcal{C}'[1]$. The image of the restriction of $\mathfrak{M}_1(S'_{(k-1)})$ to $B_k\mathcal{C}'[1]$ is in $B_1\mathcal{C}'[1]$.*

The proof is similar to the proof of Lemma 8.22 and is omitted.

We let Ψ be the restriction of $\mathfrak{M}_1(S'_{(k-1)})$ to $B_k\mathcal{C}'[1]$. Sublemma 8.38 implies

$$\Psi \in \mathrm{Hom}(B_k\mathcal{C}'[1], B_1\mathcal{C}'[1]).$$

Sublemma 8.39. $d_1 \circ \Psi - \Psi \circ d_1 = 0$.

Proof. This follows immediately from $\mathfrak{M}_1(\mathfrak{M}_1(S'_{(k-1)})) = 0$ and Sublemma 8.38. □

Sublemma 8.40. *There exists $\Phi \in \mathrm{Hom}(B_k\mathcal{C}_2, B_1\mathcal{C}_2)$. such that:*

$$\Psi = d_1 \circ \Phi + \Phi \circ d_1.$$

Proof. We define

$$T_{0*} : \mathrm{Hom}(B_k\mathcal{C}'[1], B_1\mathcal{C}[1]) \to \mathrm{Hom}(B_k\mathcal{C}'[1], B_1\mathcal{C}'[1])$$

as follows. Let $V \in \mathrm{Hom}(B_k\mathcal{C}'[1], B_1\mathcal{C}[1])$ and $c_0, \dots, c_k \in \mathfrak{Ob}(\mathcal{C}')$, $x_i \in \mathcal{C}'(c_{i-1}, c_i)$. We then put

$$T_{0*}(V)(x_1, \dots, x_k) = \mathfrak{m}_2(V(x_1, \dots, x_k), T_0(c_k)).$$

It follows from assumption that T_{0*} induces an isomorphism on cohomology. Hence, by Sublemma 8.39, it suffices to show that $T_{0*}(\Psi)$ is a boundary. By Sublemma 8.38, we find that $T_{0*}(\Psi)$ is a restriction of $\mathfrak{M}_2(\Psi, T)$ to $B_k\mathcal{C}'[1]$. By A_∞ formula of $\mathfrak{M}_k$ we have

$$\mathfrak{M}_2(\Psi, T) = \mathfrak{M}_2(\mathfrak{M}_1(T'_{(k-1)}), T) = \mathfrak{M}_1(\mathfrak{M}_2(T'_{(k-1)}), T).$$

By induction hypothesis

$$\mathfrak{M}_2(T'_{(k-1)}, T) = \mathrm{Id}^{\mathcal{G}'} - \mathfrak{M}_1(S'_{(k-1)})$$

on $B_{k-1}\mathcal{C}'[1]$. Let $-\Phi$ be the restriction of $\mathfrak{M}_2(T'_{(k-1)}, T)$ on $B_k\mathcal{C}'[1]$. Then, by induction hypothesis, $\Phi \in \mathrm{Hom}(B_k\mathcal{C}'[1], B_1\mathcal{C}'[1])$. We now have

$$\mathfrak{M}_1(\mathfrak{M}_2(T'_{(k-1)}), T) = -\mathfrak{M}_1(\Phi) = \hat{d}_1 \circ \Phi + \Phi \circ \hat{d}$$

as required. □

If we put $T'_k = \Phi$ then we have

$$\mathfrak{M}_1(T'_{(k)}) = 0.$$

We remark that Φ satisfying the conclusion of Sublemma 8.39 is not unique. Namely we may change it by any cocycle in $\mathrm{Hom}(B_k\mathcal{C}'[1], B_1\mathcal{C}'[1])$.

We next define S'_k. Using induction hypothesis, we can prove that the restriction of

$$\mathrm{Id}^{\mathcal{G}'} - \mathfrak{M}_2(T'_{(k-1)}, T) - \mathfrak{M}_1(S'_{(k-1)})$$

to $B_k\mathcal{C}'[1]$ defines an element of $\mathrm{Hom}(B_k\mathcal{C}'[1], B_1\mathcal{C}'[1])$. We denote it by Ξ. We can prove that Ξ is a cocycle in a way similar to the proof of Sublemma 8.39. Therefore, we may choose T'_k so that

$$\Xi - T_{0*}(T'_k)$$

is a coboundary. Hence we may choose S'_k such that (8.36) is satisfied. The proof of Lemma 8.35.k is now complete. □

By Lemma 8.35, we obtain T' and S'. To construct S, we proceed as follows. We apply the above construction to T' in place of T and obtain T'' and S'' such that

$$\mathrm{Id}^{\mathcal{G}} - \mathfrak{M}_2(T', T'') = \mathfrak{M}_1(S'').$$

Then

$$T'' \equiv \mathfrak{M}_2(\mathfrak{M}_2(T, T'), T'') \equiv \mathfrak{M}_2(T, \mathfrak{M}_2(T', T'')) \equiv T \mod \mathrm{Im}(\mathfrak{M}_1).$$

Therefore

$$\mathrm{Id}^{\mathcal{G}'} - \mathfrak{M}_2(T', T) \equiv \mathrm{Id}^{\mathcal{G}'} - \mathfrak{M}_2(T', T'') \equiv 0 \mod \mathrm{Im}(\mathfrak{M}_1).$$

The proof of Proposition 8.34 is complete. □

We need here some elementary properties of homotopy equivalence.

Proposition-Definition 8.41. *An A_∞ functor $\mathcal{F} : \mathcal{C}_1 \to \mathcal{C}_2$ induces A_∞ functors $\mathcal{F}_* : \mathfrak{Funk}(\mathcal{C}, \mathcal{C}_1) \to \mathfrak{Funk}(\mathcal{C}, \mathcal{C}_2)$, $\mathcal{F}^* : \mathfrak{Funk}(\mathcal{C}_2, \mathcal{C}) \to \mathfrak{Funk}(\mathcal{C}_1, \mathcal{C})$ such that $(\mathcal{F}_*)_0(\mathcal{G}) = \mathcal{F} \circ \mathcal{G}$, $(\mathcal{F}^*)_0(\mathcal{G}) = \mathcal{G} \circ \mathcal{F}$.*

Proof. Let $T^i \in \mathfrak{Funk}(\mathcal{G}^{i-1}, \mathcal{G}^i)$ be pre natural transformations such that $\mathfrak{deg}' T^i = t_i$. We put

$$(8.42)\quad \begin{aligned} ((\mathcal{F}_*)_k(T^1, \dots, T^k))(\mathbf{x}) = \sum_a (-1)^{\epsilon_a}\, \mathcal{F}(\hat{\mathcal{G}}^0(\mathbf{x}_a^{(1)}), \hat{T}^1(\mathbf{x}_a^{(2)}), \\ \dots, \hat{T}^k(\mathbf{x}_a^{(2k)}), \hat{\mathcal{G}}^k(\mathbf{x}_a^{(2k+1)})), \end{aligned}$$

where

$$\epsilon_a = \sum_{j=1}^{k} \sum_{i=1}^{2j-1} t_j \deg' \mathbf{x}_a^{(i)}.$$

We can prove that (8.42) defines an A_∞ functor in a way similar to the proof of Theorem 7.55. (We omit the detail.)

We next define

$$\begin{aligned} ((\mathcal{F}^*)_1(T^1))(\mathbf{x}) &= T^1(\hat{\mathcal{F}}(\mathbf{x})), \\ ((\mathcal{F}^*)_k(T^1, \dots, T^k))(\mathbf{x}) &= 0 \qquad k > 2. \end{aligned}$$

It is easy see that $\mathcal{F}^*$ is an A_∞ functor. □

The following two corollaries are immediate consequences.

Corollary 8.43. *If $\mathcal{F} : \mathcal{C}_1 \to \mathcal{C}_2$ is homotopic to $\mathcal{F}' : \mathcal{C}_1 \to \mathcal{C}_2$ and $\mathcal{G} : \mathcal{C}_2 \to \mathcal{C}_3$ is homotopic to $\mathcal{G}' : \mathcal{C}_2 \to \mathcal{C}_3$, then $\mathcal{G} \circ \mathcal{F}$ is homotopic to $\mathcal{G}' \circ \mathcal{F}'$.*

Corollary 8.44. *If $\mathcal{F} : \mathcal{C}_1 \to \mathcal{C}_2$ and $\mathcal{G} : \mathcal{C}_2 \to \mathcal{C}_3$ are homotopy equivalences then $\mathcal{G} \circ \mathcal{F}$ is a homotopy equivalence.*

We are now in the position to complete the proof of Proposition 8.9. We have constructed $\mathcal{F}'$. By using Proposition 8.34, we can prove that $\mathcal{F} \circ \mathcal{F}'$ is homotopic to identity.

We next prove that $\mathcal{F}' \circ \mathcal{F}$ is homotopic to identity. For this purpose, we apply the same argument to $\mathcal{F}'$ and obtain $\mathcal{F}''$ such that $\mathcal{F}' \circ \mathcal{F}''$ is homotopic to identity. It follows that $\mathcal{F}''$ is homotopic to $\mathcal{F}$. Hence by Corollary 8.43, $\mathcal{F}' \circ \mathcal{F}$ is homotopic to identity. The proof of Proposition 8.9 is complete. □

We continue the proof of Theorem 8.6. In this section we complete the proof in the case of differential graded category and postpone the proof of the general case to the next section.

Our next goal is the proof of Lemma 8.45, which is another case of Theorem 8.6. We define the notion of full subcategory of A_∞ category in an obvious way.

Lemma 8.45. *Let $\mathcal{C}_2$ be an A_∞ category such that $\mathfrak{m}_k = 0$ for $k \geq 3$. Let $\mathcal{C}_1$ be a full subcategory of $\mathcal{C}_2$ and $\mathcal{F} : \mathcal{C}_1 \to \mathcal{C}_2$ be the inclusion. We assume (8.7.2). Then $\mathcal{F} : \mathcal{C}_1 \to \mathcal{C}_2$ is a homotopy equivalece.*

Proof. Let $c \in \mathfrak{Ob}(\mathcal{C}_2) - \mathfrak{Ob}(\mathcal{C}_1)$. We choose $\mathcal{F}'_0(c) \in \mathfrak{Ob}(\mathcal{C}_1)$ which is homotopy equivalent to c. If $c \in \mathfrak{Ob}(\mathcal{C}_1)$ we put $\mathcal{F}'_0(c) = c$.

By the choice of $\mathcal{F}'_0(c)$, there exists

$$T'_0(c) \in \mathcal{C}_2(c, \mathcal{F}'_0(c)), \quad T_0(c) \in \mathcal{C}_2(\mathcal{F}'_0(c), c)$$

and

$$S'_0(c) \in \mathcal{C}_2(c, c), \quad S_0(c) \in \mathcal{C}_2(\mathcal{F}'_0(c), \mathcal{F}'_0(c))$$

such that

$$\begin{aligned} \mathbf{e}_{\mathcal{F}_0(c)} - \mathfrak{m}_2(T_0(c), T'_0(c)) &= \mathfrak{m}_1(S_0(c)) \\ \mathbf{e}_{\mathcal{F}'_0(c)} - \mathfrak{m}_2(T'_0(c), T_0(c)) &= \mathfrak{m}_1(S'_0(c)). \end{aligned}$$

For $c \in \mathfrak{Ob}(\mathcal{C}_1)$ we put $T_0(c) = T'_0(c) = \mathrm{Id}^c$ and $S_0(c) = S'_0(c) = 0$.

We next define $\mathcal{F}'_k$. We first put

$$x \circ y = (-1)^{\deg x}\mathfrak{m}_2(x, y), \quad d(x) = \mathfrak{m}_1(x).$$

Then, using the fact that $\mathfrak{m}_k = 0$ for $k \geq 3$, we have

$$\begin{aligned} d(x \circ y) &= dx \circ y + (-1)^{\deg x+1} x \circ dy, \\ (x \circ y) \circ z &= x \circ (y \circ z), \\ \mathbf{e} \circ x &= x = x \circ \mathbf{e}. \end{aligned}$$

In view of the second formula, we may write

$$x_1 \circ x_2 \circ \cdots \circ x_k$$

etc. Now let $x_i \in \mathcal{C}_2(c_{i-1}, c_i)$. We put

$$\begin{aligned}&\mathcal{F}'_k(x_1, \dots, x_k)\\ &\quad = \quad T'_0(c_0) \circ x_1 \circ S'_0(c_1) \circ x_2 \circ \cdots \circ S'_0(c_{k-1}) \circ x_k \circ T'_0(c_k).\end{aligned}$$

We remark that

$$d(S'_0(c)) = \mathbf{e}_{\mathcal{F}_0(c)} - T'_0(c) \circ T_0(c).$$

It follows from definition that

$$\begin{aligned}&d(\mathcal{F}'_k(x_1, \dots, x_k))\\ &= \sum_i (-1)^{\deg x_1 + \cdots \deg x_{i-1} + i - 1} \mathcal{F}'_k(x_1, \dots, dx_i, \dots, x_k)\\ &\quad + \sum_i (-1)^{\deg x_1 + \cdots \deg x_i + i} \mathcal{F}'_k(x_1, \dots, x_i \circ x_{i+1}, \dots, x_k)\\ &\quad + \sum_i (-1)^{\deg x_1 + \cdots \deg x_{i-1} + i - 1 + 1} \mathcal{F}'_i(x_1, \dots, x_i) \circ \mathcal{F}'_{k-i}(x_{i+1}, \dots, x_k).\end{aligned}$$

Therefore

$$\begin{aligned}&\mathfrak{m}_1(\mathcal{F}'_k(x_1, \dots, x_k))\\ &+ \sum_i (-1)^{\deg x_1 + \cdots \deg x_{i-1} + i - 1} \mathcal{F}'_i(x_1, \dots, x_i) \circ \mathcal{F}'_{k-i}(x_{i+1}, \dots, x_k)\\ &\qquad = \sum_i (-1)^{\deg x_1 + \cdots \deg x_{i-1} + i - 1} \mathcal{F}'_k(x_1, \dots, \mathfrak{m}_1(x_i), \dots, x_k)\\ &\qquad\quad + (-1)^{\deg x_1 + \cdots \deg x_{i-1} + i - 1} \mathcal{F}'_k(x_1, \dots, \mathfrak{m}_2(x_i, x_{i+1}), \dots, x_k).\end{aligned}$$

It follows that $\mathcal{F}'_k$, $k = 0, \dots$ is an A_∞ functor.

The composition $\mathcal{F}' \circ \mathcal{F}$ is an identity functor. We are going to show that $\mathcal{F} \circ \mathcal{F}'$ is homotopic to identity. For this purpose, we are going to construct $T : \mathcal{F} \circ \mathcal{F}' \to 1_{\mathcal{C}_2}$ satisfying the assumption of Proposition 8.34. We already constructed T_0. We remark that

$$(\mathcal{F} \circ \mathcal{F}')(x_1, \dots, x_k) = \mathcal{F}'_k(x_1, \dots, x_k).$$

Hence the condition $T : \mathcal{F} \circ \mathcal{F}' \to \mathrm{Id}^{\mathcal{C}_2}$ to be an A_∞ functor can be written as

$$(8.46)\quad \begin{aligned} &\sum_i (-1)^{\deg x_1 + \cdots \deg x_{i-1} + i - 1 + 1} T_0(c_0) \circ x_1 \circ S_0'(c_1) \circ x_2 \circ S_0'(c_2) \circ \cdots \\ &\qquad \cdots \circ x_i \circ T_0'(c_i) \circ T_{k-i}(x_{i+1}, \dots, x_k) \\ &\quad = \mathfrak{m}_1(T_k(x_1, \dots, x_k)) \\ &\qquad + \sum_i (-1)^{\deg x_1 + \cdots \deg x_{i-1} + i - 1} T_k(x_1, \dots, \mathfrak{m}_1(x_i), \dots, x_k) \\ &\qquad + \sum_i (-1)^{\deg x_1 + \cdots \deg x_{i-1} + i - 1} T_k(x_1, \dots, \mathfrak{m}_2(x_i, x_{i+1}), \dots, x_k). \end{aligned}$$

We put

$$T_k(x_1, \dots, x_k) = T_0(c_0) \circ x_1 \circ S_0'(c_1) \circ x_2 \circ \cdots \circ x_k \circ S_0'(c_k).$$

(8.46) can be checked easily.

Lemma 8.45 now follows from Proposition 8.34. □

Now we show:

Proposition 8.47. *Theorem* 8.6 *holds if* $\mathfrak{m}_k = 0$ *for* $k > 2$ *in* $\mathcal{C}_1$, $\mathcal{C}_2$.

Proof. Let $\mathcal{C}_2'$ be the full subcategory such that $\mathfrak{Ob}(\mathcal{C}_2')$ is the image of $\mathcal{F}_0$. Lemma 8.41 implies that the inclusion $\mathcal{C}_2' \to \mathcal{C}_2$ is a homotopy equivalence.

For each $c \in \mathfrak{Ob}(\mathcal{C}_2')$ we take and fix $\alpha(c) \in \mathfrak{Ob}(\mathcal{C}_1)$ such that $\mathcal{F}_0(\alpha(c)) = c$. Let $\mathcal{C}_1'$ be the full subcategory of $\mathcal{C}_1$ such that $\mathfrak{Ob}(\mathcal{C}_1')$ is the image of α. Proposition 8.9 implies that the restriction of $\mathcal{F}$ to $\mathcal{C}'$ induces a homotopy equivalence $\mathcal{C}_1' \to \mathcal{C}_2'$. Therefore, using Corollary 8.43 and 8.44, it suffices to show that $\mathcal{C}_1' \to \mathcal{C}_1$ is a homotopy equivalence to complete the proof of Proposition 8.47. This follows from Lemma 8.45 and the following Lemma 8.48. □

Lemma 8.48. *We assume* (8.7). *If* $\mathcal{F}_0(b) = \mathcal{F}_0(b')$ *then* b *is homotopy equivalent to* b'.

Proof. By (8.7) we have

$$\mathcal{F}_{1*} : H(\mathcal{C}_1(b, b'); \mathfrak{m}_1) \to H(\mathcal{C}_2(\mathcal{F}_0(b), \mathcal{F}_0(b')); \mathfrak{m}_1)$$

is an isomorphism. We take $[f] \in H(\mathcal{C}_1(b, b'); \mathfrak{m}_1)$ which is mapped to $[\mathbf{e}_{\mathcal{F}_0(b)}]$ by $\mathcal{F}_{1*}$. It is easy to see that

$$\begin{aligned} \mathfrak{m}_2(f, \cdot) &: \mathcal{C}_1(b', a) \to \mathcal{C}_1(b, a) \\ \mathfrak{m}_2(\cdot, f) &: \mathcal{C}_1(a, b) \to \mathcal{C}_1(a, b') \end{aligned}$$

induces isomorphism on homology. The lemma then follows from Lemma 6.21. □

We need the following proposition in the next section.

Proposition 8.49. *If $\mathcal{C}_1$ is homotopy equivalent to $\mathcal{C}_2$, then $\mathfrak{Funk}(\mathcal{C},\mathcal{C}_1)$ is homotopy equivalent to $\mathfrak{Funk}(\mathcal{C},\mathcal{C}_2)$ and $\mathfrak{Funk}(\mathcal{C}_1,\mathcal{C})$ is homotopy equivalent to $\mathfrak{Funk}(\mathcal{C}_2,\mathcal{C})$.*

Proof. Let $\mathcal{F}:\mathcal{C}_1\to\mathcal{C}_2$, $\mathcal{F}':\mathcal{C}_2\to\mathcal{C}_1$ be as in Definition 8.5 and let H, H' be natural transformations from $\mathcal{F}\circ\mathcal{F}'$ to identity functor and from identity functor to $\mathcal{F}\circ\mathcal{F}'$ respectively. Let $\mathcal{G}:\mathcal{C}\to\mathcal{C}_2$ be an A_∞ functor. We put

$$\begin{aligned}\mathfrak{H}_0(\mathcal{G}) &= (\mathcal{G}^*)_1(H)\in\mathfrak{Funk}((\mathcal{F}_*\circ\mathcal{F}'_*)_0(\mathcal{G}),\mathcal{G})\\ \mathfrak{H}_k &= 0,\quad k>0.\quad (\text{Note } (\mathcal{G}^*)_k=0 \text{ for } k>1.)\end{aligned}$$

It is easy to check that $\mathfrak{H}$ is a natural transform from $\mathcal{F}_*\circ\mathcal{F}'_*$ to the identity functor $\mathrm{Id}_{\mathfrak{Funk}(\mathcal{C},\mathcal{C}_2)}$. We define in a similar way a natural transformation $\mathfrak{H}'$ from $\mathrm{Id}_{\mathfrak{Funk}(\mathcal{C},\mathcal{C}_2)}$ to $\mathcal{F}_*\circ\mathcal{F}'_*$. We assume that

$$\mathfrak{M}_2(H,H')-\mathrm{Id}_{\mathrm{Id}_{\mathfrak{Funk}(\mathcal{C}_1,\mathcal{C}_1)}}=\mathfrak{M}_1(T).$$

(Note that the confusing symbol $\mathrm{Id}_{\mathrm{Id}_{\mathfrak{Funk}(\mathcal{C}_1,\mathcal{C}_1)}}$ denotes the identity *transform* from the identity *functor* : $\mathcal{C}_1\to\mathcal{C}_1$ to itself.) We assume also that

$$\mathfrak{M}_2(H',H)-\mathrm{Id}_{\mathcal{F}\circ\mathcal{F}'}=\mathfrak{M}_1(T').$$

We put

$$\begin{aligned}\mathfrak{T}_0(\mathcal{G}) &= (\mathcal{G}^*)_1(T)\in\mathfrak{Funk}((\mathcal{F}_*\circ\mathcal{F}'_*)_0(\mathcal{G}),\mathcal{G})\\ \mathfrak{T}_k &= 0,\quad k>0.\end{aligned}$$

And we define $\mathfrak{T}'$ in a similar way. Then we find

$$\begin{aligned}\mathrm{Id}_{\mathrm{Id}_{\mathfrak{Funk}(\mathcal{C},\mathcal{C}_1)}}-\hat{\mathfrak{M}}_2(\mathfrak{H},\mathfrak{H}') &= \hat{\mathfrak{M}}_1(\mathfrak{T}),\\ \mathrm{Id}_{(\mathcal{F}\circ\mathcal{F}')_*}-\hat{\mathfrak{M}}_2(\mathfrak{H}',\mathfrak{H}) &= \hat{\mathfrak{M}}_1(\mathfrak{T}').\end{aligned}$$

Here $\hat{\mathfrak{M}}$ is the A_∞ structure on $\mathfrak{Funk}(\mathfrak{Funk}(\mathcal{C},\mathcal{C}_1),\mathfrak{Funk}(\mathcal{C},\mathcal{C}_1))$.

Thus we proved that $(\mathcal{F}\circ\mathcal{F}')_*$ is homotopic to identity. This complete the proof of the first half of the Proposition 8.49. The second half can be proved in a similar way. □

§9. An A_∞ analogue of Yoneda's lemma

In §7 we constructed an A_∞ functor $\mathfrak{Rep}_0(c) : \mathcal{C}^o \to \mathcal{CH}$ for each object c of $\mathcal{C}$. The purpose of this section is to make it functorial. Moreover we prove that $c \mapsto \mathfrak{Rep}_0(c)$ defines a homotopy equivalence between the A_∞ category $\mathcal{C}$ and one of representable A_∞ functors : $\mathcal{C}^o \to \mathcal{CH}$. Namely we prove the following Theorem 9.1. We let $\mathfrak{Rep}(\mathcal{C}^o, \mathcal{CH})$ be the full subcategory of $\mathfrak{Funk}(\mathcal{C}^o, \mathcal{CH})$ such that $\mathfrak{Ob}(\mathfrak{Rep}(\mathcal{C}^o, \mathcal{CH}))$ is the set of all representable A_∞ functors. We define a full subcategory $\mathfrak{DRep}(\mathcal{C}^o, \mathcal{CH})$ of $\mathfrak{Funk}(\mathcal{C}^o, \mathcal{CH})$ such that $\mathfrak{Ob}(\mathfrak{DRep}(\mathcal{C}^o, \mathcal{CH}))$ is the set of all derived representable A_∞ functors.

Theorem 9.1. *There exists a homotopy equivalences of A_∞ functors $\mathfrak{Rep} : \mathcal{C} \cong \mathfrak{Rep}(\mathcal{C}^o, \mathcal{CH})$, $\mathfrak{DRep} : \mathcal{DC} \cong \mathfrak{DRep}(\mathcal{C}^o, \mathcal{CH})$, such that $\mathfrak{Rep}_0(c)$ is an in Definition* 7.28.

Remark 9.2. The first half of Theorem 9.1 was proved in [Fu4] §12, over $\mathbb{Z}_2$ coefficient. In this article we are going to put precise sign in its proof. We also improve the presentation of the proof in [Fu4].

Remark 9.3. In the case of usual category, the first half of Theorem 9.1, which is Yoneda's lemma, is well known.

We remark that actually $\mathfrak{Rep}(\mathcal{C}^o, \mathcal{CH})$ is a differential graded category, since all the higher compositions are zero. Therefore Theorem 9.1 implies the following:

Corollary 9.4. *Any A_∞ category is homotopy equivalent to a differential graded category.*

Proof of Theorem 9.1. We already defined $\mathfrak{Rep}_0$. We will define $\mathfrak{Rep}_k : B_k(\mathcal{C}) \to B_1(\mathfrak{Rep}(\mathcal{C}^o, \mathcal{CH}))$. Let

$$c_0, \dots, c_k \in \mathfrak{Ob}(\mathcal{C}),\ x_i \in \mathcal{C}(c_{i-1}, c_i).$$

We need to define a natural transformation

$$\mathfrak{Rep}_k(x_1, \dots, x_k) : \mathfrak{Rep}_0(c_0) \to \mathfrak{Rep}_0(c_k). \tag{9.5}$$

Let $b_0, \dots, b_\ell \in \mathfrak{Ob}(\mathcal{C})$, $y_i \in \mathcal{C}^o(b_{i-1}, b_i) = \mathcal{C}(b_i, b_{i-1})$. To define (9.5) we need to define

$$\mathfrak{Rep}_k(x_1, \dots, x_k)_\ell(y_1, \dots, y_\ell) \in \mathrm{Hom}(\mathfrak{Rep}_0(c_0)_0(b_0), \mathfrak{Rep}_0(c_k)_0(b_\ell)).$$

Let $z \in \mathfrak{Rep}_0(c_0)_0(b_\ell) = \mathcal{C}(b_0, c_0)$. We put $\mathbf{x} = x_1 \otimes \cdots \otimes x_k$, $\mathbf{y} = y_1 \otimes \cdots \otimes y_\ell$. We use the notations $\mathbf{x}^{op} = x_k \otimes \cdots \otimes x_1$, $\epsilon(\mathbf{x}) = \sum_{i<j}(\deg x_i + 1)(\deg x_j + 1)$ etc. introduced in §7. Now we define

Definition 9.6.

(9.7) $(\mathfrak{Rep}_k(\mathbf{x}))_\ell(\mathbf{y})(z) = (-1)^{\epsilon(\mathbf{y})+(\deg'\mathbf{x})(\deg'\mathbf{y}+\deg' z)}\mathfrak{m}_{k+\ell+1}(\mathbf{y}^{op}, z, \mathbf{x}).$

We remark that the right hand side of (9.7) is in $\mathfrak{Rep}_0(c_k)(b_\ell) = \mathcal{C}(b_\ell, c_k)$. We also remark that

$$\mathfrak{deg}'(\mathfrak{Rep}_k(\mathbf{x}))_\ell(\mathbf{y}) = \deg'\mathbf{x} + \deg'\mathbf{y}.$$

Lemma 9.8. (9.7) *defines an A_∞ functor.*

Proof. By A_∞ formula of $\mathfrak{m}$ we have

$$\begin{aligned} 0 &= \sum_a (-1)^{\deg'\mathbf{y}_a^{(3)}}\mathfrak{m}(\mathbf{y}_a^{(3)op}, \mathfrak{m}(\mathbf{y}_a^{(2)op}), \mathbf{y}_1^{(1)op}, z, \mathbf{x}) \\ &\quad + \sum_{a,b}(-1)^{\deg'\mathbf{y}_a^{\prime\prime op}}\mathfrak{m}(\mathbf{y}_a^{\prime\prime op}, \mathfrak{m}(\mathbf{y}_a^{\prime op}, z, \mathbf{x}'_b), \mathbf{x}''_b) \\ &\quad + (-1)^{\deg'\mathbf{y}+\deg' z}\mathfrak{m}(\mathbf{y}^{op}, z, \hat{d}\mathbf{x}). \end{aligned} \tag{9.9}$$

We will rewrite each term of (9.9). The first term is equal to:

(9.10) $(-1)^{\epsilon_1}\mathfrak{Rep}(\mathbf{x})(\mathbf{y}_a^{(1)}, \mathfrak{m}^o(\mathbf{y}_a^{(2)}), \mathbf{y}_a^{(3)})(z) = (-1)^{\epsilon_2}\mathfrak{Rep}(\mathbf{x})(\hat{d}^o\mathbf{y})(z)$

where $\hat{d}^{op}$ is the coderivation induced by $\mathfrak{m}^o$ on $\mathcal{C}^o$ and

$$\begin{aligned} \epsilon_1 &= \deg'\mathbf{y}_a^{(3)} + \epsilon(\mathbf{y}_a^{(2)}) + 1 + (\deg'\mathbf{x})(\deg'\mathbf{y} + \deg' z + 1) \\ &\quad + \epsilon(\mathbf{y}_a^{(1)}, \mathfrak{m}^o(\mathbf{y}_a^{(2)}), \mathbf{y}_a^{(3)}) \\ &= \deg'\mathbf{y}_a^{(1)} + (\deg'\mathbf{x})(\deg'\mathbf{y} + \deg' z + 1) + \epsilon(\mathbf{y}). \end{aligned}$$

Hence

$$\epsilon_2 = (\deg'\mathbf{x})(\deg'\mathbf{y} + \deg' z + 1) + \epsilon(\mathbf{y}). \tag{9.11}$$

We divide the second terms of (9.9) into the following five cases.
Case 1: $\mathbf{x}'_b \neq 1 \in B_0\mathcal{C}$, $\mathbf{x}''_b \neq 1 \in B_0\mathcal{C}$:

The second term of (9.9) of this case is

$$\begin{aligned} &\sum_{a,b}(-1)^{\epsilon_3(a,b)}\mathfrak{m}(\mathbf{y}_a^{\prime\prime op}, \mathfrak{Rep}(\mathbf{x}'_b)(\mathbf{y}'_a)(z), \mathbf{x}''_b) \\ &= \sum_{a,b}(-1)^{\epsilon_4(a,b)}\mathfrak{Rep}(\mathbf{x}''_b)(\mathbf{y}''_a)(\mathfrak{Rep}(\mathbf{x}'_b)(\mathbf{y}'_a)(z)) \\ &= \sum_{a,b}(-1)^{\epsilon_5(a,b)}\mathfrak{m}_2(\mathfrak{Rep}(\mathbf{x}'_b)(\mathbf{y}'_a), \mathfrak{Rep}(\mathbf{x}''_b)(\mathbf{y}''_a))(z) \end{aligned} \tag{9.12}$$

where $\mathfrak{m}_2$ in the third line is an operation in $\mathcal{CH}$ and

$$\begin{aligned}
\epsilon_3(a,b) &= \deg' \mathbf{y}''_a + (\deg' \mathbf{x}'_b)(\deg' \mathbf{y}'_a + \deg' z) + \epsilon(\mathbf{y}'_a) \\
\epsilon_4(a,b) &= \epsilon_3(a,b) + (\deg' \mathbf{x}''_b)(1 + \deg' \mathbf{y}''_a + \deg' z + \deg' \mathbf{x}'_b \\
&\quad + \deg' \mathbf{y}'_a) + \epsilon(\mathbf{y}''_a) \\
\epsilon_5(a,b) &= \epsilon_4(a,b) + (\deg' \mathbf{x}'_b + \deg' \mathbf{y}'_a + 1)(\deg' \mathbf{x}''_b + \deg' \mathbf{y}''_a) + 1.
\end{aligned}$$

We calculate (using $\mathfrak{deg}'\mathfrak{Rep}(\mathbf{x}'_b)(\mathbf{y}'_a) = \deg' \mathbf{x}'_b + \deg' \mathbf{y}'_a + 1$) that

$$\epsilon_5(a,b) = \epsilon(\mathbf{y}) + \deg' \mathbf{x} \deg' z + \deg' \mathbf{x} \deg' \mathbf{y} + \deg' \mathbf{x}''_b \deg' \mathbf{y}'_a.$$

Therefore (9.12) is equal to

$$\sum_b (-1)^{\epsilon_6} \mathfrak{M}_2(\mathfrak{Rep}(\mathbf{x}'_b), \mathfrak{Rep}(\mathbf{x}''_b))(\mathbf{y})(z) \tag{9.13}$$

where

$$\epsilon_6 = \epsilon(\mathbf{y}) + (\deg' \mathbf{x})(\deg' z + \deg' \mathbf{y}) + 1. \tag{9.14}$$

Case 2: $\mathbf{x}'_b = 1 \in B_0\mathcal{C}$, $\mathbf{y}'_a \neq 1 \in B_0\mathcal{C}$:
In this case the second term of (9.9) is:

$$\begin{aligned}
&\sum_a (-1)^{\epsilon_7(a)} \mathfrak{m}(\mathbf{y}''^{op}_a, \mathfrak{Rep}_0(c_0)(\mathbf{y}'_a)(z), \mathbf{x}) \\
&= \sum_{a,b} (-1)^{\epsilon_8(a)} \mathfrak{Rep}(\mathbf{x})(\mathbf{y}''_a)(\mathfrak{Rep}_0(c_0)(\mathbf{y}'_a)(z)) \\
&= \sum_a (-1)^{\epsilon_9(a)} \mathfrak{m}_2(\mathfrak{Rep}_0(c_0)(\mathbf{y}'_a), \mathfrak{Rep}(\mathbf{x})(\mathbf{y}''_a))(z) \\
&= (-1)^{\epsilon_6} \mathfrak{M}_2(\mathfrak{Rep}_0(c_0), \mathfrak{Rep}(\mathbf{x}))(\mathbf{y})(z)
\end{aligned} \tag{9.15}$$

where ϵ_6 is as in (9.14). In fact

$$\begin{aligned}
\epsilon_7(a) &= \deg' \mathbf{y}''_a + \epsilon(\mathbf{y}'_a), \\
\epsilon_8(a) &= \epsilon_7(a) + (\deg' \mathbf{x})(\mathbf{y}''_a + \mathbf{y}'_a + \deg' z + 1) + \epsilon(\mathbf{y}''_a) \\
\epsilon_9(a) &= \epsilon_8(a) + (\deg' \mathbf{y}'_a + 1)(\deg' \mathbf{x} + \deg' \mathbf{y}''_a) \\
&\equiv \epsilon_6 + \deg' \mathbf{x} \deg' \mathbf{y}'_a \mod 2.
\end{aligned}$$

Case 3: $\mathbf{x}'_b = 1 \in B_0\mathcal{C}$, $\mathbf{y}'_a = 1 \in B_0\mathcal{C}$:
In a similar way, we have

$$(-1)^{\epsilon_6 + \deg' \mathbf{x} + \deg' \mathbf{y}} \mathfrak{Rep}(\mathbf{x})(\mathbf{y})(\mathfrak{m}_1(z)). \tag{9.16}$$

In face we have

$$\deg' \mathbf{y} + (\deg' \mathbf{x})(\deg' z + \deg' \mathbf{y}) + \epsilon(\mathbf{y}) = \epsilon_6 + \deg' \mathbf{x} + \deg' \mathbf{y}.$$

Case 4: $\mathbf{x}''_b = 1 \in B_0\mathcal{C}$, $\mathbf{y}''_a \neq 1 \in B_0\mathcal{C}$:

In a similar way we have

$$(-1)^{\epsilon_6}\mathfrak{M}_2(\mathfrak{Rep}(\mathbf{x}), \mathfrak{Rep}_0(c_k))(\mathbf{y})(z). \tag{9.17}$$

Case 5: $\mathbf{x}''_b = 1 \in B_0\mathcal{C}$, $\mathbf{y}''_a = 1 \in B_0\mathcal{C}$:

In a similar way we have

$$(-1)^{\epsilon_6}\mathfrak{m}_1(\mathfrak{Rep}(\mathbf{x})(\mathbf{y})(z)). \tag{9.18}$$

We remark that the sum of (9.10), (9.16) and (9.18) is

$$(-1)^{\epsilon_6}\mathfrak{M}_1(\mathfrak{Rep}(\mathbf{x}))(\mathbf{y})(z) \tag{9.19}$$

Finally the third term of (9.9) is:

$$(-1)^{\epsilon_6+1}\mathfrak{Rep}(\hat{d}\mathbf{x})(\mathbf{y})(z). \tag{9.20}$$

Thus $(-1)^{\epsilon_6}$ times (9.9) implies

$$\begin{aligned}\sum_{\mathbf{x}'_b\neq 1,\ \mathbf{x}''_b\neq 1} &\mathfrak{M}_2(\mathfrak{Rep}(\mathbf{x}'_b), \mathfrak{Rep}(\mathbf{x}''_b))(\mathbf{y}) + \mathfrak{M}_2(\mathfrak{Rep}_0(c_0), \mathfrak{Rep}(\mathbf{x}))(\mathbf{y}) \\ &+ \mathfrak{M}_2(\mathfrak{Rep}(\mathbf{x}), \mathfrak{Rep}_0(c_k))(\mathbf{y}) + \mathfrak{M}_1(\mathfrak{Rep}(\mathbf{x}))(\mathbf{y}) \\ = \mathfrak{Rep}(\hat{d}\mathbf{x})(\mathbf{y}).\end{aligned}$$

(We remark that we need over all minus sign in the definition of $\mathfrak{M}_k$ to show this formula.) The proof of Lemma 9.8 is complete. □

Using Lemma 9.8 and Proposition 8.9, it suffices to check (8.7.1) for $\mathfrak{Rep}$ to complete the proof of Theorem 9.1. Namely we need to show

$$\mathfrak{Rep}_1 : \mathcal{C}(c_1, c_2) \to \mathfrak{Func}(\mathfrak{Rep}_0(c_1), \mathfrak{Rep}_0(c_2))$$

induces an isomorphism on homology. We define

$$\Pi : \mathfrak{Func}(\mathfrak{Rep}_0(c_1), \mathfrak{Rep}_0(c_2)) \to \mathcal{C}(c_1, c_2)$$

by

$$\Pi(T) = (-1)^{\mathfrak{deg}' T}T_0(c_1)(\mathbf{e}_{c_1}).$$

(Note $\mathbf{e}_{c_1} \in \mathcal{C}(c_1, c_1) = \mathfrak{Rep}_0(c_1)_0(c_1)$. Hence $T_0(c_1)(\mathbf{e}_{c_1}) \in \mathfrak{Rep}_0(c_2)_0(c_1) = \mathcal{C}(c_1, c_2)$.)

Lemma 9.21. *Π is a chain map.*

Proof. Let $T \in \mathfrak{Func}(\mathfrak{Rep}_0(c_1), \mathfrak{Rep}_0(c_2))$. We have

$$\begin{aligned} \Pi(\mathfrak{M}_1(T)) &= (-1)^{\mathfrak{deg}' T}(\mathfrak{M}_1(T))_0(c_1)(\mathbf{e}_{c_1}) \\ &= (-1)^{\mathfrak{deg}' T+1}\mathfrak{m}_1(T_0(c_1)(\mathbf{e}_{c_1})) = \mathfrak{m}_1(\Pi(T)). \end{aligned}$$

(We remark that overall minus sign in the definition of $\mathfrak{M}_1$ is essential here.) □

We have

$$(\Pi \circ \mathfrak{Rep}_1)(x) = (-1)^{\deg' x}(\mathfrak{Rep}_1(x)_0(c_1))(\mathbf{e}_{c_1}) = \mathfrak{m}_2(\mathbf{e}_{c_1}, x) = x.$$

So, to complete the proof of Theorem 9.1, it suffices to show that $\mathfrak{Rep}_1 \circ \Pi$ is homotopic to identity. We define an operator

$$\mathcal{H} : \mathfrak{Func}(\mathfrak{Rep}_0(c_1), \mathfrak{Rep}_0(c_2)) \to \mathfrak{Func}(\mathfrak{Rep}_0(c_1), \mathfrak{Rep}_0(c_2))$$

of degree +1 by

$$(\mathcal{H}(T))_k(\mathbf{y})(z) = (-1)^{\deg' z \deg' \mathbf{y} + \deg' z + \deg' \mathbf{y} + \mathfrak{deg}' T} T_{k+1}(z, \mathbf{y})(\mathbf{e}_{c_1}).$$

Here $\mathbf{y} \in B_k\mathcal{C}^j(b_1, b_k)$, $z \in \mathcal{C}(b_1, c_1) = \mathfrak{Rep}_0(c_1)(b_1)$. (Then $T_{k+1}(\mathbf{y}, z)$ $(\mathbf{e}_{c_1}) \in \mathfrak{Rep}_0(c_2)(b_k)$.)

Lemma 9.22.

$$T - (\mathfrak{Rep}_1 \circ \Pi)(T) = (\mathfrak{M}_1 \circ \mathcal{H} + \mathcal{H} \circ \mathfrak{M}_1)(T). \tag{9.23}$$

Proof. We have

$$\begin{aligned} &(-1)^{\mathfrak{deg}' T}\mathfrak{M}_1(\mathcal{H}(T)(\mathbf{y})(z)) \\ &\quad = \mathfrak{m}_1(\mathcal{H}(T)(\mathbf{y})(z)) + (-1)^{\mathfrak{deg}' T}\mathcal{H}(T)(\hat{d}^{op}\mathbf{y})(z) \\ &\qquad + (-1)^{\mathfrak{deg}' T + \deg' \mathbf{y}}\mathcal{H}(T)(\mathbf{y})(\mathfrak{m}_1(z)) \\ &\qquad + \sum_a \mathfrak{m}_2(\mathcal{H}(T)(\mathbf{y}'_a), \mathfrak{Rep}_0(\mathbf{y}''_a))(z) \\ &\qquad + \sum_a (-1)^{(\mathfrak{deg}' T+1)\deg' \mathbf{y}'_a}\mathfrak{m}_2(\mathfrak{Rep}_0(\mathbf{y}'_a), \mathcal{H}(T)(\mathbf{y}''_a))(z) \\ &\quad = (-1)^{\epsilon_1}\mathfrak{m}_1(T(z, \mathbf{y})(\mathbf{e})) \\ &\qquad + \sum_a (-1)^{\epsilon_2} T(z, \mathbf{y}^{(1)}_a, \mathfrak{m}(\mathbf{y}^{(2)op}_a), \mathbf{y}^{(3)}_a)(\mathbf{e}) \\ &\qquad + (-1)^{\epsilon_3} T(\mathfrak{m}_1(z), \mathbf{y})(\mathbf{e}) \\ &\qquad + \sum_a (-1)^{\epsilon_4}\mathfrak{m}(\mathbf{y}^{\prime\prime op}_a, T(z, \mathbf{y}'_a)(\mathbf{e})) \\ &\qquad + \sum_a (-1)^{\epsilon_5} T(\mathfrak{m}(\mathbf{y}^{\prime op}_a, z), \mathbf{y}''_a)(\mathbf{e}). \end{aligned}$$

Here

$$\begin{aligned}
\epsilon_1 &= \deg' \mathbf{y} \deg' z + \deg' \mathbf{y} + \deg' z, \\
\epsilon_2 &= \mathfrak{deg}' T + (\deg' \mathbf{y} + 1)(\deg' z) + \deg' \mathbf{y} + 1 + \deg' z + \deg' \mathbf{y}_a^{(1)} \\
&\quad + \epsilon(\mathbf{y}_a^{(2)}), \\
\epsilon_3 &= \mathfrak{deg}' T + \deg' \mathbf{y} + (\deg' \mathbf{y})(\deg' z + 1) + \deg' \mathbf{y} + \deg' z + 1, \\
\epsilon_4 &= (\mathfrak{deg}' T + 1 + \deg' \mathbf{y}'_a)(\deg' \mathbf{y}''_a) + \deg' z \deg' \mathbf{y}'_a + \epsilon(\mathbf{y}''_a) + \deg' \mathbf{y}'_a \\
&\quad + \deg' z, \\
\epsilon_5 &= (\mathfrak{deg}' T + 1)(\deg' \mathbf{y}'_a) + (\deg' \mathbf{y}'_a + 1)(\mathfrak{deg}' T + \deg' \mathbf{y}''_a + 1) + \epsilon(\mathbf{y}'_a) \\
&\quad + \deg' \mathbf{y}''_a(\deg' z + \deg' \mathbf{y}'_a + 1) + \deg' \mathbf{y}''_a + \deg' z + \deg' \mathbf{y}'_a + 1.
\end{aligned}$$

We also have

$$\begin{aligned}
&(-1)^{\mathfrak{deg}' T + 1} \mathcal{H}(\mathfrak{M}_1(T))(\mathbf{y})(z) \\
&= (-1)^{\deg' \mathbf{y} + \deg' z + \deg' z \deg' \mathbf{y}} \mathfrak{m}_1(T(z, \mathbf{y})(\mathbf{e})) \\
&\quad + \sum_a (-1)^{\epsilon_6} T(\mathfrak{m}(\mathbf{y}_a'^{op}, z), \mathbf{y}''_a)(\mathbf{e}) \\
&\quad + \sum_a (-1)^{\epsilon_7} T(z, \mathbf{y}_a^{(1)}, \mathfrak{m}(\mathbf{y}_a^{(2)op}), \mathbf{y}_a^{(3)})(\mathbf{e}) \qquad (9.24) \\
&\quad + (-1)^{\deg' \mathbf{y} + \deg' z + \deg' z \deg' \mathbf{y}} \sum_{\deg \mathbf{y}''_a \neq 0} \mathfrak{m}_2(T(z, \mathbf{y}'_a), \mathfrak{Rep}(\mathbf{y}''_a))(\mathbf{e}) \\
&\quad + (-1)^{\deg' \mathbf{y} + \deg' z + \deg' z \deg' \mathbf{y}} \mathfrak{m}_2(T_0, \mathfrak{Rep}_0(z, \mathbf{y}))(\mathbf{e}) \\
&\quad + (-1)^{\deg' \mathbf{y} + \deg' z + \deg' z \deg' \mathbf{y} + \mathfrak{deg}' T \deg' z} \mathfrak{m}_2(\mathfrak{Rep}_0(z), T(\mathbf{y}))(\mathbf{e}).
\end{aligned}$$

Here

$$\begin{aligned}
\epsilon_6 &= \deg' \mathbf{y} + \deg' z + \deg' z \deg' \mathbf{y} + \deg' z \deg' \mathbf{y}'_a + \epsilon(\mathbf{y}'_a) \\
&\quad + \mathfrak{deg}' T + 1 = \epsilon_5, \\
\epsilon_7 &= \deg' \mathbf{y} + \deg' z + \deg' z \deg' \mathbf{y} + \mathfrak{deg}' T + 1 + \deg' z + \deg' \mathbf{y}_a^{(1)} \\
&\quad + \epsilon(\mathbf{y}^{(2)}) = \epsilon_2.
\end{aligned}$$

The fourth term of (9.24) is:

$$(-1)^{\epsilon_8} \mathfrak{m}(\mathbf{y}_a''^{op}, T(z, \mathbf{y}'_a)(\mathbf{e})),$$

where

$$\begin{aligned}
\epsilon_8 &= \deg' \mathbf{y} + \deg' z + \deg' z \deg' \mathbf{y} + (\mathfrak{deg}' T + \deg' z + \deg' \mathbf{y}'_a) \deg' \mathbf{y}''_a \\
&\quad + \epsilon(\mathbf{y}''_a) = \epsilon_4.
\end{aligned}$$

The fifth term of (9.24) is

$$\begin{aligned}
(-1)^{\epsilon_9} \mathfrak{m}(\mathbf{y}^{op}, z, T(\mathbf{e})) &= \mathfrak{Rep}_1(T(\mathbf{e}))(\mathbf{y})(z) \\
&= (-1)^{\mathfrak{deg}' T}((\mathfrak{Rep}_1 \circ \Pi)(T))(\mathbf{y})(z),
\end{aligned}$$

where

$$\begin{aligned}\epsilon_9 &= \deg' \mathbf{y} + \deg' z + \deg' z \deg' \mathbf{y} + (\mathfrak{deg}' T)(\deg' \mathbf{y} + \deg' z) \\ &\quad + \deg' \mathbf{y} \deg' z + \epsilon(\mathbf{y}).\end{aligned}$$

Finally the sixth term of (9.24) is:

$$\begin{aligned}&(-1)^{\deg' \mathbf{y} + \deg' z + \deg' z \deg' \mathbf{y} + \mathfrak{deg}' T \deg' z + (\deg' z + 1)(\mathfrak{deg}' T + \deg' \mathbf{y})} \\ &\qquad\qquad T(\mathbf{y})(\mathfrak{m}_2(\mathbf{e}, z)), \\ &= (-1)^{\mathfrak{deg}' T + 1} T(\mathbf{y})(z).\end{aligned}$$

The lemma follows. □

The proof of the first half of Theorem 9.1 is now complete. We omit the proof of the second half. (Which is analogous to the argument of the last section of [Fu7].) □

We are now in the position to complete the proof of Theorem 8.6. We consider the following Diagram. We can easily see that the diagram

$$\begin{array}{ccc} \mathcal{C}_1 & \xrightarrow{\mathcal{F}} & \mathcal{C}_2 \\ \downarrow & & \downarrow \\ \mathcal{R}ep(\mathcal{C}_1^o, \mathcal{CH}) & \xrightarrow{\mathcal{F}_*} & \mathcal{R}ep(\mathcal{C}_2^o, \mathcal{CH}) \end{array}$$

Diagram 9.1.

commutes. Moreover

$$\mathfrak{F}_* : \mathfrak{Rep}(\mathcal{C}_1^o, \mathcal{CH}) \to \mathfrak{Rep}(\mathcal{C}_2^o, \mathcal{CH})$$

satisfies the assumption of Theorem 8.6, by the proof of Proposition 8.49. Hence by Proposition 8.47 it is a homotopy equivalence. Moreover by Theorem 9.1 the vertical allows are homotopy equivalence. Therefore by Corollary 8.44 $\mathfrak{F}$ is a homotopy equivalence. The proof of Theorem 8.6 is now complete. □

References

[Ak] M. Akaho, *An Intersection theory of Lagrangian Immersion*, Thesis, Kyoto University, 2000.

[AG] V. Arnold and A. Givental, *Symplectic geometry*, Encyclopaedia of Mathematical Sciences, Dynamical system IV (V. Arnold and S. Novikov, ed.), Springer, Berlin, 1980.

[APS] M. Atiyah, V. Patodi and I. Singer, *Spectral asymmetry and Riemannian geometry III*, Math. Proc. Cambridge Phil. Soc., **79** (1976), 71–99.

[BaK] S. Barannikov and M. Kontsevich, *Frobenius manifolds and formality of Lie algebras of polyvector fields*, Internat. Math. Res. Notices, **4** (1998), 201–215.

[BBS] K. Becker, M. Becker and A. Strominger, *Fivebranes, membranes and nonperturbative string theory*, Nucl. Phys., **B456** (1995), 130–152.

[BoK] A. Bondal and M. Kapranov, *Framed triangulated categories*, Math. USSR. Sbornik, **1** (1991 (1990 in Russian)), 93–107.

[Do] S. Donaldson, *A lecture at University of Warwick*, 1992.

[Fl1] A. Floer, *Morse theory for Lagrangian intersections*, J. Diff. Geom., **28** (1988), 513–547.

[Fl2] A. Floer, *Witten's complex and infinite dimensional Morse theory*, J. Diff. Geom., **30** (1989), 207–221.

[Fl3] A. Floer, *Symplectic fixed point and holomorphic sphere*, Commun. Math. Phys., **120** (1989), 575–611.

[Fu1] K. Fukaya, *Morse homotopy, A^{∞}-category and Floer homologies*, Proceedings of GARC Workshop on GEOMETRY and TOPOLOGY (H. J. Kim, ed.), Seoul National University, 1993.

[Fu2] K. Fukaya, *Morse homotopy and Chern-Simons Perturbation theory*, Comm. Math. Phys., **181** (1996), 37–90.

[Fu3] K. Fukaya, *Floer homology of connected sum of homology* 3*-spheres*, Topology, **35** (1996), 89–136.

[Fu4] K. Fukaya, *Floer homology for* 3 *manifolds with boundary I*, preprint, never to appear, (in http://www.kusm.kyoto-u.ac.jp/~fukaya /fukaya.html) (1995).

[Fu5] K. Fukaya, *Floer homology for families - report of a project in progress* -, preprint, 2000.

[Fu6] K. Fukaya, *Floer homology of Lagrangian foliations and noncommutative mirror symmetry*, preprint, 1998.

[Fu7] K. Fukaya, *Mirror symmetry of Abelian variety and multi theta functions*, J. Alg. Geom. (to appear).

[Fu8] K. Fukaya, *Floer homology for oriented* 3*-manifolds*, Aspects of Low dimensional manifolds, (Y. Matumoto and S. Morita, ed.), 1992, pp. 1–92.

[FOOO] K. Fukaya, Y. Oh, H. Ohta and K. Ono, *Lagrangian intersection Floer theory - anomaly and obstruction* -, preprint.

[FOh] K. Fukaya and Y. Oh, *Zero-loop open strings in the cotangent bundle and Morse homotopy*, Asian J. Math., **1** (1997), 96–180.

[FOn1] K. Fukaya and K. Ono, *Arnold conjecture and Gromov-Witten invariants for general symplectic manifolds*, Fields Inst. Comm., **24** (1999), 173–190.

[FOn2] K. Fukaya and K.Ono, *Arnold conjecture and Gromov-Witten invariants*, Topology, **38** (1999), 933–1048.

[GJ] E. Getzler and J. Jones, *A_∞ algebra and cyclic bar complex*, Illinois J. Math., **34** (1990), 256–283.

[GM] S. Gelfand and Y. Manin, *Methods of Homological algebra*, Springer, Berlin, 1991.

[Gr] M. Gromov, *Pseudoholomorhpic curves in symplectic manifolds*, Invent. Math., **82** (1985), 307–347.

[KaS] M. Kashiwara and P. Schapira, *Sheaves on Manifolds*, Springer, Berlin, 1994.

[Ha] R. Hartshone, *Residues and duality*, Lecture Notes Math., **20**, Springer, Berlin, 1966.

[Ko1] M. Kontsevitch, *A_∞-algebras in mirror symmetry*, preprint.

[Ko2] M. Kontsevitch, *Homological algebra of Mirror symmetry*, International congress of Mathematics, Birkhäuser, Züich, 1995.

[Ko3] M. Kontsevitch, *Deformation quantization of Poisson manifolds I*, `q-alg/9709040`.

[KoS1] M. Kontsevitch and Y. Soibelman, *Homological mirror symmetry and torus fibration*, Symplectic Geometry and Mirror Symmetry (K. Fukaya, Y.-G. Oh, K. Ono and G. Tian, ed.), World Sci. Press, Singapore, 2001.

[KoS2] M. Kontsevitch and Y. Soibelman, *Deformation Theory*, a book anounced in [KoS1].

[KZ] M. Khovanov and P. Seidel, *Quivers, Floer cohomology, and braid group actions*, `QA/0006056`.

[No] S. Novikov, *Multivalued functions and functionals - an analogue of the Morse theory*, Soviet Math. Dokl., **24** (1981), 222–225.

[Oh1] Y.-G. Oh, *Floer cohomology of Lagrangian intersections and pseudo-holomorphic disks I & II*, Comm. Pure Appl. Math., **46** (1993), 949–994 & 995–1012.

[Oh2] Y.-G. Oh, *Floer cohomology of Lagrangian intersections and pseudo-holomorphic disks III*, Floer Memorial Volume, Birkhäuser, Basel, 1995, pp. 555–573.

[Ot] H. Ohta, *Obstraction to and Deformation of Lagrangian Intersection Floer Cohomology*, Symplectic Geometry and Mirror Symmetry (K. Fukaya, Y.-G. Oh, K. Ono and G. Tian, ed.), World Sci. Press, Singapore, 2001.

[PZ] A. Polishchuk and E. Zaslow, *Categorical mirror symmetry: the elliptic curve*, Adv. Theor. Math. Phys., **2** (1998), 443–470.

[Po] M. Pozniak, *Floer homology, Novikov ring, and clean intersections*, Thesis, Univ. of Warwick, 1994.

[Sc] V. Schechtman, *Remarks on formal deformations and Batalin-Vilkovsky algebras*, `math/9802006`.

[Se1] P. Seidel, *Graded Lagrangian Submanifolds*, preprint, 1999.

[Se2] P. Seidel, *Vanishing cycles and mutation*, `math.SG/0007115` (2000).

[Se3] P. Seidel, *More about vanishing cycles and mutation*, Symplectic Geometry and Mirror Symmetry (K. Fukaya, Y.-G. Oh, K. Ono and G. Tian, ed.), World Sci. Press, Singapore, 2001.

[ST] P. Seidel and R. Thomas, *Braid group actions on derived categories of coherent sheaves*, `math.AG/0001043` (2000).

[SS] M. Schlessinger and J. Stasheff, *The Lie algebra structure on tangent cohomology and deformation theory*, J. Pure Appl. Algebra, **38** (1985), 313–322.

[Si] V. D. Silva, *Products on Symplectic Floer homology*, Thesis, Oxford Univ. ,1997.

[Sm] V. Smilnov, *Simplicial and Operad Methods in Algebraic Topology*, Translations of Amer. Math. Soc., **198**, Amer. Math. Soc., Providence, 2000.

[St] J. Stasheff, *Homotopy associativity of H spaces I & II*, Trans. Amer. Math. Soc., **108** (1963), 275–292 & 293–312.

[SYZ] A. Strominger, S. Yau and E. Zaslow, *Mirror symmetry is T-duality*, Nucl. Phys., **B476** (1996), 243–259.

Department of Mathematics
Faculty of Science
Kyoto University
Kitashirakawa, Kyoto, 606-8224
Japan
`fukaya@kusm.kyoto-u.ac.jp`

Advanced Studies in Pure Mathematics 34, 2002
Minimal Surfaces, Geometric Analysis and Symplectic Geometry
pp. 129–142

Solution to the Shadow Problem in 3-Space

Mohammad Ghomi

Abstract.

If a convex surface, such as an egg shell, is illuminated from any given direction, then the corresponding shadow cast on the surface forms a connected subset. The shadow problem, first studied by H. Wente in 1978, asks whether a converse of this phenomenon is true as well. In this report it is shown that the answer is yes provided that each shadow is *simply* connected; otherwise, the answer is no. Further, the motivations behind this problem, and some ramifications of its solution for studying constant mean curvature surfaces in 3-space (soap bubbles) are discussed.

§1. Introduction

Let $M \subset \mathbf{R}^3$ be a smooth *convex surface*, i.e., the boundary of a convex body; let $n\colon M \to \mathbf{S}^2$ denote the outward unit normal vectorfield, which we also refer to as the *Gauss map*, of M; and let $u \in \mathbf{S}^2$ be a unit vector. Suppose that M is illuminated by parallel rays of light flowing in the direction of u, see Figure 1. Then, the *shadow* cast on M, i.e., the set of points in M not reached by the rays of light, is given by

$$S_u := \{\, p \in M \mid \langle n(p), u \rangle > 0 \,\}, \tag{1.1}$$

where $\langle \cdot, \cdot \rangle$ denotes the standard inner product in $\mathbf{R}^3$. It is intuitively clear, and not too difficult to show [Ghm], that if M is convex, then, for every $u \in \mathbf{S}^2$, S_u is a connected subset of M.

It is natural to ask whether the connectedness of the shadows characterizes convex surfaces, i.e., whether the converse of the above phenomenon holds as well. More precisely, let M be a *closed* (i.e., compact and connected) surface immersed in $\mathbf{R}^3$. Suppose that M is oriented, so that the gauss map is globally well-defined. Then, for every unit vector

2000 *Mathematics Subject Classification.* Primary 53A05; Secondary 52A15, 53A02, 53C45.

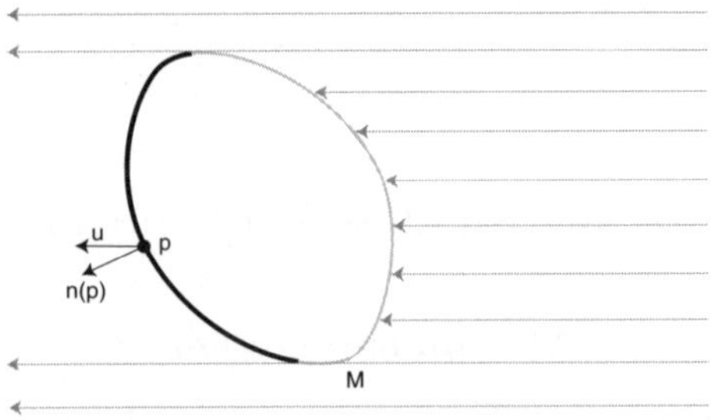

Fig. 1. The shadow cast on a surface M, when illuminated by light rays parallel to u corresponds to points p in M where the inner product between u and the unit normal $n(p)$ is positive.

$u \in \mathbf{S}^2$, let the corresponding shadow, S_u, be defined as in (1.1). Suppose that for every u, S_u is a connected subset of M. Does it then follow that M is convex?

In 1978, motivated by problems concerning the stability of constant mean curvature surfaces, H. Wente appears to have been the first person to have studied the above question [Wnt1], see Section 4, which has since become known as the *shadow problem* (a.k.a. the *illumination conjecture*). Recently, the author has proved that this problem has a positive solution provided that the shadows are *simply* connected:

Theorem 1.1. *Let M be an oriented compact surface immersed in $\mathbf{R}^3$. Suppose that for every $u \in \mathbf{S}^2$, the corresponding shadow, S_u, is* simply *connected. Then M is convex. In particular, M is embedded and homeomorphic to $\mathbf{S}^2$.*

A proof of the above theorem is outlined in Section 2. Furthermore, in Section 3, we will show that the additional condition in Theorem 1.1 (the word *simply*) is in fact necessary, as there exist embedded closed surfaces of genus one all of whose shadows are connected. Some ramifications for studying constant mean curvature surfaces will be discussed in Section 4.

Note 1.2. If M is assumed to be *simply* connected, then the assumption, in Theorem 1.1, that the shadows be 'simply connected' may be weakened to 'connected' [Ghm].

Note 1.3. The compactness assumption in Theorem 1.1 cannot be removed. Suppose, for instance, that M is a hyperbolic paraboloid, such as the one given by the graph of the equation $z = xy$. Then all the shadows of M are simply connected, even though M is not a convex surface. To see this, let H_u denote the (open) hemisphere in $\mathbf{S}^2$ determined by the unit vector u, i.e., let $H_u := \{x \in \mathbf{S}^2 \mid \langle x, u\rangle > 0\}$. A direct computation shows that the Gauss map of M, n, is a homeomorphism into $H_{(0,0,1)}$ (the Northern hemisphere). Further, note that $S_u = n^{-1}(H_u) = n^{-1}(H_u \cap H_{(0,0,1)})$. Thus, since $H_u \cap H_{(0,0,1)}$ is simply connected, it follows that S_u is simply connected as well.

§2. Outline of the Proof

The proof is by contradiction, and is organized into three steps described below. The first two steps employ techniques from Morse theory [Mil], and the third step, which is the main part of the proof, introduces a topological invariant for shadows by permuting the critical points of height functions. For a full treatment of all the details, we refer the reader to [Ghm].

2.1. Critical points of height functions

Suppose that M satisfies the hypothesis of Theorem 1.1, but is not convex. Then there exists a unit vector $v \in \mathbf{S}^2$ such that the corresponding height function $h_v \colon M \to \mathbf{R}$, defined by

$$h_v(p) := \langle p, v\rangle,$$

has at least three nondegenerate critical points, see Figure 2. This fol-

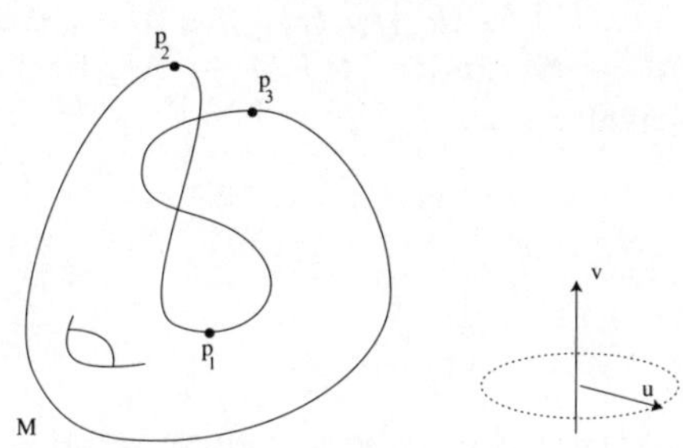

Fig. 2. If a closed surface, M, immersed in $\mathbf{R}^3$, is not convex, then there exists a direction, v, such that the corresponding height function, h_v, is a Morse function with at least three critical points.

lows from basics of the theory of *tight immersions* [CC] going back to the works of Chern and Lashof [CL]: let $\#C(h_u)$ denote the number of critical points of a Morse height function h_u, and let K be the Gauss curvature of M; then one has the following formula

$$\frac{1}{2}\int_{\mathbf{S}^2} \#C(h_u)\,du = \int_M |K(x)|\,dx.$$

Note that the integral on the left is well-defined, because for almost every $u \in S^2$, h_u is a Morse function (this is an easy application of Sard's theorem), and consequently $\#C(h_u)$ is finite. The integral on the right is known as the *total absolute curvature*, which is bounded below by 4π, because the Gauss map is surjective. $\int_M |K(x)|dx$ attains its minimum only when M is convex [CL, Thm 3]. Thus, assuming that M is not convex, $\int_M |K(x)|dx > 4\pi$. Consequently, by the above formula, there has to exist a Morse function, h_v, with more than two critical points.

2.2. Regularity of the boundary of shadows

Let $v^\perp := \{u \in \mathbf{S}^2 \mid \langle u, v\rangle = 0\}$. Using Sard's theorem, it can be shown that, after a perturbation of v, we can assume that there exists a vector $u_0 \in v^\perp$, such that the boundary of the corresponding shadow, ∂S_{u_0}, is a regular submanifold of M. This is a consequence of the fact that, for almost every $u \in \mathbf{S}^2$, ∂S_u is regular, which, briefly, may be proved as follows: define the *shadow function* $f_u \colon M \to \mathbf{R}$ by

$$f_u(p) := \langle n(p), u\rangle,$$

and observe that $\partial S_u \subset f_u^{-1}(0)$. Further, let UTM denote the unit tangent bundle of M, i.e., $UTM := \{(p, t_p) \mid p \in M,\ t_p \in T_pM, \text{ and } \|t_p\| = 1\}$. Define $\tau\colon UTM \to \mathbf{S}^2$ and $\pi\colon UTM \to M$, by $\tau(p, t_p) := t_p$ and $\pi(p, t_p) := p$ respectively.

$$\begin{array}{ccc} UTM & \xrightarrow{\ \tau\ } & \mathbf{S}^2 \\ \Big\downarrow{\scriptstyle \pi} & & \\ M & & \end{array}$$

Then $f_u^{-1}(0) = \pi(\tau^{-1}(u))$. Let u be a regular value of τ. Then $\tau^{-1}(u)$ is a regular curve in UTM. Further, it is not too difficult to show that π is an embedding on $\tau^{-1}(u)$. Hence, by Sard's theorem, for almost every u, $f_u^{-1}(0)$, and consequently ∂S_u, is a regular curve (for more results of this type and an introduction to studying geometry of the shadow boundaries on illuminated surfaces see [Hwd2]).

After a rotation of the coordinate axis, and for the sake of convenience, we assume from now on that $v = (0, 0, 1)$ and $u_0 = (1, 0, 0)$. Further, we parameterize $v^\perp$ by $u(\theta) := (\cos\theta, \sin\theta, 0)$, $\theta \in [0, 2\pi]$.

2.3. Induced permutations on the critical points

Let p_i, $i = 1, 2, 3$, denote three critical points of h_v. For every $\theta \in [0, 2\pi]$, we define a permutation $\sigma(\theta) \in \mathrm{Sym}(p_1, p_2, p_3)$, the symmetric group of three elements, as follows.

Fix $\theta \in [0, 2\pi]$. Note that $p_i \in \partial S_{u(\theta)}$; furthermore, since p_i is a *nondegenerate* critical point of the height function h_v, it follows that $\partial S_{u(\theta)}$ is regular in a neighborhood of p_i, see Figure 3. This together

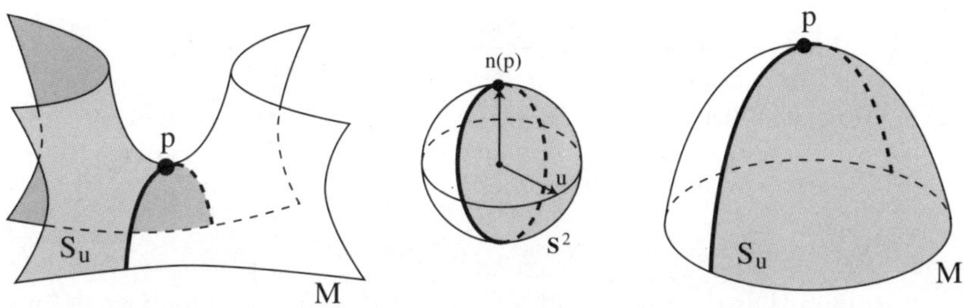

Fig. 3. If p is a regular critical point of the height function h_v, then $n(p) = \pm v$, and for every $u \in \mathbf{S}^2$ orthogonal to $n(p)$ the boundary of the corresponding shadow ∂S_u is regular in a neighborhood of p; because, n is a local diffeomorphism at p, and ∂S_u is the pull-back via n of a great circle in $\mathbf{S}^2$.

with the simply-connectedness of $S_{u(\theta)}$ implies that there exists a simple closed curve T in the closure of $S_{u(\theta)}$ such that: (i) T is composed of three smooth arcs which end at p_i, (ii) each arc meets $\partial S_{u(\theta)}$ transversally, and (iii) the interior of each arc lies in $S_{u(\theta)}$. We say that such a curve is a *standard triangle* for $S_{u(\theta)}$, see Figure 4. Since $S_{u(\theta)}$ is simply connected, T bounds a unique region in $S_{u(\theta)}$. This region inherits an orientation from M (recall that M is, by assumption, oriented), which in turn induces a preferred sense of direction on T. The induced direction on T determines a permutations for p_i in a natural way; for instance, suppose that as we move along T away from p_1 we encounter p_2 before reaching p_3, then we say that the induced permutation is the cycle $(p_1 p_2 p_3)$. Finally, note that the induced permutation on p_i does not depend on the choice of the standard triangle; because, if T' is any other standard triangle in $S_{u(\theta)}$, then T' and T are homotopic in $S_{u(\theta)}$ by the simply-connectedness of $S_{u(\theta)}$. So we conclude that each shadow $S_{u(\theta)}$ determines a unique permutation on $\{p_1, p_2, p_3\}$, which we denote by $\sigma(\theta)$.

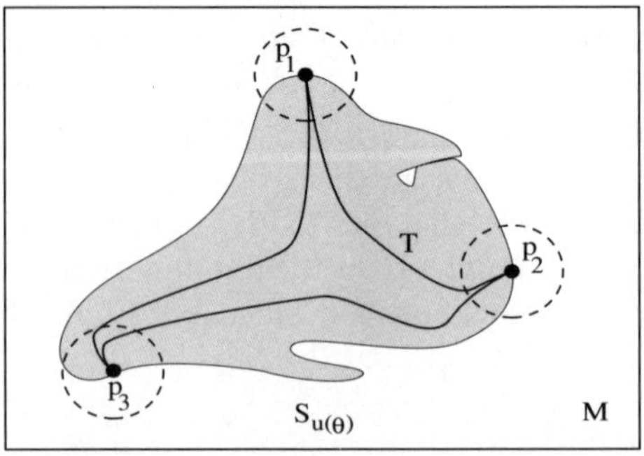

Fig. 4. Each shadow, $S_{u(\theta)}$, contains a *standard triangle.* Note that the boundary of the shadow is a regular curve in a neighborhood of the critical points p_i.

We claim that the map $\sigma\colon [0, 2\pi] \to \mathrm{Sym}(p_1, p_2, p_3)$ which we defined above is constant. To this end, since $[0, 2\pi]$ is connected, it suffices to show that σ is locally constant. This follows from the fact that whenever θ and θ' are sufficiently close, then $S_{u(\theta)}$ and $S_{u(\theta')}$ have a standard triangle in common, see Figure 5. The proof of this is based on the

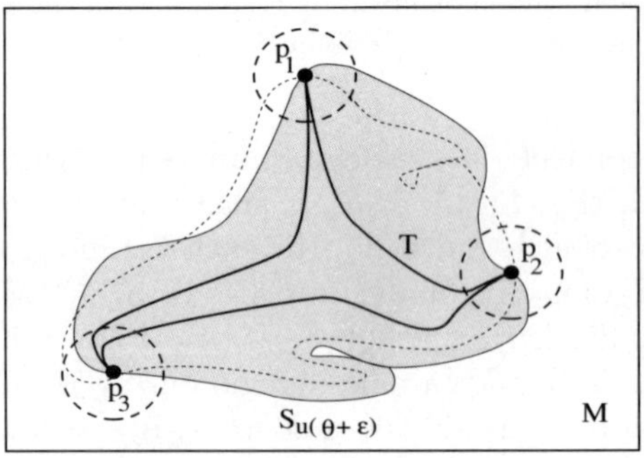

Fig. 5. For every θ there exists an $\epsilon > 0$ such that the shadows $S_{u(\theta)}$ and $S_{u(\theta+\epsilon)}$ have a standard triangle in common. This shows that the induced permutation on $\{p_1, p_2, p_3\}$ is locally constant.

compactness of T, the assumption that T meets $\partial S_{u(\theta)}$ only at p_i and does so transversally, and the observation that in a neighborhood of p_i $\partial S_{u(\theta)}$ depends continuously on θ.

On the other hand, it is not difficult to show that $\sigma(0) \neq \sigma(\pi)$. To see this, recall that $\partial S_{u(0)}$ is regular by construction. This implies that $\partial S_{u(\pi)} = \partial S_{u(0)}$. Further, since $S_{u(0)}$ is, by assumption, simply connected, $\partial S_{u(0)}$ is connected. In particular, $\partial S_{u(0)}$ is a simple closed curve passing through p_i. Suppose that $\partial S_{u(0)}$ is given the orientation induced by $S_u(0)$, and note that the corresponding permutation induced on p_i coincides with $\sigma(0)$, because all standard triangles in $S_{u(0)}$ are homotopic to $\partial S_{u(0)}$. Similarly, if $\partial S_{u(\theta)}$ is oriented by $S_{u(\pi)}$, then this gives rise to a permutation of p_i which is identical with $\sigma(\pi)$. $S_{u(0)}$ and $S_{u(\pi)}$ induce opposite orientations on $\partial S_{u(\theta)}$. Hence $\sigma(0) = -\sigma(\pi)$, which produces the desired contradiction and completes the proof.

§3. A counterexample

In this section we show that Theorem 1.1 does not remain valid if the condition that the shadows be 'simply connected' is replaced by 'connected'. More specifically, we show that there exists a smooth embedded surface of genus one all of whose shadows are connected. This surface is given by building a tube around a closed curve without any pairs of parallel tangent lines. An explicit example of such a curve, formulated by Ralph Howard, is given by $\gamma(t) := (x(t), y(t), z(t))$, where

$$
\begin{aligned}
x(t) &:= -\cos(t) - \frac{1}{20}\cos(4t) + \frac{1}{10}\cos(2t), \\
y(t) &:= +\sin(t) + \frac{1}{10}\sin(2t) + \frac{1}{20}\sin(4t), \\
z(t) &:= -\frac{46}{75}\sin(3t) - \frac{2}{15}\cos(3t)\sin(3t),
\end{aligned}
\tag{3.1}
$$

$t \in [0, 2\pi]$. Figure 6 shows the pictures of a small tube built around the above curve. Let Γ denote the trace of γ. Since Γ is a regular submanifold, it follows from the tubular neighborhood theorem that there exists an $r > 0$ such that

$$M := \{x \in \mathbf{R}^3 \mid \operatorname{dist}(x, \Gamma) = r\}$$

is a smooth surface, where $\operatorname{dist}(x, \Gamma) := \inf_{y \in \Gamma} \|x - y\|$. We claim that, since Γ has no pair of parallel tangent lines, each shadow of M is a connected subset. Before proving this, however, we describe a general procedure for constructing Γ.

Let $T \subset \mathbf{S}^2$ be a smooth simple closed curve such that (i) the origin is contained in the interior of the convex hull of T, $(0,0,0) \in \operatorname{int}\operatorname{conv} T$, and (ii) T does not contain any pair of antipodal points. Although it

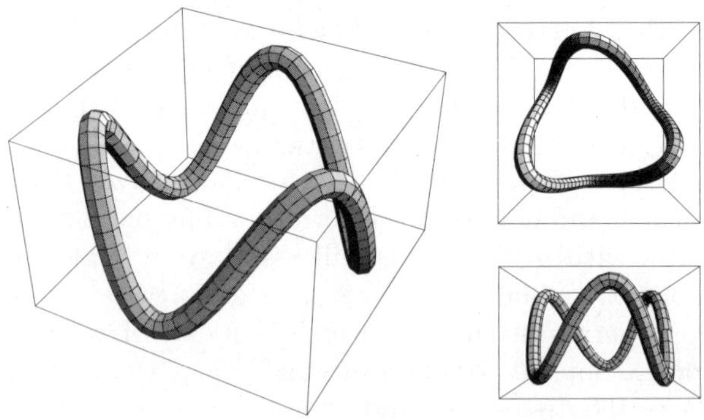

Fig. 6. Three different views of a nonconvex surface all of whose shadows are connected. This surface is constructed by building a tube around a curve with no pair of parallel tangents.

is not immediately clear that such curves exist, they are not difficult to construct. Figure 7 shows an example, which is perhaps, qualitatively speaking, the simplest. Let $T(s)$, $s \in \mathbf{R}$, denote a periodic parameter-

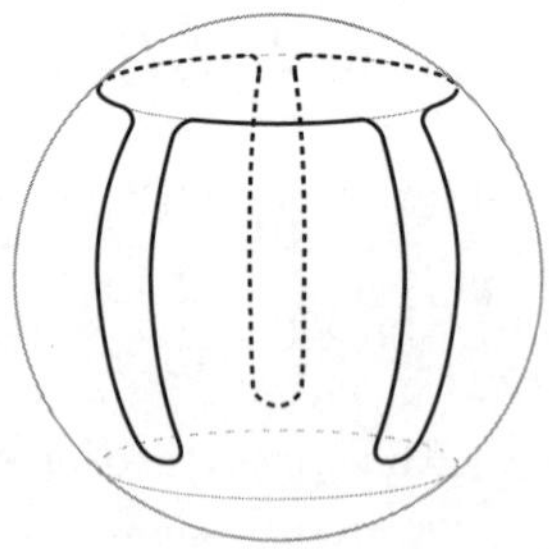

Fig. 7. A simple closed curve on the sphere which contains the origin in the interior of its convex and is disjoint from its antipodal reflection. An appropriate integration of the above yields a space curve with no parallel tangents.

ization of T by arclength. So, assuming T has total length L, we have $T(s+L) = T(s)$. Since $(0,0,0) \in \operatorname{int}\operatorname{conv} T$, there exists a (density) function $v(s)$ with period L such that $\int_0^L v(s)T(s)\,ds = 0$; or, intuitively speaking, it is possible to distribute mass along T so that the center of gravity of the resulting object coincides with the origin. Now set

$$\gamma(t) := \int_0^t v(s)T(s)\,ds.$$

Then $\gamma(t+L) = \gamma(t)$. Further, $\gamma'(t)/\|\gamma'(t)\| = T(t)$. Thus γ is a closed curve whose tangential spherical image coincides with T. In particular, γ has no parallel tangent lines. Hence Γ (the trace of γ) is the desired curve.

Next we show that M, given by a small tube around Γ, has connected shadows. To see this, let $\pi\colon M \to \Gamma$ be the obvious projection, i.e., the nearest point map. For every $x \in \Gamma$, let $F_x := \pi^{-1}(x)$ be the corresponding fiber. Note that (i) each fiber, F_x, is a circle, (ii) the image of each fiber under the Gauss map, $n(F_x)$, is the great circle in $\mathbf{S}^2$ which lies in the plane perpendicular to $T(x)$, and (iii) n is one-to-one on each F_x. Let $u \in S^2$, and let S_u be the corresponding shadow cast on M. Recall that $S_u = n^{-1}(H_u)$, where $H_u := \{x \in S^2 \mid \langle x, u\rangle > 0\}$ is an open hemisphere, see Figure 8. Thus, for each fiber, F_x, we have

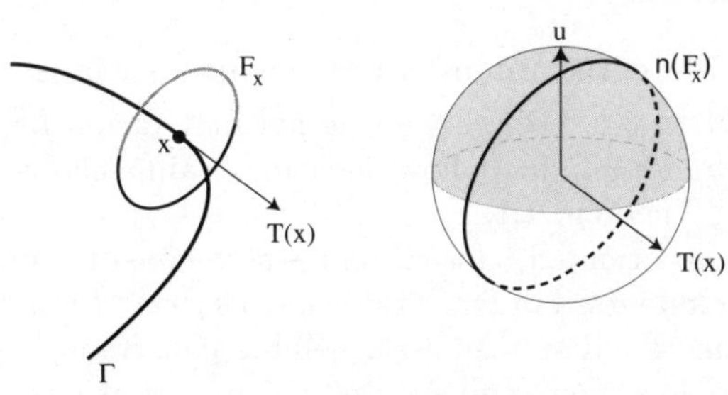

Fig. 8. Unless $T(x)$ and u are parallel, the fiber F_x of the tube around Γ intersects the shadow S_u along an open semicircle.

only two possibilities: either F_x intersects S_u in an open half-circle, or F_x is disjoint from S_u. But, by construction of Γ, the latter occurs for at most for one $x \in \Gamma$. Hence, it follows that each shadow, S_u, is either homeomorphic to a disk or an annulus. In particular, S_u is connected for every $u \in \mathbf{S}^2$.

Note 3.1. It is an elementary and well-known fact that if Γ is a closed curve, then its tangential spherical image (a.k.a. tangent indicatrix or *tantrix*), contains the origin in the interior of its convex hull. Here we showed that a converse of this phenomenon holds as well. This observation is also known, and has been attributed to Löwner; but it is not clear if it had ever been published by him. See [Hwd3] for detailed proofs and historical comments. A proof may also be found in [Gmv, p. 168]

Note 3.2. It is possible to construct a simple closed curve without parallel tangents which lies on a cylinder with a convex base. In fact, the equations (3.1) give one such example. So a loop without parallel tangents may lie on the boundary of a convex body. Interestingly enough, however, no such curve may be constructed on an ellipsoid. This follows from recent results of Joel Weiner [Wne] or Bruce Solomon [Slm] who showed that the tantrix of a spherical curve, if embedded, divides the sphere into equal areas. Consequently, any loop on a sphere has to have a pair of parallel tangents. Further, ellipsoids must have this property as well, because they are equivalent to the sphere up to a linear transformation. It would be interesting to know if ellipsoids are the only closed surfaces which admit no loops without parallel tangent.[1]

§4. Applications

4.1. Stable constant mean curvature surfaces

In this section we discuss the original motivation for studying the shadow problem, and indicate how one can obtain a classical isoperimetric result using Theorem 1.1.

Let M be an oriented, closed, and stable constant mean curvature (CMC) surface immersed in $\mathbf{R}^3$. Stable means that M is a critical surface for the area functional subject to a volume constraint. In 1978, when the shadow problem seems to have first originated, it was not yet known that M is necessarily a (round) sphere. Motivated by this question, one might make the following observation: M, much like a sphere, has connected shadows. This is based on a variation argument, described below, which the author first learned from Henry Wente [Wnt1].

For all $u \in S^2$, the shadow function $f_u \colon M \to \mathbf{R}$, $f_u(p) := \langle n(p), u \rangle$, is a Jacobi field on M, i.e., for the perturbation

$$p \longmapsto p + t f_u(p) n(p),$$

the first variation of volume and the first and second variation of area are all zero; because, the variations corresponds to a rigid motion of

[1]Note added in proof: since this paper was first written, the author and Bruce Solomon have proved that the property of having no loops without parallel tangent lines (*skew loops*), does indeed characterize ellipsoids amongst all closed surfaces immersed in 3-space [GS].

M in the direction u. Consider the nodal regions of f_u on M. These are the sets where f_u is either positive or negative, and correspond, therefore, to the shadows S_u and S_{-u}, respectively. Suppose, towards a contradiction, that S_u is not connected, then there are at least three nodal regions A_i, $i = 1, 2, 3$. Consequently, one can form three functions f_i by setting $f_i := f$ on A_i and $f_i := 0$ elsewhere. One can then take a suitable linear combination $\sum_{i=1}^{3} \lambda_i f_i$, to obtain a function for which the first variation of volume is zero but the second variation of area is negative, contradicting the stability assumption. Hence, we conclude that all shadows of M are connected.

Suppose now that M is simply connected, then, see Note 1.2, the connectedness of the shadows of M imply that each shadow is simply connected. Hence, by Theorem 1.1, it follows that M is convex. In particular, M is embedded. Consequently, by applying the maximum principle together with the reflection technique introduced by Aleksandrov [Akv], it follows that M is a sphere.

The above result is well-known, and may be regarded as a weak version of a theorem of Hopf [Hpf, p. 138], or a theorem of Barbosa and do Carmo [BC]. Hopf showed, without assuming stability, that any closed CMC surface of genus zero must be a sphere, and Barbosa and do Carmo proved that a closed oriented surface of higher genus must also be a sphere provided it is stable (for an elementary proof of this result, see [Wnt2]). Finally, Wente showed that the stability assumption in higher genus is not superfluous [Wnt3] by constructing a CMC torus in $\mathbf{R}^3$; thus, settling a famous and long standing question of Hopf [Hpf, p. 131].

In closing this section, we should also point out that a number of results concerning the connection between the number of components of nodal regions of the shadow function (the *vision number*) and the stability index of complete minimal surfaces in $\mathbf{R}^3$ have been obtained by Jaigyoung Choe [Cho].

4.2. Convexity of the level sets of H-graphs

Recently, shadows on illuminated surfaces have been studied within the context of another problem involving constant mean curvature. This problem, unlike those mentioned in the previous subsection, is still open. Let $\Omega \subset \mathbf{R}^2$ be a convex domain, and $f \in C^2(\Omega) \cap C^0(\overline{\Omega})$ be a solution to the following boundary value problem:

$$\operatorname{Div}\left(\frac{\operatorname{grad} f}{\sqrt{1 + \|\operatorname{grad} f\|^2}}\right) = 2H \quad \text{on } \Omega, \text{ and } f = 0 \quad \text{on } \partial\Omega.$$

Let M denote the graph of f. Then M has constant mean curvature H. Intuitively, one may think of M as the membrane of least area, spanned by $\partial\Omega$, which traps a given volume above the xy-plane. It has been a well-known and long standing problem [Kwl] to show that the level sets of M, and those given by equations of similar type, are convex. Recently, John McCuan [Mcn1] has obtained a number of results on this problem. In particular, he has shown that for every unit vector $u(\theta) := (\cos\theta, \sin\theta, 0)$, the set $X_{u(\theta)} := \{x \in \overline{\Omega} \mid \langle \operatorname{grad} f(x), u(\theta)\rangle = 0\}$ is a connected regular curve, assuming that $\partial\Omega$ has strictly positive curvature. This implies that the shadow $S_{u(\theta)}$ is a simply connected subset of M, because $X_{u(\theta)}$ is the projection of $\partial S_{u(\theta)}$ into the xy-plane. One is then led to consider the following question [Mcn2]: does the simply-connectedness of the shadows $S_{u(\theta)}$ imply that the level sets of M are convex? The answer is negative, see Figure 9. At present it is not

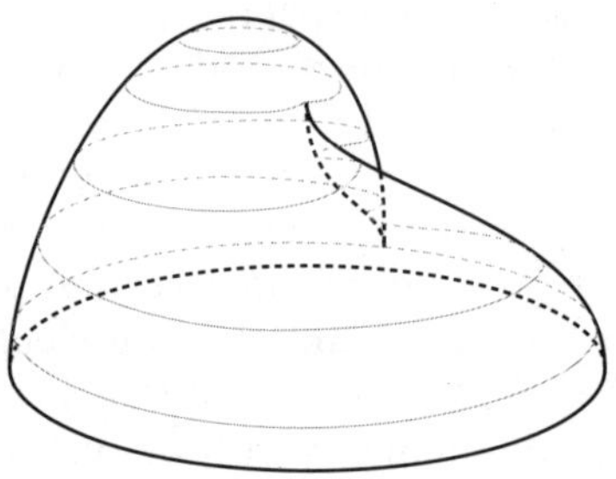

Fig. 9. A graph with zero boundary values over a convex domain which has nonconvex level sets, even though the shadows $S_{u(\theta)}$ are simply connected. $S_{u(\theta)}$ is simply connected because the graph has a unique critical point.

clear what shadow property, if any, would characterize the convexity of level sets.

Acknowledgments. The author is grateful to Ralph Howard for his interest in this work and help and advice. Further, the author wishes to thank John McCuan for bringing the shadow problem to his attention, Henry Wente for his encouragement, and Bruce Solomon for his interest in aspects of this work. Finally, the author thanks the Department of Mathematics at University of California at Santa Cruz for their hospitality while working on this project.

References

[Akv] A. D. Aleksandrov, *Uniqueness theorems for surfaces in the large,* I, Amer. Math. Soc. Transl., **21** (1962), 341–354.

[BC] J. L. Barbosa and M. do Carmo, *Stability of hypersurfaces with constant mean curvature*, Math. Z., **185** (1984), 339–353.

[Cho] J. Choe, *Index, vision number and stability of complete minimal surfaces*, Arch. Rational Mech. Anal., **109** (1990), 195–212.

[CC] T. Cecil and S. S. Chern, *Tight and taut submanifolds*, Cambridge Univ. Press, Cambridge, 1997.

[CL] S. S. Chern and R. K. Lashof, *On the total curvature of immersed manifolds*, Amer. J. Math., **79** (1957), 306–318; *On the total curvature of immersed manifolds,* II, Michigan Math. J., **5** (1958), 5–12.

[Ghm] M. Ghomi, *Shadows and Convexity of Surfaces*, Ann. of Math., (2002), to appear.

[Gmv] M. Gromov, *Partial differential relations*, Ergebnisse der Mathematik und ihrer Grenzgebiete (3) [Results in Mathematics and Related Areas (3)], **9**, Springer-Verlag, Berlin-New York, 1986.

[GS] M. Ghomi and B. Solomon, *Skew loops and quadric surfaces*, to appear, preprint available at `www.math.sc.edu/~ghomi`.

[Hpf] H. Hopf, *Differential geometry in the large*, Springer-Verlag, Berlin-New York, 1983.

[Hwd1] R. Howard, *Mohammad Ghomi's Solution to the Shadow Problem*, Lecture Notes, available at `www.math.sc.edu/~howard`.

[Hwd2] ——, *The Geometry of Shadow Boundaries on Surfaces in Space*, Lecture Notes, available at `www.math.sc.edu/~howard`.

[Hwd3] ——, *Characterization of Tantrix Curves*, Lecture Notes, available at `www.math.sc.edu/~howard`.

[Kwl] B. Kawohl, *Rearrangements and convexity of level sets in PDE*, Springer-Verlag, Berlin-New York, 1985.

[Mil] J. Milnor, *Morse theory*, Princeton University Press, Princeton, N. J., 1963.

[Mcn1] J. McCuan, *Continua of H-graphs: convexity and isoperimetric stability*, Calc. Var. Partial Differential Equations, **9** (1999), 297–325, 246.

[Mcn2] ——, *Personal e-mail*, June 23, 1998.

[Slm] B. Solomon, *Tantrices of spherical curves*, Amer. Math. Month., **103** (1996), 30–39.

[Wnt1] H. C. Wente, *Personal e-mail*, January 9, 1999.

[Wnt2] ——, *A note on the stability theorem of J. L. Barbosa and M. Do Carmo for closed surfaces of constant mean curvature*, Pacific J. Math., **147** (1991), 375–379.

[Wnt3] ——, *Counterexample to a conjecture of H. Hopf*, Pacific J. Math., **121** (1986), 193–243.

[Wne] J. Weiner, *Flat tori in S^3 and their Gauss maps*, Proc. London Math. Soc., **62** (1991), 54–76.

Department of Mathematics
University of South Carolina
Columbia, SC 29208
U.S.A.
`ghomi@math.sc.edu`
`http://www.math.sc.edu/~ghomi/`

Advanced Studies in Pure Mathematics 34, 2002
Minimal Surfaces, Geometric Analysis and Symplectic Geometry
pp. 143–154

On 4-dimensional CR-Submanifolds of a 6-dimensional Sphere

Hideya Hashimoto, Katsuya Mashimo and Kouei Sekigawa

Abstract.

We prove several fundamental properties of 4-dimensional CR-submanifolds of a nearly Kähler 6-dimensional sphere and construct explicit examples of such submanifolds.

§1. Introduction

Let S^6 be the 6-dimensional unit sphere centered at the origin of a 7-dimensional Euclidean space $\mathbf{R}^7$. We denote by $\mathbf{O}$ the normed algebra of octonions (or Cayley algebra) and identify the set of pure imaginary octonions $\operatorname{Im}\mathbf{O}$ with $\mathbf{R}^7$. An almost complex structure on S^6 is defined as follows:

$$JX = X \times x, \quad x \in S^6, \ X \in T_x(S^6),$$

where $\times$ denotes the cross product of octonions. The almost complex structure J is compatible with the canonical metric $\langle\ ,\ \rangle$ and the almost Hermitian structure $(J, \langle,\rangle)$ on S^6 is nearly Kähler ([F-I]).

In this paper, we shall study 4-dimensional CR-submanifolds of the nearly Kähler manifold $(S^6, J, \langle,\rangle)$. Let M be a submanifold of S^6. We put $\mathcal{H}_x = T_xM \cap J(T_xM)$ for $x \in M$ and denote by $\mathcal{H}_x^\perp$ the orthogonal complement of $\mathcal{H}_x$ in T_xM. If the dimension of $\mathcal{H}_x$ is constant and $J(\mathcal{H}_x^\perp) \subset T_x^\perp M$ for any $x \in M$, the submanifold M is called a *CR submanifold.*

Concerning the existence of almost complex submanifolds and totally real submanifolds of $(S^6, J, \langle,\rangle)$, many results have been obtained (see, [Gr], [Se]). On the other hand, about the existence of CR-submanifolds, only a result by Sekigawa was known before our previous paper ([H-M]), in which the first and the second authors proved that there exist many 3-dimensional CR-submanifolds.

One aim of this paper is to give some topological restrictions on the existence of compact 4-dimensional CR-submanifolds of S^6. For

2000 *Mathematics Subject Classification.* Primary 53B25; Secondary 53C15.

example, we prove that the Euler number of a compact 4-dimensional CR-submanifold is equal to zero. We also consider the integrability of the distributions $\mathcal{H}$ and $\mathcal{H}^{\perp}$. Many examples of 4-dimensional CR-submanifolds of S^6 will be given in the last section.

The authors wish to express their gratitude to Professor Yasuo Matsushita for his many valuable comments on characteristic classes.

§2. Preliminaries

Let $\mathbf{Q}$ be the skew field of all quaternions. The algebra of octonions $\mathbf{O}$ is the direct sum $\mathbf{O} = \mathbf{Q} \oplus \mathbf{Q}$ with the following multiplication:

$$(q, r) \cdot (s, t) = (qs - t^{\iota}\, r, tq + r\, s^{\iota}), \quad q, r, s, t \in \mathbf{Q},$$

where $^{\iota}$ means the conjugation in $\mathbf{Q}$. We define a conjugation in $\mathbf{O}$ by $(q, r)^{\iota} = (q^{\iota}, -r)$, $q, r \in \mathbf{Q}$, and an inner product $\langle , \rangle$ by

$$\langle x, y \rangle = \frac{(x \cdot y^{\iota} + y \cdot x^{\iota})}{2}, \quad x, y \in \mathbf{O}.$$

We denote by $\mathbf{G}_2$ the group of automorphisms of $\mathbf{O}$, that is,

$$\mathbf{G}_2 = \{g \in \mathbf{GL}(8, \mathbf{R});\ g(uv) = g(u)g(v) \text{ for any } u,\ v\ \in \mathbf{O}\}.$$

Each element of $\mathbf{G}_2$ leaves invariant the identity element $(1, 0)$ and its orthogonal complement $\operatorname{Im}\mathbf{O}$. Thus we may regard $\mathbf{G}_2$ as a subgroup of $\mathbf{GL}(7, \mathbf{R}) = \mathbf{GL}(\operatorname{Im}\mathbf{O})$.

Now, we define a basis of $\mathbf{C} \otimes \operatorname{Im}\mathbf{O}$,

$$(\varepsilon, E, \overline{E}) = (\varepsilon, E_1, E_2, E_3, \overline{E}_1, \overline{E}_2, \overline{E}_3)$$

as follows:

$$\varepsilon = (0, 1) \in \mathbf{Q} \oplus \mathbf{Q},$$

$$E_1 = iN,\ E_2 = jN,\ E_3 = -kN,$$

$$\overline{E}_1 = i\overline{N},\ \overline{E}_2 = j\overline{N},\ \overline{E}_3 = -k\overline{N},$$

where $N = (1 - \sqrt{-1}\varepsilon)/2$, $\overline{N} = (1 + \sqrt{-1}\varepsilon)/2 \in \mathbf{C} \otimes \mathbf{O}$. We denote also by g the complex linear extension of $g \in \mathbf{G}_2$. A basis $(u,\ f,\ \overline{f})$ of $\mathbf{C} \otimes \operatorname{Im}\mathbf{O}$ is said to be *admissible*, if there exists an element g of $\mathbf{G}_2$ such that $(u,\ f,\ \overline{f}) = (\varepsilon,\ E,\ \overline{E})g$. We identify an element of $\mathbf{G}_2$ with an admissible basis by the injection

$$\iota\ :\ \mathbf{G}_2 \to \mathbf{GL}(7, \mathbf{C})\ ;\ g \mapsto (\varepsilon,\ E,\ \overline{E})g.$$

We denote by $M_{p\times q}(\mathbf{C})$ the set of $p \times q$ complex matrices. Let $[a]$ be the element given by

$$[a] = \begin{pmatrix} 0 & a_3 & -a_2 \\ -a_3 & 0 & a_1 \\ a_2 & -a_1 & 0 \end{pmatrix} \in M_{3\times 3}(\mathbf{C})$$

for $a = {}^t(a_1\ a_2\ a_3\ ) \in M_{3\times 1}(\mathbf{C})$. Then we have

$$[a]b + [b]a = 0,$$

where $a, b \in M_{3\times 1}(\mathbf{C})$. We adopt the matrix representation of elements of $\mathbf{GL}(7,\mathbf{C})$ with respect to $(\varepsilon,\ E,\ \overline{E})$.

Proposition 2.1 (cf. Bryant [Br]). *The pull-back Φ of the Maurer-Cartan form of $\mathbf{GL}(7,\mathbf{C})$ is of the form*

$$\Phi = \begin{pmatrix} 0 & -\sqrt{-1}\ {}^t\overline{\theta} & \sqrt{-1}\ {}^t\theta \\ -2\sqrt{-1}\ \theta & \kappa & [\overline{\theta}] \\ 2\sqrt{-1}\ \overline{\theta} & [\theta] & \overline{\kappa} \end{pmatrix} \tag{2.1}$$

where $\kappa = ({\kappa_j}^i)$ $(1 \le i, j \le 3)$ (resp. $\theta = {}^t(\theta^1\ \theta^2\ \theta^3)$) is an $\mathfrak{su}(3)$-valued (resp. $M_{3\times 1}(\mathbf{C})$-valued) left invariant 1-forms. The Maurer-Cartan equation $d\Phi = -\Phi \wedge \Phi$ reduces to

$$d\theta = -\kappa \wedge \theta + [\overline{\theta}] \wedge \overline{\theta}, \tag{2.2}$$

$$d\kappa = -\kappa \wedge \kappa + 3\theta \wedge {}^t\overline{\theta} - ({}^t\theta \wedge \overline{\theta})\, I_3. \tag{2.3}$$

§3. Structure equations

Let $\varphi : M \to S^6$ be a 4-dimensional submanifold of S^6. We denote by ∇ (resp. D) the Levi Civita connection of M (resp. S^6) and by $\nabla^\perp$ the induced connection on the normal bundle of M in S^6. We denote by σ the second fundamental form and A_ν the shape operator in the direction of ν. The Gauss and the Weingarten formulas are given respectively by

$$\begin{aligned} D_X(\varphi_*(Y)) &= \varphi_*(\nabla_X Y) + \sigma(X, Y), \\ D_X\nu &= -\varphi_*(A_\nu(X)) + {\nabla^\perp}_X \nu, \end{aligned}$$

where X, Y are tangent vector fields and ν is a normal vector field.

Let $\varphi : M \to S^6$ be an oriented 4-dimensional CR-submanifold of S^6. Define an orientation on $\mathcal{H}^\perp$ in such a way that an orthonormal base $\{\xi_1, \xi_2\}$ of $\mathcal{H}_p^\perp$ for $p \in M$ is oriented if and only if $\{v,\ J(v),\ \xi_1,\ \xi_2\}$ is oriented for some unit vector $v \in \mathcal{H}_p$.

Lemma 3.1. *Take an oriented orthonormal base $\{\xi_1, \xi_2\}$ of $\mathcal{H}_p^\perp$ for $p \in M$. The vector $\xi_1 \times \xi_2$ is an element of $\mathcal{H}_p$ and is independent of the choice of the base.*

We denote by $\mathcal{F}$ the bundle of unit vectors of $\mathcal{H}^\perp$. For a vector $\xi \in \mathcal{F}$ we denote by ξ' the vector such that $\{\xi, \xi'\}$ is an oriented orthonormal frame of $\mathcal{F}$. We define a mapping $\psi : \mathcal{F} \to \mathbf{GL}(7, \mathbf{C})$ by

$$\psi(\xi) = (\varphi \circ \pi(\xi), f, \overline{f})$$

where

$$\begin{aligned} f_1 &= \frac{1}{2}(\xi - \sqrt{-1}J\xi), \\ f_2 &= \frac{1}{2}(\xi' - \sqrt{-1}J\xi'), \\ f_3 &= -\overline{f_1 \times f_2} = -\frac{1}{2}\left(\xi \times \xi' - \sqrt{-1}J(\xi \times \xi')\right). \end{aligned}$$

Define $\mathbf{C} \otimes \operatorname{Im} \mathbf{O}$-valued functions f_3, Ξ_1 and Ξ_2 on $\mathcal{F}$ as follows:

$$\begin{aligned} \mathbf{f}_3\left((\varphi \circ \pi(\xi), f, \overline{f})\right) &= f_3, \\ \Xi_1\left((\varphi \circ \pi(\xi), f, \overline{f})\right) &= \xi, \\ \Xi_2\left((\varphi \circ \pi(\xi), f, \overline{f})\right) &= \xi'. \end{aligned}$$

Note that the image of the mapping ψ is contained in $\iota(\mathbf{G}_2)$. Also any element of the fibre is expressed as $\cos(\theta)\ \xi + \sin(\theta)\ \xi'$.

Proposition 3.2. *Restricting the 1-forms $\kappa_i{}^j$ and θ^i given in Proposition 2.1 to $\mathcal{F}$, we have the following:*

$$\begin{aligned} (3.1)\quad d\varphi \circ \pi_* &= \mathbf{f}_3 \otimes (-2\sqrt{-1}\ \theta^3) + \overline{\mathbf{f}_3} \otimes (2\sqrt{-1}\ \overline{\theta^3}) \\ &\quad + \Xi_2 \otimes \mu_2 + \Xi_1 \otimes \mu_1, \\ (3.2)\quad \theta^3(\tilde{X}) &= \sqrt{-1}\left\langle \pi^* d\varphi(\tilde{X}), \overline{\mathbf{f}_3} \right\rangle, \\ \theta^1(\tilde{X}) &= \frac{\sqrt{-1}}{2}\left\langle \pi^* d\varphi(\tilde{X}), \Xi_1 \right\rangle = \frac{\sqrt{-1}}{2}\mu_1(\tilde{X}), \\ (3.3)\quad \theta^2(\tilde{X}) &= \frac{\sqrt{-1}}{2}\left\langle \pi^* d\varphi(\tilde{X}), \Xi_2 \right\rangle = \frac{\sqrt{-1}}{2}\mu_2(\tilde{X}), \\ (3.4)\quad d\mathbf{f}_3 &= \pi \circ \psi \otimes (-\sqrt{-1}\ \overline{\theta^3}) + \mathbf{f}_3 \otimes \kappa_3{}^3 \\ &\quad + \Xi_2 \otimes \frac{1}{2}\left(\frac{\sqrt{-1}}{2}\mu_1 + \kappa_3{}^2\right) \end{aligned}$$

$$
\begin{aligned}
& - \Xi_1 \otimes \frac{1}{2}\left(\frac{\sqrt{-1}}{2}\mu_2 - {\kappa_3}^1\right) \\
& - J\Xi_2 \otimes \frac{1}{2}\left(\frac{1}{2}\mu_1 + \sqrt{-1}{\kappa_3}^2\right) \\
& + J\Xi_1 \otimes \frac{1}{2}\left(\frac{1}{2}\mu_2 - \sqrt{-1}{\kappa_3}^1\right),
\end{aligned}
$$

$$
\begin{aligned}
(3.5) \qquad d\Xi_2 = {} & \pi \circ \psi \otimes (-\mu_2) + \mathbf{f}_3 \otimes \left({\kappa_2}^3 + \frac{\sqrt{-1}}{2}\mu_1\right) \\
& + \overline{\mathbf{f}_3} \otimes \left(\overline{{\kappa_2}^3} - \frac{\sqrt{-1}}{2}\mu_1\right) \\
& + \Xi_1 \otimes \frac{1}{2}({\kappa_2}^1 + \overline{{\kappa_2}^1} + \theta^3 + \overline{\theta^3}) \\
& - J\Xi_2 \otimes (\sqrt{-1}{\kappa_2}^2) \\
& + J\Xi_1 \otimes \frac{\sqrt{-1}}{2}(-{\kappa_2}^1 + \overline{{\kappa_2}^1} + \theta^3 - \overline{\theta^3}),
\end{aligned}
$$

$$
\begin{aligned}
(3.6) \qquad d\Xi_1 = {} & \pi \circ \psi \otimes (-\mu_1) + \mathbf{f}_3 \otimes \left({\kappa_1}^3 - \frac{\sqrt{-1}}{2}\mu_2\right) \\
& + \overline{\mathbf{f}_3} \otimes \left(\overline{{\kappa_1}^3} + \frac{\sqrt{-1}}{2}\mu_2\right) \\
& + \Xi_2 \otimes \frac{1}{2}({\kappa_1}^2 + \overline{{\kappa_1}^2} - \theta^3 - \overline{\theta^3}) \\
& + J\Xi_2 \otimes \frac{\sqrt{-1}}{2}(-{\kappa_1}^2 + \overline{{\kappa_1}^2} - \theta^3 + \overline{\theta^3}) \\
& + J\Xi_1 \otimes (-\sqrt{-1}{\kappa_1}^1),
\end{aligned}
$$

$$
\begin{aligned}
(3.7) \quad d(J\Xi_2) = {} & \mathbf{f}_3 \otimes \sqrt{-1}\left({\kappa_2}^3 - \frac{\sqrt{-1}}{2}\mu_1\right) \\
& - \overline{\mathbf{f}_3} \otimes \sqrt{-1}\left(\overline{{\kappa_2}^3} + \frac{\sqrt{-1}}{2}\mu_1\right) + \Xi_2 \otimes \sqrt{-1}{\kappa_2}^2 \\
& + \Xi_1 \otimes \frac{\sqrt{-1}}{2}({\kappa_2}^1 - \overline{{\kappa_2}^1} + \theta^3 - \overline{\theta^3}) \\
& + J\Xi_1 \otimes \frac{1}{2}({\kappa_2}^1 + \overline{{\kappa_2}^1} - \theta^3 - \overline{\theta^3}),
\end{aligned}
$$

$$
\begin{aligned}
(3.8) \quad d(J\Xi_1) = {} & \mathbf{f}_3 \otimes \sqrt{-1}\left({\kappa_1}^3 + \frac{\sqrt{-1}}{2}\mu_2\right) \\
& - \overline{\mathbf{f}_3} \otimes \sqrt{-1}\left(\overline{{\kappa_1}^3} - \frac{\sqrt{-1}}{2}\mu_2\right)
\end{aligned}
$$

$$
\begin{aligned}
&+ \Xi_2 \otimes \frac{\sqrt{-1}}{2}({\kappa_1}^2 - \overline{{\kappa_1}^2} - \theta^3 + \overline{\theta^3}) \\
&+ \Xi_1 \otimes \sqrt{-1}{\kappa_1}^1 \\
&+ J\Xi_2 \otimes \frac{1}{2}({\kappa_1}^2 + \overline{{\kappa_1}^2} + \theta^3 + \overline{\theta^3}).
\end{aligned}
$$

Remark 3.3. From Lemma 3.1, there exists a complex valued global 1-form Θ on M^4 such that $\pi^*\Theta = \theta^3$.

Next we give the explicit expression of the integarability conditions (2.2) and (2.3).

Lemma 3.4. *On $\mathcal{F}$, we have the following:*

$$
\begin{aligned}
&(3.9) & d\mu^1 &= -{\kappa_1}^1 \wedge \mu^1 - {\kappa_2}^1 \wedge \mu^2 \\
& & &\quad - {\kappa_3}^1 \wedge (-2\sqrt{-1}\,\theta^3) + 2\mu^2 \wedge \overline{\theta^3}, \\
&(3.10) & d\mu^2 &= -{\kappa_1}^2 \wedge \mu^1 - {\kappa_2}^2 \wedge \mu^2 \\
& & &\quad - {\kappa_3}^2 \wedge (-2\sqrt{-1}\,\theta^3) - 2\mu^1 \wedge \overline{\theta^3}, \\
&(3.11) & d\theta^3 &= -\frac{\sqrt{-1}}{2}({\kappa_1}^3 \wedge \mu^1 + {\kappa_2}^3 \wedge \mu^2) \\
& & &\quad - {\kappa_3}^3 \wedge \theta^3 + \frac{1}{2}\mu^1 \wedge \mu^2, \\
&(3.12) & d{\kappa_3}^3 &= -\sum_{j=1}^{3} {\kappa_j}^3 \wedge {\kappa_3}^j + 2\theta^3 \wedge \overline{\theta^3}, \\
&(3.13) & d{\kappa_i}^i &= -\sum_{j=1}^{3} {\kappa_j}^i \wedge {\kappa_i}^j - \theta^3 \wedge \overline{\theta^3} \quad (i = 1, 2), \\
&(3.14) & d{\kappa_2}^1 &= -\sum_{j=1}^{3} {\kappa_j}^1 \wedge {\kappa_2}^j + \frac{4}{3}\mu^1 \wedge \mu^2, \\
&(3.15) & d{\kappa_3}^1 &= -\sum_{j=1}^{3} {\kappa_j}^1 \wedge {\kappa_3}^j + \frac{3\sqrt{-1}}{2}\mu^1 \wedge \overline{\theta^3}, \\
&(3.16) & d{\kappa_3}^2 &= -\sum_{j=1}^{3} {\kappa_j}^2 \wedge {\kappa_3}^j + \frac{3\sqrt{-1}}{2}\mu^2 \wedge \overline{\theta^3}.
\end{aligned}
$$

Finally we shall represent the connection 1-form $\left\langle (d\Xi_1)(\tilde{X}), \Xi_2 \right\rangle$ of the S^1 bundle $\mathcal{F}$ explicitely, in terms of the local data. We put

$$\partial_\theta = \left.\frac{d}{d\theta}\right|_{\theta=0} (\cos(\theta)\xi + \sin(\theta)\xi') = \xi',$$

and denote by $d\theta$ its dual 1-form. By (3.6), we obtain

$$\left\langle (d\Xi_1)(\tilde{X}), \Xi_2 \right\rangle = -\frac{1}{2}({\kappa_2}^1 + \overline{{\kappa_2}^1} + \theta^3 + \overline{\theta^3})(\tilde{X}) = \left\langle \nabla_{d\pi(\tilde{X})}\xi_1, \xi_2 \right\rangle + d\theta(\tilde{X}).$$

In particular, we have $(1/2)({\kappa_2}^1 + \overline{{\kappa_2}^1})(\partial_\theta) = 1$.

§4. Topological restrictions

In this section we prove several topological properties of 4-dimensional CR-submanifolds of S^6. From Lemma 3.1 and Hopf's Index theorem, we immediately obtain the following

Proposition 4.1. *Let $\varphi : M^4 \to S^6$ be an oriented 4-dimensional CR-submanifold of S^6. Then both of the Euler class of M^4 and the Euler class of the complex subbundle $\mathcal{H}$ over M vanish. If M^4 is compact, then the Euler number $\chi(M^4)$ is equal to zero. In particular, $S^4, S^2 \times S^2$ and $\mathbf{C}P^2$ can not be immersed into S^6 as a CR-submanifold.*

Next we shall establish the relations of the various characteristic classes of the bundles $\mathcal{H}$, $\mathcal{H}^\perp$ and $T^\perp M^4$ over M^4. We denote by $J_\mathcal{H}$ the restriction to $\mathcal{H}$ of the almost complex structure of S^6, and J' the almost complex structure on $\mathcal{H}^\perp$ such that the orientaion determined by the almost complex structure $J_1 = J_\mathcal{H} \oplus J'$ on M coincides with that given on M. We denote by J_2 the opposite almost complex structure: $J_2 = J_\mathcal{H} \oplus (-J')$. We also denote by $J^\perp$ the almost complex structure of $T^\perp M^4$ which is compatible with the orientation of $T^\perp M^4$. Recall that

$$\varphi^*(TS^6)|_{M^4} = \mathcal{H} \oplus \mathcal{H}^\perp \oplus T^\perp M^4. \tag{4.1}$$

Let V be the direct sum $V = \mathcal{H}^\perp \oplus T^\perp M^4$. We denote by J_V the restriction to V of the almost complex structure J of S^6. We denote by $V^{(1,0)}$ (resp. $V^{(0,1)}$) the set of vectors of type (1,0) (resp. (0,1)) in the complexification $V \otimes \mathbf{C}$.

Proposition 4.2. *Let $\varphi : M \to S^6$ be an oriented 4-dimensional CR-submanifold of S^6. Then we have in $H^*(M; \mathbf{Z})$*

(1) $e(\mathcal{H}) = c_1(\mathcal{H}^{(1,0)}) (\equiv c_1(\mathcal{H}^{(1,0)}, J_\mathcal{H})) = 0,$

(2) $p_1(TM^4) = \{c_1(\mathcal{H}^{\perp(1,0)}, J')\}^2 = -\{c_1(T^{\perp(1,0)}M^4, J^\perp)\}^2$,
(3) $p_1(V) = 0$,
(4) $c_1(V^{(1,0)}) = 0$,

where we denote by $p_1(\ )$ (resp. $c_1(\ )$) the first Pontrjagin (resp. Chern) class and by $e(\ )$ the Euler class of the respective bundles.

Proof. By Lemma 3.1, we get (1) immediately. For (2), we calculate the second Chern class of the complexified tangent bundle $TM^4 \otimes \mathbf{C}$ by making use of the above decomposition. Then, we have

$$\begin{aligned} c(TM^4 \otimes \mathbf{C}) &= c(\mathcal{H}^{(1,0)} \oplus \mathcal{H}^{(0,1)} \oplus \mathcal{H}^{\perp(1,0)} \oplus \mathcal{H}^{\perp(0,1)}) \\ &= (1 - \{c_1(\mathcal{H}^{(1,0)})\}^2)(1 - \{c_1(\mathcal{H}^{\perp(1,0)})\}^2). \end{aligned}$$

Therefore we have $c_2(TM^4 \otimes \mathbf{C}) = -\{c_1(\mathcal{H}^{(1,0)})\}^2 - \{c_1(\mathcal{H}^{\perp(1,0)})\}^2$, from which we get $p_1(TM^4) = \{c_1(\mathcal{H}^{(1,0)})\}^2 + \{c_1(\mathcal{H}^{\perp(1,0)})\}^2$. Hence we have (2).

Next, we prove (3) and (4). From the decomposition $\varphi^*(T^{(1,0)}S^6)|_{M^4} = \mathcal{H}^{(1,0)} \oplus V^{(1,0)}$ and $c(T^{(1,0)}S^6) = 1$, we have

$$\begin{aligned} 1 &= 1 + c_1(\mathcal{H}^{(1,0)}) + c_1(V^{(1,0)}) \\ &\quad + c_1(\mathcal{H}^{(1,0)})c_1(V^{(1,0)}) + c_2(V^{(1,0)}) + c_1(\mathcal{H}^{(1,0)})c_2(V^{(1,0)}). \end{aligned}$$

Thus we obtain (4). Since $c_2(V^{(1,0)}) = 0$, we have $p_1(V) = -c_2(V \otimes C) = c_1(V^{(1,0)})^2 - 2c_2(V^{(1,0)}) = 0$. □

Theorem 4.3. *Let $\varphi : M^4 \to S^6$ be an oriented 4-dimensional CR-submanifold of S^6. Then the first Portrjagin class of M^4 vanishes. In particular, if M^4 is compact, its Hirzebruch signature is equal to zero.*

Proof. First we can show that the structure group of the vector bundle V reduces to $Sp(1) \simeq SU(2)$. The vector bundle $V = \mathcal{H}^\perp \oplus T^\perp M^4$ admits two different orthogonal almost complex structures $J' \oplus J^\perp$ and J_V. We may easily check that the composition $(J' \oplus J^\perp) \circ J_V$ is also an orthogonal almost complex structure on V. Furthermore, these three orthogonal almost complex structures satisfy the quaternionic relations. Thus we get $c_1(V, (J' \oplus J^\perp)) = c_1(V, -(J' \oplus J^\perp)) = -c_1(V, (J' \oplus J^\perp))$ (see [p.46; Theorem (5.11); Kob]). Therefore, we have

$$c_1(V, (J' \oplus J^\perp) = c_1(\mathcal{H}^\perp, J') + c_1(T^\perp M^4, J^\perp) = 0,$$

from which we get immediately $c_1(\mathcal{H}^{\perp(1,0)}) + c_1(T^{\perp(1,0)}M^4) = 0$. Therefore, by Proposition 4.2 (2), we obtain the desired result. □

§5. Distributions $\mathcal{H}$ and $\mathcal{H}^\perp$

Proposition 5.1. *The totally real distribution $\mathcal{H}^\perp$ of an oriented 4-dimensional CR-submanifold $\varphi : M \to S^6$ is not involutive.*

Proof. By Frobenius' theorem, $\mathcal{H}^\perp$ is involutive if and only if

$$d\theta^3 \equiv 0 \mod \{\theta^3, \overline{\theta^3}, d\theta\}. \tag{5.1}$$

From 3.12, we have

$$d\theta^3 \equiv \frac{\sqrt{-1}}{2}\left(-\sqrt{-1} + {\kappa_1}^3(E_2) - {\kappa_2}^3(E_1)\right) \mu_1 \wedge \mu_2 \mod \{\theta^3, \overline{\theta^3}, d\theta\},$$

where $\{E_1, E_2\}$ is the dual basis of $\{\mu_1, \mu_2\}$. Thus (5.1) is equivalent to

$$-\sqrt{-1} + {\kappa_1}^3(E_2) - {\kappa_2}^3(E_1) = 0.$$

On the other hand, taking account of (3.5), (3.6) and $\pi^* d\varphi(E_i) = \Xi_i$ for $i = 1, 2$, we get

$$\begin{aligned} {\kappa_1}^3(E_2) &= \sqrt{-1}\left(2\left\langle \sigma(\Xi_2, \overline{f_3}), J\Xi_1\right\rangle - \frac{1}{2}\right), \\ {\kappa_2}^3(E_1) &= \sqrt{-1}\left(2\left\langle \sigma(\Xi_1, \overline{f_3}), J\Xi_2\right\rangle + \frac{1}{2}\right). \end{aligned}$$

Finally, by (3.6) and (3.7), we have

$$\begin{aligned} &-\sqrt{-1} + {\kappa_1}^3(E_2) - {\kappa_2}^3(E_1) \\ &\quad= -2\sqrt{-1} + 2\sqrt{-1}\left(\left\langle \sigma(\Xi_2, \overline{f_3}), J\Xi_1\right\rangle - \left\langle \sigma(\Xi_1, \overline{f_3}), J\Xi_2\right\rangle\right) \\ &\quad= -2\sqrt{-1} + 2\sqrt{-1}\left(\left\langle d\Xi_2(\overline{\mathbf{f}_3}), J\Xi_1\right\rangle - \left\langle d\Xi_1(\overline{\mathbf{f}_3}), J\Xi_2\right\rangle\right) \\ &\quad= -2\sqrt{-1} - 2\overline{\theta^3}(\overline{\mathbf{f}_3}) \\ &\quad= -3\sqrt{-1}, \end{aligned}$$

which is a contradaiction. □

As an immediate consequence of Proposition 4.2 (1), we have the following lemma on the involutivity of the distibution $\mathcal{H}$.

Lemma 5.2. *Let $\varphi : M^4 \to S^6$ be an oriented 4-dimensional CR-submanifold of S^6. If the distribution $\mathcal{H}$ is involutive, then each compact leaf of $\mathcal{H}$ is homeomorphic to a torus.*

Let $\varphi : M \to S^6$ be an oriented 4-dimensional CR-submanifold of S^6. Take a (locally defined) oriented orthonormal frame $\{\xi_1, \xi_2\}$ of $\mathcal{H}^\perp$. We put $e_1 = \xi_1 \times \xi_2$, $e_2 = J(e_1)$ and denote by ω_1, ω_2, ω_3, ω_4 the dual 1-forms of e_1, e_2, ξ_1, ξ_2, respectively. From Lemma 3.1, ω_1, ω_2 are independent of the choice of the frame, and it is easily seen that so is the 2-form $\omega_3 \wedge \omega_4$.

Proposition 5.3. *Let $\varphi : M \to S^6$ be an oriented 4-dimensional CR-submanifold of S^6. The pull-back by $\pi : \mathcal{F} \to M$ of the complex valued 3-form*

$$(\omega_1 + \sqrt{-1}\omega_2) \wedge \omega_3 \wedge \omega_4$$

is equal to $2\sqrt{-1}\theta^3 \wedge \mu_1 \wedge \mu_2$ and is a closed form.

Proof. By (3.10), (3.11) and (3.12), we have

$$d(\theta^3 \wedge \mu_1 \wedge \mu_2) = -({\kappa_3}^3 + {\kappa_2}^2 + {\kappa_1}^1) \wedge \theta^3 \wedge \mu_1 \wedge \mu_2 = 0.$$

□

Remark 5.4. The proposition 5.3 is equivalent to the fact that $\mathrm{div}(e_1) = \mathrm{div}(J(e_1)) = 0$.

§6. Examples

In this section, we give two kinds of 4-dimensional CR-submanifolds of S^6. A 4-dimensional submanifold M of S^6 is a CR-submanifold if and only if the normal bundle $T^\perp M$ of M is a totally real subbundle (namely, $\Omega(T^\perp M) = \Omega \wedge \Omega(TM) = 0$, where Ω is the fundamental 2-form of S^6 defined by $\Omega(X, Y) = \langle JX, Y\rangle$ for $X, Y \in \mathfrak{X}(S^6)$).

Proposition 6.1. *Let $\gamma : I \to S^2 \subset \mathrm{Im}\,\mathbf{Q}$ be a regular curve in the unit 2-sphere. Then the following immersion $\psi : I \times Sp(1) \to S^6$ is a 4-dimensional CR-submanifold of S^6:*

$$\psi(t, q) = a\gamma(t) + bq^\iota\varepsilon,$$

where a, b are positive real numbers satisfying $a^2 + b^2 = 1$.

Proof. It is easy to verify that the vector fields

$$\nu_1 = \dot\gamma(t) \times \gamma(t), \quad \nu_2 = b\gamma(t) - aq^\iota\varepsilon$$

form an orthonormal frame field of the normal bundle and satisfy $\langle \nu_1, J(\nu_2)\rangle = 0$. □

For an element (z, q) of $U(1) \times Sp(1)$, we have an automorphism $\tau(z, q)$ of the Cayley algebra defined by

$$(6.1) \qquad (\tau(z, q))(r + s\varepsilon) = (qrq^\iota) + (zsq^\iota)\varepsilon, \quad r, s \in \mathbf{Q}, r + r^\iota = 0.$$

We denote by L the image of the Lie group homomorphism $\tau : U(1) \times Sp(1) \to \mathrm{Aut}(\mathbf{O}) = \mathbf{G}_2$.

It is easily verified that on each orbit of the action of L on S^6, there exists a point of the form $ai + (b + cj)\varepsilon$ with $a \geq 0,\ b \geq 0,\ c \geq 0$ and $a^2 + b^2 + c^2 = 1$.

Proposition 6.2. *For any positive numbers a, b, c satisfying $a^2 + b^2 + c^2 = 1$, the orbit*

$$a(qiq^{\iota}) + (z(b + cj)q^{\iota})\,\varepsilon, \quad z \in U(1),\ q \in Sp(1),$$

is a 4-dimensional CR-submanifold of S^6.

Proof. We denote by X^* a Killing vector field on S^6 induced by $X \in T_1(U(1) \times Sp(1))$. If we denote by X_0, X_1, X_2, X_3 the vectors $(i, 0), (0, i), (0, j), (0, k)$ of $T_1(U(1) \times Sp(1))$ respectively, then the tangent space $T_{p_0}(L(p_0))$ of the orbit $L(p_0)$ through the point $p_0 = ai + (b + cj)\varepsilon$ is spanned by the vectors

$$\begin{aligned} {X_0}^*(p_0) &= (bi + ck)\varepsilon, & {X_1}^*(p_0) &= (-bi + ck)\varepsilon, \\ {X_2}^*(p_0) &= -2ak + (c - bj)\varepsilon, & {X_3}^*(p_0) &= 2aj - (ci + bk)\varepsilon. \end{aligned}$$

From

$$\Omega(X_i^*(p_0), X_j^*(p_0)) = \begin{cases} 6abc, & \text{if } i = 0,\ j = 2, \\ a(5 - 9a^2), & \text{if } i = 2,\ j = 3, \\ 0, & \text{otherwise,} \end{cases}$$

we easily obtain

$$\Omega \wedge \Omega({X_0}^*(p_0), {X_1}^*(p_0), {X_2}^*(p_0), {X_3}^*(p_0)) = 0.$$

□

Proposition 6.3. *The orbit of L through the point $p = ai + (b + cj)\varepsilon$ $(a, b, c \geq 0,\ a^2 + b^2 + c^2 = 1)$ is a minimal submanifold of S^6 if and only if*

$$a = \sqrt{\frac{3 + \sqrt{57}}{24}}, \quad b = c = \sqrt{\frac{21 - \sqrt{57}}{48}}.$$

Proof. With respect to the basis $\{X_0(p_0), X_1(p_0), X_2(p_0), X_3(p_0)\}$, the induced metric g is represented as follows:

$$g = \begin{pmatrix} b^2 + c^2 & c^2 - b^2 & 0 & -2bc \\ c^2 - b^2 & b^2 + c^2 & 0 & 0 \\ 0 & 0 & 3a^2 + 1 & 0 \\ -2bc & 0 & 0 & 3a^2 + 1 \end{pmatrix}.$$

Since the orbit of the action (6.1) through a point $p = (ai) + (b + cj)\varepsilon$ $(a,\ b,\ c\ > 0)$ is diffeomorphic to $U(2)$, the volume of the orbit is equal to

$$\text{const.} \times \det(g) = \text{const.} \times 4abc\sqrt{1 + 3a^2}.$$

Considering the extremal of the volume under the condition $a^2+b^2+c^2 = 1$, we obtain the result. □

References

[Br] R. L. Bryant, *Submanifolds and special structures on the octonians*, J. Diff. Geom., **17** (1982), 185–232.

[E] N. Ejiri, *Totally real submanifolds in a 6-sphere*, Proc. Amer. Math. Soc., **83** (1981), 759–763.

[F-I] T. Fukami and S. Ishihara, *Almost Hermitian structure on S^6*, Tohoku Math. J., **7** (1955), 151–156.

[Gr] A. Gray, *Almost complex submanifolds of six sphere*, Proc. Amer. Math. Soc., **20** (1969), 277–279.

[H-M] H. Hashimoto and K. Mashimo, *On some 3-dimensional CR submanifolds in S^6*, Nagoya Math. J., **156** (1999), 171–185.

[Kob] S. Kobayashi, Differential geometry of complex vector bundles, Iwanami Shoten, Publishers and Princeton University Press, Tokyo and Princeton, 1987.

[Se] K. Sekigawa, *Some CR-submanifolds in a 6-dimensioanl sphere*, Tensor, N. S., **6** (1984), 13–20.

Hideya Hashimoto
Nippon Institute of Technology
4-1, Gakuendai, Miyashiro
Minami-Saitama Gun, Saitama, 345-8501
Japan.
hideya@nit.ac.jp

Katsuya Mashimo
Department of Mathematics
Tokyo University of Agriculture and Technology
Koganei, Tokyo 184-8588
Japan
mashimo@cc.tuat.ac.jp

Kouei Sekigawa
Department of Mathematics
Niigata University
Niigata 950-2181
Japan

Advanced Studies in Pure Mathematics 34, 2002
Minimal Surfaces, Geometric Analysis and Symplectic Geometry
pp. 155–171

On Isotropic Minimal Surfaces in Euclidean Space

Masatoshi Kokubu

Abstract.

We investigate a certain class of minimal surfaces in Euclidean space, which are constructed from a generalization of the Weierstrass formula. We also show a characterization of the catenoid.

§1. Introduction

Let f be a conformal minimal immersion from a Riemann surface M into Euclidean N-space $\mathbb{E}^N$. It is given (at least locally) by the real part of an isotropic holomorphic immersion F from M into complex Euclidean space $\mathbb{C}^N$. We say that f is *m-isotropic* if the derivatives $f^{(k)}$ of f of order k ($k = 1, 2, \dots, m$) are isotropic vectors in $\mathbb{C}^N$. (Note that f is necessarily 1-isotropic, which is equivalent to the conformality of f.) In other words, an m-isotropic minimal surface is locally the projection from $\mathbb{C}^N$ of an m-isotropic curve to $\mathbb{E}^N$.

The m-isotropic curves fully immersed in $\mathbb{C}^{2m+1}$ have a remarkable representation formula (cf. [4]), which is a generalization of the integral-free form of the Weierstrass formula for minimal surfaces. In the first half of this paper, applying it, we present some examples of complete minimal surfaces in $\mathbb{E}^{2m+1}$. They are based on Enneper's surface and the catenoid.

In the latter half of this paper, we study the total curvature of m-isotropic complete minimal surfaces. Several interesting inequalities concerning the total curvature of complete minimal surfaces in $\mathbb{E}^N$ have been known (cf. [1], [5], [6]). Among those, we focus our attention on the following two inequalities.

Given an m-isotropic complete minimal immersion $f\colon M \to \mathbb{E}^N$, we denote the Gaussian curvature by K, the area element by dA, the genus by g, and the number of ends by r, respectively.

- **(Chern-Osserman's inequality)**

$$\int_M K dA \le 4(1 - g - r)\pi. \tag{1}$$

2000 *Mathematics Subject Classification.* Primary 53A10; Secondary 53A07, 53C42.

- **(Ejiri's inequality)**
  If the immersion f is full and k-degenerate, then

$$\int_M K dA \leq 2(1-g-N+k)\pi. \tag{2}$$

Here, we say that an immersion $f\colon M \to \mathbb{E}^N$ is *full* if the image $f(M)$ is not contained in any hyperplanes of $\mathbb{E}^N$, and that f is *k-degenerate* if its Gauss image $\nu(M)$ is contained in an $(N-1-k)$-dimensional subspace of complex projective $(N-1)$-space $\mathbb{C}P^{N-1}$. (By definition, the Gauss map ν of f is given by $\nu = [\partial f/\partial z]\colon M \to \mathbb{C}P^{N-1}$, where $[\partial f/\partial z]$ denotes the complex line spanned by the vector $\partial f/\partial z \in \mathbb{C}$.

Recall that the catenoid is a complete minimal surface in $\mathbb{E}^3$, which is of genus zero, with two ends and of total curvature -4π. So it satisfies the equality in (1).

Jorge and Meeks [7] showed the formula

$$\int_M K dA = 2\left(2(1-g) - r - \sum_{j=1}^{r} I_j\right)\pi,$$

where $I_1, \ldots, I_r$ are positive integers that describe the behaviours of ends $p_1, \ldots, p_r$, respectively. In particular, they proved that an end p_j is embedded if and only if $I_j = 1$, and hence, that the equality in (1) holds if and only if all ends of M are embedded. Indeed, the catenoid has embedded ends. In [7], they also constructed examples with arbitrary number of embedded ends, which are now called Jorge-Meeks' n-noid (n is an integer greater than 1). Note that Jorge-Meeks' 2-noid is the catenoid.

On the other hand, the catenoid also satisfies the equality in (2). So the catenoid is an example satisfying the equality both in (1) and in (2). Then it is natural to ask if there are any other examples with the same property. We can answer this question for strictly m-isotropic complete minimal surfaces as follows:

Main Theorem (A characterization of the catenoid). *The catenoid in $\mathbb{E}^3$ is the only strictly m-isotropic complete minimal surface in $\mathbb{E}^{2m+1}$, which attains the equality both in Chern-Osserman's inequality and in Ejiri's inequality.*

The author is grateful to Professor Masaaki Umehara for directing his attention to this subject. He also thanks Professor Katsuei Kenmotsu and the referee for their helpful comments.

§2. Preliminaries

We denote by M a Riemann surface, and by $F\colon M \to \mathbb{C}^N$ a meromorphic curve. Let $\langle\ ,\ \rangle$ denote the standard inner product of $\mathbb{E}^N$, and the quadratic form on $\mathbb{C}^N$ which is the $\mathbb{C}$-linear extension of itself as well. A linear subspace V of $\mathbb{C}^N$ is said to be *isotropic* if $V \subset V^\perp := \{w \in \mathbb{C}^N \mid \langle v, w\rangle = 0 \quad \text{for all } v \in V\}$ holds. Note that if V is isotropic then $\overline{V}$ is also isotropic, $V \cap \overline{V} = \{0\}$ and $2\dim V \leq N$. Here, we denote by $\overline{V}$ the set of all complex conjugate vectors in V.

Definition 1. $F\colon M \to \mathbb{C}^N$ is called an *m-isotropic* curve if $\langle F^{(k)}, F^{(k)}\rangle = 0\,(1 \leq k \leq m)$ hold except at the poles. Here, $F^{(k)}$ denotes the derivative of F of order k with respect to a local coordinate z of M. For simplicity, a 1-isotropic curve is said to be isotropic. An m-isotropic curve that is not $(m+1)$-isotropic is called a *strictly m-isotropic* curve.

The following two lemmas can be easily checked.

Lemma 1. *If $F\colon M \to \mathbb{C}^N$ is m-isotropic, then the following equations hold except for poles of $F^{(k)}$:*

$$\langle F^{(i)}, F^{(j)}\rangle = 0, \quad i + j \leq 2m + 1.$$

Lemma 2. *$F\colon M \to \mathbb{C}^N$ is full if and only if at each point $p \in M$, the vectors $F', F'', \ldots, F^{(N)}$ are linearly independent except for isolated points.*

Note that Lemma 1 implies that Definition 1 is well-defined.

Proposition 1. *If $F\colon M \to \mathbb{C}^N$ is strictly m-isotropic, then $2m + 1 \leq N$. Namely, $N = 2m + 1$ is the minimum dimension of $\mathbb{C}^N$ for which an m-isotropic curve exists.*

Proof. It is enough to prove this under the assumption that F is full. By Lemma 2, $F', \ldots, F^{(N)}$ are linearly independent almost everywhere on M. At such a point p, the subspace $V = \mathrm{Span}\{F'(p), \ldots, F^{(m)}(p)\}$ is an m-dimensional isotropic subspace of $\mathbb{C}^N$ by Lemma 1. Since F is strictly m-isotropic, $\langle F^{(m+1)}(p), F^{(m+1)}(p)\rangle \neq 0$. Hence, $F^{(m+1)}(p) \notin V \oplus \overline{V}$. In fact, suppose that $F^{(m+1)}(p) \in V \oplus \overline{V}$. Then we may write

$$F^{(m+1)}(p) = \sum \lambda_i F^{(i)}(p) + \sum \mu_i \overline{F^{(i)}(p)}. \tag{3}$$

The inner product of (3) and $F^{(j)}(p)$ then implies

$$\sum \mu_i \langle \overline{F^{(i)}(p)}, F^{(j)}(p)\rangle = 0. \tag{4}$$

Here, by the linearly independency of $F^{(k)}$, the matrix $(\langle\overline{F^{(i)}(p)}, F^{(j)}(p)\rangle)$ is nonsingular. Hence, each μ_i must be zero by (4). Substituting these into (3), we have $F^{(m+1)}(p) = \sum \lambda_i F^{(i)}(p)$. This contradicts to the linearly independency of $F^{(k)}$.

Therefore, $\mathbb{C}^N$ contains a $(2m+1)$-dimensional subspace $V \oplus \overline{V} \oplus \{F^{(m+1)}\}$. It implies that $2m+1 \leq N$. □

Lemma 3. *Let $F\colon M \to \mathbb{C}^{2m+1}$ be an m-isotropic curve. Then F is strictly m-isotropic if and only if F is full.*

Proof. It is obvious from Proposition 1 that the strictness implies the fullness.

Suppose now that the m-isotropicity of F is not strict. It implies that F is $(m+1)$-isotropic. By Lemma 1, the equations

$$\langle F^{(i)}, F^{(j)}\rangle = 0,\ i+j \leq 2m+3$$

holds, which implies that

$$\begin{pmatrix} {}^tF' \\ \vdots \\ {}^tF^{(2m+1)} \end{pmatrix} \begin{pmatrix} F' & \cdots & F^{(2m+1)} \end{pmatrix} = \begin{pmatrix} 0 & \cdots & 0 \\ \vdots & * & \\ 0 & & \end{pmatrix}.$$

(Here, we regard F as a column vector and denote by ${}^tF$ its transpose.) Hence, $\det(F' \cdots F^{(2m+1)}) = 0$. Therefore, F is not full. □

Now we are in a position to state an important and fundamental theorem concerning full m-isotropic curves in $\mathbb{C}^{2m+1}$, which will be needed in Section 3.

Theorem 1 (Weierstrass-Ejiri formula). *Let $G\colon M \to \mathbb{C}^{2m-1}$ be a full $(m-1)$-isotropic curve. Suppose that g is a meromorphic function on M which is not of the form $a\langle G, G\rangle + \langle B, G\rangle + c$, where a and c are complex numbers and B is a constant vector in $\mathbb{C}^{2m-1}$. Then the following system of equations*

$$\langle G^{(k)}, H\rangle = g^{(k)}, \quad k = 1, 2, \ldots, 2m-1,$$

has a unique solution $H\colon M \to \mathbb{C}^{2m-1}$.

Moreover, if we define a function h by $h = \langle G, H'\rangle / \langle G, G'\rangle$, then the curve defined by

$$\left(\frac{1}{2}\{1 - \langle G, G\rangle\}h + \langle H, G\rangle - g,\ \sqrt{-1}\left(\frac{1}{2}\{1 + \langle G, G\rangle\}h - \langle H, G\rangle + g\right),\right.$$
$$\left. hG - H\right)$$

is full and m-isotropic in $\mathbb{C}^{2m+1}$.

Conversely, any full m-isotropic curve in $\mathbb{C}^{2m+1}$ *can be represented in this form.*

In the case of $m = 1$, Theorem 1 is the integral-free version of the Weierstrass formula for minimal surfaces (cf. [2]). For general m, this formula was proved by Ejiri [4].

§3. Applications of Theorem 1

Recall that a minimal surface in $\mathbb{E}^n$ is given by the real part of an isotropic curve in $\mathbb{C}^n$ (at least locally). Namely, for a conformal minimal immersion $f\colon M \to \mathbb{E}^N$, there exists an isotropic curve $F\colon \tilde{M} \to \mathbb{C}^N$ such that $f \circ \pi = \mathrm{Re} F$. Here, $\pi\colon \tilde{M} \to M$ denotes the universal covering of M. In other words, there exists a multi-valued isotropic curve $F\colon M \to \mathbb{C}^N$ such that $f = \mathrm{Re} F$. We call F the *lift* of f.

First, let us recall well-known minimal surfaces in $\mathbb{E}^3$.

Example 1 (Enneper's surface). $M = \mathbb{C}$, and

$$f(z) = \mathrm{Re}\left(3z - z^3,\ \sqrt{-1}(3z + z^3),\ 3z^2\right). \tag{5}$$

Example 2 (the catenoid). $M = \mathbb{C} \setminus \{0\}$, and

$$f(z) = \mathrm{Re}\left(\frac{1}{2}\left(-\frac{1}{z} - z\right),\ \frac{\sqrt{-1}}{2}\left(-\frac{1}{z} + z\right),\ \log z\right). \tag{6}$$

Example 3 (Jorge-Meeks' n-noid). $M = (\mathbb{C} \cup \{\infty\}) \setminus \{z^n = 1\}$, and

$$f(z) = \mathrm{Re}\left(\int \frac{1 - z^{2n-2}}{2(z^n - 1)^2} dz,\ \int \frac{\sqrt{-1}(1 + z^{2n-2})}{2(z^n - 1)^2} dz,\ \int \frac{z^{n-1}}{(z^n - 1)^2} dz\right). \tag{7}$$

In the case of $n = 3$, integrating (7), we have

$$f(z) = \mathrm{Re}\begin{pmatrix} \dfrac{z}{6\,(1 + z + z^2)} - \dfrac{2\,\log(-1 + z)}{9} + \dfrac{\log(1 + z + z^2)}{9} \\ \sqrt{-1}\left\{\dfrac{z\,(1 + z)}{6 - 6\,z^3} + \dfrac{2\,\arctan((1 + 2z)/\sqrt{3}))}{3\,\sqrt{3}}\right\} \\ \dfrac{1}{3\,(z^3 - 1)} \end{pmatrix}. \tag{8}$$

We define the m-isotropicity for minimal surfaces in $\mathbb{E}^N$ as well as for curves in $\mathbb{C}^N$.

Definition 2 ([3]*). A conformal minimal immersion $f\colon M \to \mathbb{E}^N$ is said to be *m-isotropic* if it satisfies the condition that $\langle f^{(k)}, f^{(k)} \rangle = 0$ for $1 \leq k \leq m$. Here, $f^{(k)}$ denotes the partial derivative $\partial^k f / \partial z^k$ with respect to a local coordinate z of M. An m-isotropic minimal surface that is not $(m+1)$-isotropic is called a *strictly m-isotropic* minimal surface.

A conformal minimal immersion is necessarily 1-isotropic, because the conformality is nothing but the 1-isotropicity. Assume that F is a lift of f, that is, $f = \mathrm{Re}F$. It then follows from $2f^{(k)} = F^{(k)}$ that the (strictly) m-isotropicity of f is equivalent to that of F.

We will construct examples of strictly m-isotropic minimal surface in $\mathbb{E}^{2m+1}$ by making use of the Weierstrass-Ejiri formula. Our examples are based on Enneper's surface and the catenoid.

First, we recall Theorem 1, the Weierstrass-Ejiri formula. It asserts that a full m-isotropic curve $F\colon M \to \mathbb{C}^{2m+1}$ is constructed from a full $(m-1)$-isotropic curve $G\colon M \to \mathbb{C}^{2m-1}$ and a meromorphic function g. We denote the curve F constructed with these data by $\mathrm{WE}(M, G, g)$.

With this notation, Enneper's surface can be written as the real part of $F = \mathrm{WE}(\mathbb{C}, z, z^3)$. Namely, (5) is constructed from $G(z) = z$ and $g(z) = z^3$ through the Weierstrass-Ejiri formula. It is also easily verified that the catenoid (6) is given by the real part of $F = \mathrm{WE}(\mathbb{C} \setminus \{0\}, z, z \log z)$.

We note that the data of Enneper's surface are given by polynomials, and Enneper's surface is also given by polynomials.† We can construct a series of m-isotropic minimal surfaces ($m = 1, 2, \ldots$) which are given by polynomials.

Proposition 2. *Consider the following recurrence formula:*

$$F_0(z) = z, \ \ F_m = \mathrm{WE}(\mathbb{C}, F_{m-1}, z^{2m+1}) \ (m \geq 1)$$

Then it inductively defines strictly m-isotropic polynomials $F_m\colon \mathbb{C} \to \mathbb{C}^{2m+1}$ *of degree* $2m+1$. *The real part* $\mathrm{Re}F_m\colon \mathbb{C} \to \mathbb{E}^{2m+1}$ *is a simply-connected, complete minimal surface of total curvature* $-4m\pi$. *In particular,* $\mathrm{Re}F_1$ *is Enneper's surface.*

For the proof, we need the following lemma.

---

*In [3], a full m-isotropic minimal surface in $\mathbb{E}^{2m+1}$ is simply called an *isotropic minimal surface.*

†We say that $F = (F_1, \ldots, F_N)$ is a polynomial if each component F_i is a polynomial. By the degree of F we mean the maximum of $\deg F_i$.

Lemma 4. *Let $F\colon \mathbb{C} \to \mathbb{C}^{2m+1}$ be an m-isotropic polynomial of degree $2m+1$. Then $\langle F, F\rangle$ is a polynomial of degree smaller than or equal to $2m+2$. Moreover, if F is full, then the degree of $\langle F, F\rangle$ is equal to $2m+2$.*

Proof. It follows from the m-isotropicity that

$$\langle F^{(i)}, F^{(j)}\rangle = 0, \quad i+j \le 2m+1. \tag{9}$$

Since F is a polynomial of degree $2m+1$,

$$F^{(k)} = 0, \quad k \ge 2m+2. \tag{10}$$

Equations (9) and (10) then imply

$$\langle F, F\rangle^{(2m+2)} = 2\langle F^{(2m+1)}, F'\rangle, \tag{11}$$

$$\langle F, F\rangle^{(2m+3)} = 2\langle F^{(2m+1)}, F''\rangle. \tag{12}$$

In particular, we consider the case of $i+j = 2m+1$ in (9), that is,

$$\langle F^{(i)}, F^{(2m-i+1)}\rangle = 0, \quad i = 1, 2, \ldots, 2m. \tag{13}$$

Differentiating (13) twice, we have for i=1,2,...,2m,

$$\langle F^{(i+2)}, F^{(2m-i+1)}\rangle + 2\langle F^{(i+1)}, F^{(2m-i+2)}\rangle + \langle F^{(i)}, F^{(2m-i+3)}\rangle = 0. \tag{14}$$

We write the cases $i = 1, \ldots, m$ in (14) into the matrix form:

$$\begin{pmatrix} 2 & 1 & & & \\ 1 & 2 & 1 & & 0 \\ & 1 & \ddots & \ddots & \\ & 0 & \ddots & 2 & 1 \\ & & & 1 & 3 \end{pmatrix} \begin{pmatrix} \langle F^{(2)}, F^{(2m+1)}\rangle \\ \langle F^{(3)}, F^{(2m)}\rangle \\ \vdots \\ \langle F^{(m)}, F^{(m+3)}\rangle \\ \langle F^{(m+1)}, F^{(m+2)}\rangle \end{pmatrix} = \begin{pmatrix} 0 \\ \vdots \\ \vdots \\ \vdots \\ 0 \end{pmatrix}. \tag{15}$$

Here, the $m \times m$ matrix on the left-hand side is nonsingular, in fact, its determinant is equal to $2m+1$. It then follows that $\langle F^{(2)}, F^{(2m+1)}\rangle = 0$. This implies that $\langle F, F\rangle^{(2m+3)} = 0$ by (12). So we can conclude that the degree of $\langle F, F\rangle$ is smaller than or equal to $2m+2$.

Suppose now that F is full. By Lemma 3 it is strictly m-isotropic. So $\langle F^{(m+1)}, F^{(m+1)}\rangle \neq 0$. Hence,

$$\begin{aligned} \langle F^{(2m+1)}, F'\rangle &= \langle F^{(2m)}, F'\rangle' - \langle F^{(2m)}, F''\rangle = -\langle F^{(2m)}, F''\rangle \\ &\;\;\vdots \\ &= (-1)^m \langle F^{(m+1)}, F^{(m+1)}\rangle \neq 0. \end{aligned}$$

Therefore, by (11), we may conclude that the degree of $\langle F, F\rangle$ is $2m+2$. □

Proof of Proposition 2. We prove this by an induction.

First, note that it is trivial in the case of $m=1$.

Assuming that the assertion is true up to $m-1$, we are going to show the case m.

The data for constructing F_m are $G=F_{m-1}$ and $g(z)=z^{2m+1}$. By our induction assumption and Lemma 4, the function $a\langle G,G\rangle+\langle B,G\rangle+c$ is a polynomial of degree $2m$, and hence, g is not identical with it. Therefore, it is assured that F_m can be constructed.

We show that F_m is a polynomial of degree $2m+1$. For this, it suffices to prove that H and h are also polynomials and that the following inequalities hold:

$$\deg H \le 2m+1,\ \deg\langle H, G\rangle \le 2m+1,\ \deg h \le 1,$$

because $\deg G = 2m-1$ and $\deg g = 2m+1$.

First, we prove that H is a polynomial. Recall that H is determined by $\langle G^{(k)}, H\rangle = g^{(k)}$. Note that the determinant of the matrix $(G^{(k)})$ satisfies

$$|G'\cdots G^{(2m-1)}|' = |G'\cdots G^{(2m-2)}\ G^{(2m)}| = 0,$$

and hence, it is constant. This implies that the inverse of $(G^{(k)})$ has components consisting of polynomials. Therefore, H is also a polynomial.

In the following, we calculate the degree of H. Differentiating

$$(16)\qquad \langle G^{(k)}, H\rangle = g^{(k)},\quad k=1,\dots,2m-1$$

we have

$$(17)\qquad \langle G^{(k+1)}, H\rangle + \langle G^{(k)}, H'\rangle = g^{(k+1)},\quad k=1,\dots,2m-1.$$

Substituting (17) into (16), we have

$$(18)\qquad \begin{cases} \langle G^{(k)}, H'\rangle = 0, & k=1,\dots,2m-2,\\ \langle G^{(2m-1)}, H'\rangle = g^{(2m)}. \end{cases}$$

Moreover, if we differentiate (18) and carry out the calculation similar to the above, then we have

$$(19)\qquad \begin{cases} \langle G^{(k)}, H''\rangle = 0, \quad k=1,\dots,2m-3\\ g^{(2m)} + \langle G^{(2m-2)}, H''\rangle = 0,\\ \langle G^{(2m-1)}, H''\rangle = g^{(2m+1)}. \end{cases}$$

Similarly, it follows from (19) that

$$\begin{cases} \langle G^{(k)}, H''' \rangle = 0, \quad k = 1, \dots, 2m-4, \\ -g^{(2m)} + \langle G^{(2m-3)}, H''' \rangle = 0, \\ 2g^{(2m+1)} + \langle G^{(2m-2)}, H''' \rangle = 0, \\ \langle G^{(2m-1)}, H''' \rangle = 0. \end{cases}$$

Proceeding successively, we obtain

$$\begin{cases} \langle G', H^{(2m)} \rangle = (2m-1) g^{(2m+1)} \neq 0, \\ \langle G^{(k)}, H^{(2m)} \rangle = 0, \quad k = 2, \dots, 2m-1, \end{cases}$$

and

$$\langle G^{(k)}, H^{(2m+1)} \rangle = 0, \quad k = 1, \dots, 2m-1.$$

Hence, we have $H^{(2m)} \neq 0$ and $H^{(2m+1)} = 0$, since $G', \dots, G^{(2m-1)}$ are linearly independent. Hence the degree of H is $2m$.

Since $G', \dots, G^{(2m-1)}$ form a basis of $\mathbb{C}^{2m-1}$ at every point $p \in \mathbb{C}$, we can write $H' = a_1 G' + \cdots + a_{2m-1} G^{(2m-1)}$. Hence, for $k = 1, \dots, 2m-2$,

$$\begin{aligned} \langle G^{(k)}, H' \rangle &= \langle G^{(k)}, a_1 G' + \cdots + a_{2m-1} G^{(2m-1)} \rangle \\ &= a_1 \langle G^{(k)}, G' \rangle + \cdots + a_{2m-1} \langle G^{(k)}, G^{(2m-1)} \rangle. \end{aligned}$$

It follows from the isotropicity of G and (18) that $a_2 = \cdots = a_{2m-1} = 0$. So, $H' = a_1 G'$. It is easy to see that $a_1 = h$. Hence, $H' = hG'$. This implies that h is a rational function P_1/P_2 and $\deg P_1 - \deg P_2 = 1$. Furthermore, taking the inner product of $H' = hG'$ with $G^{(2m-1)}$, we conclude by (18) that $h \langle G', G^{(2m-1)} \rangle = g^{(2m)}$. Since $g^{(2m)}$ has degree 1, it follows from the above fact that $\deg h = 1$ and $\deg \langle G', G^{(2m-1)} \rangle = 0$.

Finally, it follows from

$$\langle H, G \rangle' = \langle H', G \rangle + \langle H, G' \rangle = h \langle G', G \rangle + g' = \frac{h}{2} \langle G, G \rangle' + g'$$

that $\deg \langle H, G \rangle'$ is at most $2m$. □

In Proposition 2, We have constructed examples F_m as a generalization of Enneper's surface. We also construct a generalized catenoid by applying the Weierstrass-Ejiri formula.

Let F_{m-1} be an $(m-1)$-isotropic curve obtained in Proposition 2. Then F_{m-1} and the multi-valued function $g(z) = z^m \log z$ on $\mathbb{C} \setminus \{0\}$ satisfy the assumption of Theorem 1. So, $C_m := \mathrm{WE}(\mathbb{C} \setminus \{0\}, F_{m-1}, z^m \log z)$

is a multi-valued strictly m-isotropic curve $C_m\colon \mathbb{C}\setminus\{0\} \to \mathbb{C}^{2m+1}$. Explicit computations according to Theorem 1 shows that $\mathrm{Re}C_m\colon \mathbb{C}\setminus\{0\} \to \mathbb{E}^{2m+1}$ is single-valued for $m = 1, 2, 3$. Indeed, they are given by

$$
\begin{aligned}
F_0(z) &= z,\\
C_1(z) &= \left(\frac{1}{2}\left(-\frac{1}{z} - z\right),\ \frac{\sqrt{-1}}{2}\left(-\frac{1}{z} + z\right),\ \log z\right),\\
F_1(z) &= \left(3z - z^3,\ \sqrt{-1}(3z + z^3),\ 3z^2\right),\\
C_2(z) &= \left(\frac{1}{72}\left(\frac{1}{z^2} - 3z^2\right),\ \frac{\sqrt{-1}}{72}\left(\frac{1}{z^2} + 3z^2\right),\ \frac{1}{18}\left(\frac{3}{z} + z\right),\right.\\
&\qquad \left.\frac{\sqrt{-1}}{18}\left(\frac{3}{z} - z\right),\ \frac{1}{12}(1 - 2\log z)\right),\\
F_2(z) &= \left(\frac{-1}{3}z(3z^4+5), \frac{\sqrt{-1}}{3}z(3z^4-5), \frac{5}{6}z^2(z^2-6),\right.\\
&\qquad \left.\frac{-5}{6}\sqrt{-1}z^2(z^2+6), \frac{-10}{3}z^3\right),\\
C_3(z) &= \left(\frac{1}{3600}\left(\frac{9}{z^3} + 10z^3\right), \frac{\sqrt{-1}}{3600}\left(\frac{9}{z^3} - 10z^3\right), -\frac{1}{400}\left(\frac{5}{z^2} - 3z^2\right),\right.\\
&\qquad -\frac{\sqrt{-1}}{400}\left(\frac{5}{z^2} + 3z^2\right),\ -\frac{1}{80}\left(\frac{6}{z} + z\right),\ -\frac{\sqrt{-1}}{80}\left(\frac{6}{z} - z\right),\\
&\qquad \left.\frac{1}{120}(6\log z - 5)\right).
\end{aligned}
$$

Hence, $\mathrm{Re}C_1$, $\mathrm{Re}C_2$ and $\mathrm{Re}C_3$ are single-valued.‡

In the cases of $m = 1, 2, 3$, explicit formulas of C_m also show that $\mathrm{Re}C_m$ is a complete minimal surface of genus zero, with two ends and of total curvature $-4m\pi$. In particular, $\mathrm{Re}C_1$ is the catenoid.

If we represent Jorge-Meeks' trinoid (8) by the Weierstrass-Ejiri Formula, then it is given by

$$
\begin{aligned}
G(z) &= z^2,\\
g(z) &= \frac{z^2}{6} + \frac{1}{3\sqrt{3}}\arctan\left(\frac{1+2z}{\sqrt{3}}\right) + \frac{1}{18}(z^4-1)\log\frac{z^2+z+1}{(z-1)^2}.
\end{aligned}
$$

Also, the following is an example similar to the trinoid.

---

‡For general m, it is still open whether $\mathrm{Re}C_m$ is single-valued or not.

Example 4. $M = (\mathbb{C} \cup \{\infty\}) \setminus \{z^3 = 1\}$, and

$$f(z) = \mathrm{Re} \int \left(\frac{3 - 3z^{10}}{(z^3-1)^4}, \frac{\sqrt{-1}\,(3 + 3z^{10})}{(z^3-1)^4}, \frac{z^2\,(5 + 5z^6)}{(z^3-1)^4}, \frac{\sqrt{-1}z^2\,(5 - 5z^6)}{(z^3-1)^4}, \frac{8\sqrt{-1}z^5}{(z^3-1)^4} \right) dz,$$

which is given by

$$G(z) = \left(\frac{5}{6}z^2(1+z^6), \frac{5}{6}\sqrt{-1}z^2(1-z^6), \frac{4}{3}\sqrt{-1}z^5 \right)$$

$$g(z) = \frac{4}{243}\left\{ 2z^2(10 - 41z^3 + 40z^6) + 30\sqrt{3}(z^{10}+1)\arctan\left(\frac{1+2z}{\sqrt{3}}\right) + 15(z^{10}-1)\log\frac{z^2+z+1}{(z-1)^2} \right\}$$

through the Weierstrass-Ejiri formula.

This is a strictly 2-isotropic complete minimal surface with three ends, of genus zero and of total curvature -20π.

§4. Total curvature

First, we recall some fundamental facts needed later (see [1], [5] etc.).

If a complete minimal surface $f\colon M \to \mathbb{E}^N$ has finite total curvature, then M is biholomorphic to a compact Riemann surface $\overline{M}$ punctured at a finite number of points $p_1, \ldots, p_r$, i.e., $M \cong \overline{M} \setminus \{p_1, \ldots, p_r\}$. A sufficiently small neighborhood of each p_s is called an *end* of M. The Gauss map $[\partial f/\partial z]\colon M \to \mathbb{C}P^{N-1}$ can extend to a holomorphic map from $\overline{M}$ to $\mathbb{C}P^{N-1}$. In other words, $\partial f/\partial z$ has a pole at each end. It is known that because of the completeness the order of pole at any end is greater than or equal to 2, that is, the Laurent expansion of $\partial f/\partial z$ centered at p_s $(s = 1, \ldots, r)$

$$(20) \quad \frac{\partial f}{\partial z} = \frac{1}{z^{l_s}}a^s_{-l_s} + \cdots + \frac{1}{z}a^s_{-1} + \text{holomorphic part}, \quad a^s_{-l_s} \neq 0 \in \mathbb{C}^N$$

has the property that

$$(21) \qquad l_s \geq 2, \quad s = 1, \ldots, r.$$

Note that Chern-Osserman's inequality is an immediate conclusion of (21).

We also have

$$a^s_{-1} \in \mathbb{R}^N, \quad s = 1, \dots, r, \tag{22}$$

since Re $\int(\partial f/\partial z)dz$ is single-valued.

Let V be a complex vector subspace of $\mathbb{C}^N$ spanned by

$$a^s_{-l_s}, \dots, a^s_{-1} \quad (1 \le s \le r),$$

and $\tilde{V}$ a real vector subspace of $\mathbb{E}^N$ spanned by

$$\mathrm{Re}a^s_{-l_s}, \mathrm{Im}a^s_{-l_s}, \dots, \mathrm{Re}a^s_{-1}, \mathrm{Im}a^s_{-1} \quad (1 \le s \le r).$$

If f is a full immersion, then it holds that

$$\dim \tilde{V} = N. \tag{23}$$

On the other hand, it is known that the following equality, which is called the *Balancing formula* (see [5]), holds:

$$\sum_s a^s_{-1} = 0. \tag{24}$$

Hence, we have

$$\dim_{\mathbb{C}} V \le \sum_s l_s - 1. \tag{25}$$

Note that the inequality (25) is one of the reason why Ejiri's inequality holds.

In the following, we investigate what surface attains the equality both in Chern-Osserman's inequality and in Ejiri's inequality.

Recall that Jorge-Meeks' n-noid attains the equality in Chern-Osserman's inequality for any n and that Jorge-Meeks' 2-noid is the catenoid. On the other hand, the equality in Ejiri's inequality is attained by ReF_m ($m = 1, 2, \dots$) or ReC_m ($m = 1, 2, 3$) obtained in Section 3. This is verified by proving the following lemma.

Lemma 5. *A strictly m-isotropic minimal surface in $\mathbb{E}^{2m+1}$ is nondegenerate.*

Proof. Let $f\colon M \to \mathbb{E}^{2m+1}$ be a strictly m-isotropic minimal surface, and $F\colon M \to \mathbb{C}^{2m+1}$ its lift. Then F is also strictly m-isotropic, and hence is full by Lemma 3. The Gauss map $[f']$ is equal to $[F']$.

Assume that f is degenerate. Then there exists a constant vector $\xi \in \mathbb{C}^{2m+1}$ such that $\langle F', \xi\rangle = 0$. Hence, $\langle F, \xi\rangle =$ constant, which contradicts to the fullness of F. □

Recall that ReC_1 is also the catenoid. So, the catenoid is an example of complete minimal surfaces which attain the equality both in Chern-Osserman's inequality and in Ejiri's inequality. Conversely, Main Theorem in Section 1 asserts that the catenoid can be characterized as an m-isotropic surface in $\mathbb{E}^{2m+1}$ with these properties.

We now give a proof of Main Theorem in what follows.

Lemma 6. *If a strictly m-isotropic complete minimal surface M in $\mathbb{E}^{2m+1}$ attains the equality in Chern-Osserman's inequality, then the number of ends of M is greater than m.*

Proof. The equality implies that the order of pole at each end is exactly 2. Hence, the Laurent expansion (20) leads to

$$\frac{\partial f}{\partial z} = \frac{1}{z^2}\overset{s}{a}_{-2} + \frac{1}{z}\overset{s}{a}_{-1} + \text{holomorphic part}, \; \overset{s}{a}_{-2} \neq 0 \in \mathbb{C}^N. \tag{26}$$

If $m \geq 2$, then it is verified from the 2-isotropicity and (22) that

$$\overset{s}{a}_{-1} = 0 \tag{27}$$

in (26). Hence, $\dim \tilde{V} \leq 2r$. Therefore, $2m+1 \leq 2r$ by (23). Since m and r are integers, we conclude that $m+1 \leq r$.

If $m = 1$, then by the Balancing formula (24), we have

$$3 = \dim \tilde{V} \leq 2r + (r-1).$$

Hence, $4 \leq 3r$, which means that $2 \leq r$, since r is an integer. □

Lemma 7. *If a strictly m-isotropic complete minimal surface M in $\mathbb{E}^{2m+1}$ attains the equality both in Chern-Osserman's inequality and in Ejiri's inequality, then the genus of M is zero and the number of ends of M is $m+1$.*

Proof. By our assumption, we have

$$\int_M K dA = 4(1-g-r)\pi = 2(1-g-(2m+1))\pi,$$

which implies that

$$2(m+1-r) = g. \tag{28}$$

The left-hand side of (28) is smaller than or equal to 0 by Lemma 6 and the right-hand side is greater than or equal to 0. Therefore, both side must be 0. □

Proof of Main Theorem. The equality in Chern-Osserman's inequality implies that

$$\sum_s l_s = 2r = 2(m+1) \tag{29}$$

by Lemma 7. Moreover, the equality in Ejiri's inequality implies that the equality must hold in (25). It follows from (29) that

$$\dim_{\mathbb{C}} V = 2(m+1) - 1 = 2m+1. \tag{30}$$

On the other hand, if we assume $m \geq 2$, then $\dim_{\mathbb{C}} V \leq r = m+1$ holds because of (27), and hence the equality (30) cannot occur.

Therefore, we have $m = 1$. In this case, $g = 0$, $r = 2$ and the total curvature is -4π. So, it is the catenoid. □

Next, we consider only Chern-Osserman's inequality for strictly m-isotropic complete minimal surfaces in $\mathbb{E}^{2m+1}$.

Assume now that the equality is attained by a strictly m-isotropic complete minimal surface $f\colon M \to \mathbb{E}^{2m+1}$. By Lemma 6, the number of ends of M is greater than m. Hence, in the case of $m = 1$, the possibility of the number of ends is $2, 3, 4, \ldots$. Indeed, Jorge Meeks' n-noids realize these values. In the case of $m = 2$, the possibility of the number of ends is $3, 4, 5, \ldots$. However, this case is not quite similar to the case of $m = 1$.

Proposition 3. *A strictly 2-isotropic complete minimal surface of genus zero with three ends in $\mathbb{E}^5$ never attain the equality in Chern-Osserman's inequality.*

Proof. Assume that there exists a strictly 2-isotropic complete minimal surface of genus zero with three ends in $\mathbb{E}^5$ which attains the equality in Chern-Osserman's inequality.

By our assumption, the surface is biholomorphic to $\mathbb{C} \cup \{\infty\}$ punctured at three points. Without loss of generality, we may assume that these three points are cubic roots of 1, i.e., $\{z^3 = 1\}$ (if necessary, three punctured points can be mapped to $\{z^3 = 1\}$ by a linear transformation of $\mathbb{C} \cup \{\infty\}$).

Since the equality is attained in Chern-Osserman's inequality, the $\mathbb{C}^N$-valued one-form $(\partial f/\partial z)dz$ has a pole of order 2 at each end and the other points are regular. Hence, $\Omega := (z^3-1)^2(\partial f/\partial z)$ has a pole only at $z = \infty$. This implies that Ω is a polynomial of z. The degree of Ω is 4, since the induced metric

$$\left(\frac{\partial f}{\partial z}dz\right)\left(\overline{\frac{\partial f}{\partial z}dz}\right)$$

determines a positive definite inner product at $z = \infty$.

Now, we put

$$(31) \qquad \frac{\partial f}{\partial z} = \frac{a_0 z^4 + a_1 z^3 + a_2 z^2 + a_3 z + a_4}{(z^3-1)^2}, \qquad a_j \in \mathbb{C}^5.$$

Since $\mathrm{Re}\int(\partial f/\partial z)dz$ is single-valued, the residue at each pole takes a value in $\mathbb{R}$.

Indeed,

$$\mathrm{Res}_{z=1} = \frac{1}{9}(2a_0 + a_1 - a_3 - 2a_4),$$

$$\mathrm{Res}_{z=(-1)^{2/3}} = -\frac{1}{18}(2a_0 + a_1 - a_3 - 2a_4) - \frac{\sqrt{3}}{18}i(2a_0 - a_1 - a_3 + 2a_4),$$

$$\mathrm{Res}_{z=(-1)^{4/3}} = -\frac{1}{18}(2a_0 + a_1 - a_3 - 2a_4) + \frac{\sqrt{3}}{18}i(2a_0 - a_1 - a_3 + 2a_4),$$

where i denotes the imaginary unit $\sqrt{-1}$. Hence, the following holds.

$$(32) \qquad \begin{cases} 2a_0 + a_1 - a_3 - 2a_4 \in \mathbb{R}^5, \\ 2a_0 - a_1 - a_3 + 2a_4 \in \sqrt{-1}\mathbb{R}^5. \end{cases}$$

In the following, we show that the equation (32) contradicts the strictly 2-isotropicity of the surface.

By (31), the 1-isotropicity $\langle \partial f/\partial z, \partial f/\partial z\rangle = 0$ implies that

$$\langle a_0,a_0\rangle=0,\ \langle a_0,a_1\rangle=0,\ \langle a_1,a_1\rangle+2\langle a_0,a_2\rangle=0,\ \langle a_1,a_2\rangle+\langle a_0,a_3\rangle=0,$$
$$\langle a_2,a_2\rangle+2\langle a_1,a_3\rangle+2\langle a_0,a_4\rangle=0,\ \langle a_2,a_3\rangle+\langle a_1,a_4\rangle=0,$$
$$\langle a_3,a_3\rangle+2\langle a_2,a_4\rangle=0,\ \langle a_3,a_4\rangle=0,\ \langle a_4,a_4\rangle=0,$$

and the condition $\langle \partial^2 f/\partial z^2, \partial^2 f/\partial z^2\rangle = 0$ implies that

$$\langle a_0,a_0\rangle=0,\ \langle a_0,a_1\rangle=0,\ 9\langle a_1,a_1\rangle+16\langle a_0,a_2\rangle=0,$$
$$3\langle a_1,a_2\rangle+2\langle a_0,a_3\rangle=0,\ 2\langle a_2,a_2\rangle+3\langle a_1,a_3\rangle=0,\ \langle a_2,a_3\rangle=0,$$
$$\langle a_3,a_3\rangle=0.$$

Summing up these, we have

$$(33) \quad \langle a_0,a_0\rangle = \langle a_0,a_1\rangle = \langle a_0,a_2\rangle = \langle a_0,a_3\rangle = \langle a_1,a_1\rangle = \langle a_1,a_2\rangle$$
$$= \langle a_1,a_4\rangle = \langle a_2,a_3\rangle = \langle a_2,a_4\rangle = \langle a_3,a_3\rangle = \langle a_3,a_4\rangle = \langle a_4,a_4\rangle = 0,$$

$$(34) \quad \langle a_2,a_2\rangle + 2\langle a_1,a_3\rangle + 2\langle a_0,a_4\rangle = 0,\ 2\langle a_2,a_2\rangle + 3\langle a_1,a_3\rangle = 0.$$

Equations in (34) imply that

$$\langle a_1, a_3\rangle + 4\langle a_0, a_4\rangle = 0. \tag{35}$$

It then follows from (33) and (35) that $2a_0 + a_1 - a_3 - 2a_4$ and $2a_0 - a_1 - a_3 + 2a_4$ are both isotropic vectors. However, they are real-valued and purely imaginary-valued, respectively. Hence, they must be zero, i.e.,

$$2a_0 + a_1 - a_3 - 2a_4 = 0,\ 2a_0 - a_1 - a_3 + 2a_4 = 0.$$

Therefore, we have

$$2a_0 = a_3,\ a_1 = 2a_4.$$

It follows that

$$\langle a_1, a_3\rangle = 4\langle a_0, a_4\rangle,$$

which implies from (35) that

$$\langle a_1, a_3\rangle = \langle a_0, a_4\rangle = 0.$$

By (34), we also have $\langle a_2, a_2\rangle = 0$. Consequently, we have for all $j, k = 0, 1, 2, 3, 4$,

$$\langle a_j, a_k\rangle = 0.$$

Therefore, $\langle \partial^3 f/\partial z^3, \partial^3 f/\partial z^3\rangle = 0$, which is a contradiction to the strictness of the surface. □

Finally of this paper, we propose a problem related to Proposition 3.

Problem. Is there an inequality sharper than Chern-Osserman's inequality for strictly m-isotropic complete minimal surfaces in $\mathbb{E}^{2m+1}$ $(m \geq 2)$? Namely, is there a constant $C(m, g, r)$ depending only on m, g, r such that

$$\int_M K dA \leq C(m, g, r) \leq 4(1 - g - r)\pi$$

holds for all strictly m-isotropic complete minimal surfaces in $\mathbb{E}^{2m+1}$ of genus g and with r ends?

Proposition 3 means that

$$\int_M K dA < 4(1 - g - r)\pi$$

in the case of $m = 2$, $g = 0$ and $r = 3$. Hence, $C(2, 0, 3)$ is at most $4(1 - 0 - 3)\pi - 2\pi = -10\pi$, because the total curvature of complete minimal surface takes value in $2\pi\mathbb{Z}$. On the other hand, there exists a strictly 2-isotropic complete minimal surface with three ends, of genus zero and of total curvature -20π, which is stated in Example 4. Therefore we conclude that $-20\pi \leq C(2,0,3) \leq -10\pi$.

References

[1] S. Chern and R. Osserman, *Complete minimal surfaces of in euclidean n-space*, J. d'Analyse Math., **19** (1967), 15–34.

[2] G. Darboux, Leçons sur la théorie générale des surfaces, third edition, Chelsea, New York, 1972.

[3] N. Ejiri, *Isotropic harmonic maps of Riemann surfaces into the de Sitter space*, Quart. J. Oxford, **43** (1992), 421–440.

[4] N. Ejiri, *A Darboux theorem for null curves in* C^{2m+1}, Lecture Notes Vol. 2, Geometry and Global Analysis, Tohoku Univ., 1993.

[5] N. Ejiri, *Degenerate minimal surfaces of finite total curvature in* R^N, Kobe J. Math., **14** (1997), 11–22.

[6] F. Gackstatter, *Über die Dimension einer Minimalfläche und zur Ungleichung von St. Cohn-Vossen*, Arch. Rational Mech. Anal., **61** (1976), 141–152.

[7] L. P. M. Jorge and W. H. Meeks III, *The topology of complete minimal surfaces of finite total curvature*, Topology, **22** (1983), 203–221.

Department of Natural Science
School of Engineering
Tokyo Denki University
2-2, Kanda-Nishiki-Cho
Chiyoda-Ku, Tokyo 101-8457
Japan
kokubu@cck.dendai.ac.jp

Advanced Studies in Pure Mathematics 34, 2002
Minimal Surfaces, Geometric Analysis and Symplectic Geometry
pp. 173–184

The Topology of Toric HyperKähler Manifolds

Hiroshi Konno[1]

Abstract.

The topology of hyperKähler quotients of quaternionic vector spaces by tori is studied. We discuss the relation between their topology and a combinatorial property of some polyhedral complexes. As its simple application we compute their Chern classes.

§1. Introduction

The topology of symplectic quotients has been intensively studied in the last two decades. Especially, Kirwan's theory enables us to compute the Betti numbers of symplectic quotients [9], and thanks to the theory of Jeffrey and Kirwan [8] we can investigate their cohomology rings. On the other hand, various classes of hyperKähler quotients were introduced and studied in detail by many authors, but their topology has not yet been studied well. Recently, in this regard Bielawski and Dancer studied hyperKähler quotients of quaternionic vector spaces $\mathbf{H}^N$ by subtori of T^N, which they call toric hyperKähler manifolds [2].

Being influenced by their work, we intend to study the topology of toric hyperKähler manifolds. It should be remarked that every toric hyperKähler manifold, if we deform its hyperKähler structure appropriately, contains a union of projective toric manifolds as its deformation retract. Because of this fact we call it the core of the toric hyperKähler manifold. Generally speaking, the topology of projective toric manifolds is well-known [4]. However, since they intersect in a complicated way, it is not easy to study the topology of the core. Concerning this, in [10] we determined their cohomology rings.

In this note we also study the topology of toric hyperKähler manifolds. The structure of the core is described by a polyhedral complex

[1]2000 *Mathematics Subject Classification.* Primary 53C26; Secondary 53D20.

Supported in part by the Grant-in-Aid for Scientific Research (C), No. 09640124.

associated to it. We discuss the relation between the topology of toric hyperKähler manifolds and a combinatorial property of the associated polyhedral complex. As its simple application we compute the total Chern class of toric hyperKähler manifolds.

In Section 2 we define toric hyperKähler manifolds and describe their cohomology rings, which is proved in [10]. The relation of the topology of toric hyperKähler manifolds and their associated polyhedral complexes is studied in Section 3. In Section 4 we compute their Chern classes.

The author would like to thank T. Gocho for suggesting a proof of Lemma 3.3. He also would like to thank for the organizers of the conference and JAMI for their hospitality.

§2. Toric hyperKähler manifolds

In this section we define toric hyperKähler manifolds and describe their cohomology rings.

First, let us recall the hyperKähler structure on the quaternionic vector space $\mathbf{H}^N$. Let $\{1, I_1, I_2, I_3\}$ be the standard basis of $\mathbf{H}$. On $\mathbf{H}^N$ we define three complex structures by the multiplication of I_1, I_2, I_3 from the left, respectively. We denote these complex structures also by I_1, I_2, I_3. The real torus $T^N = \{\alpha = (\alpha_1, \dots, \alpha_N) \in \mathbf{C}^N \| \alpha_i| = 1\}$ acts on $\mathbf{H}^N$ from the right diagonally, and preserves its hyperKähler structure. If we identify $\xi \in \mathbf{H}^N$ with $(z, w) \in \mathbf{C}^N \times \mathbf{C}^N$ by $\xi = z + wI_2$, then the action is given by

$$(z, w)\alpha = (z\alpha, w\alpha^{-1}).$$

Let K be a subtorus of T^N with Lie algebra $k \subset t^N$. Then we have the torus $T^n = T^N/K$ with Lie algebra $t^n = t^N/k$. Moreover, we have the following exact sequences:

$$\begin{array}{ccccccccc} 0 & \longrightarrow & k & \xrightarrow{\iota} & t^N & \xrightarrow{\pi} & t^n & \longrightarrow & 0, \\ 0 & \longleftarrow & k^* & \xleftarrow{\iota^*} & (t^N)^* & \xleftarrow{\pi^*} & (t^n)^* & \longleftarrow & 0. \end{array}$$

Since the action of K on $\mathbf{H}^N$ preserves its hyperKähler structure, we obtain the hyperKähler moment map

$$\mu_K = (\mu_{K,1}, \mu_{K,2}, \mu_{K,3})\colon \mathbf{H}^N \to k^* \otimes \mathbf{R}^3,$$

which is given by

$$\mu_{K,1}(z,w) = \pi \sum_{i=1}^{N} (|z_i|^2 - |w_i|^2)\iota^* u_i,$$

$$(\mu_{K,2} + \sqrt{-1}\mu_{K,3})(z,w) = -2\pi\sqrt{-1} \sum_{i=1}^{N} z_i w_i \iota^* u_i,$$

where $\{u_1, \ldots, u_N\} \subset (t^N)^*$ is the dual basis of the standard basis $\{X_1, \ldots, X_N\} \subset t^N$. Now we define toric hyperKähler manifolds.

Definition. If $\nu \in k^* \otimes \mathbf{R}^3$ is a regular value of the hyperKähler moment map μ_K and if the action of K on $\mu_K^{-1}(\nu)$ is free, we call the hyperKähler quotient

$$X(\nu) = \mu_K^{-1}(\nu)/K$$

a toric hyperKähler manifold.

Note that $X(\nu)$ is a $4n$ dimensional hyperKähler manifold. We denote its hyperKähler structure by $(g_\nu, I_{\nu,1}, I_{\nu,2}, I_{\nu,3})$. The torus $T^n = T^N/K$ acts on $X(\nu)$, preserving its hyperKähler structure. This action gives the hyperKähler moment map

$$\mu_{T^n} = (\mu_{T^n,1}, \mu_{T^n,2}, \mu_{T^n,3})\colon X(\nu) \to (t^n)^* \otimes \mathbf{R}^3.$$

The terminology 'a toric hyperKähler manifold' is due to Bielawski and Dancer [2]. One of their results is the following:

Fact 2.1. *The diffeomorphism type of a toric hyperKähler manifold $X(\nu)$ is independent of the choice of ν.*

In [10], for each $h \in (t^N_{\mathbf{Z}})^* = \sum_{i=1}^{N} \mathbf{Z}u_i$, we constructed a holomorphic line budle L_h on $X(\nu)$ with respect to the complex structure $I_{\nu,1}$. The equation $z_i = 0$ defines a divisor D_{u_i} on $X(\nu)$, and we showed that the holomorphic line bundle defined by the divisor D_{u_i} is L_{u_i}. Moreover, we showed that the dual line bundle $L^*_{u_i}$ corresponds to the divisor defined by the equation $w_i = 0$. In [10] we described the cohomology ring of $X(\nu)$ in terms of the subtorus K as follows.

Theorem 2.2. *Let $\Phi\colon \mathbf{Z}[u_1, \ldots, u_N] \to H^*(X(\nu);\mathbf{Z})$ be a ring homomorphism defined by $\Phi(u_i) = c_1(L_{u_i})$. Then the following holds:*
(1) *The map Φ is surjective. Therefore we have an isomorphism as a ring:*

$$H^*(X(\nu);\mathbf{Z}) \cong \mathbf{Z}[u_1, \ldots, u_N]/\ker\Phi.$$

(2) $\ker\Phi$ *is an ideal generated by all*

1. $\sum_{i=1}^N a_i u_i \in \ker \iota^* \cap (t_{\mathbf{Z}}^N)^*$, *and*
2. $\prod_{b_i\neq 0} u_i$ *for* $\sum_{i=1}^N b_i X_i \in k\setminus\{0\}$.

Example. Let $\pi\colon t^5 \to t^3$ be a surjevtive map such that $\pi(X_4) = -\pi(X_1) - \pi(X_2)$ and $\pi(X_5) = -\pi(X_1) - \pi(X_3)$. Then we have a toric hyperKähler manifold $X(\nu)$ for $\nu \in k^* \otimes \mathbf{R}^3$ satisfying the condition mentioned above. Since k is spanned by $\{X_1+X_2+X_4, X_1+X_3+X_5\}$, there are 4 types of elements in k as follows:

$$X_1+X_2+X_4, X_1+X_3+X_5, X_2-X_3+X_4-X_5,$$
$$\sum_{i=1}^5 a_i X_i \quad \text{where } a_i \neq 0 \text{ for } i=1,\dots,5.$$

Moreover, since $\ker\iota^*$ is spanned by $\{u_2-u_4, u_3-u_5, u_1-u_2-u_3\}$, Theorem 2.2 implies that in this case $\ker\Phi$ is generated by

$$\{u_2-u_4, u_3-u_5, u_1-u_2-u_3, u_1u_2u_4, u_1u_3u_5, u_2u_3u_4u_5, u_1u_2u_3u_4u_5\}.$$

§3. The associated polyhedral complex

In this section we associate a polyhedral complex $\mathcal{C}(X(\nu))$ to a toric hyperKähler manifold $X(\nu)$ with $\nu = (\nu_1, 0, 0) \in k^* \otimes \mathbf{R}^3$. We also discuss the relation between the topology of $X(\nu)$ and the associated polyhedral complex. Throughout this section, we assume that $\nu = (\nu_1, 0, 0) \in k^* \otimes \mathbf{R}^3$. We also fix an element $h \in (t^N)^*$ such that $\iota^* h = \nu_1$.

First, let us recall the notion of a polyhedral complex. A polyhedral complex $\mathcal{C}$ is by definition a family of polyhedra in the fixed $\mathbf{R}^n$ satisfying the following conditions:

1. If σ is an element of $\mathcal{C}$, then every face of σ belongs to $\mathcal{C}$.
2. If σ and τ are elements of $\mathcal{C}$ and the intersection $\sigma\cap\tau$ is not empty, then $\sigma\cap\tau$ is a face of both σ and τ.

We define the support of $\mathcal{C}$ by $|\mathcal{C}| = \bigcup_{\sigma\in\mathcal{C}} \sigma$.

Now we associate a polyhedral complex $\mathcal{C}(X(\nu))$ to a toric hyperKähler manifold $X(\nu)$ with $\nu = (\nu_1, 0, 0) \in k^* \otimes \mathbf{R}^3$. Recall that we fixed $h \in (t^N)^*$ such that $\iota^* h = \nu_1$. We define hyperplanes F_i in $(t^n)^*$ by

$$F_i = \{p \in (t^n)^* \mid \langle \pi^* p + h, X_i\rangle = 0\} \quad \text{for } i=1,\dots,N.$$

Then these hyperplanes devide $(t^n)^*$ into a finite number of closed convex polyhedra $\{\Delta_\epsilon | \epsilon \in \Theta\}$, where Θ is the set consisting of all maps from

$\{1, \ldots, N\}$ to $\{1, -1\}$, and $\Delta_\epsilon \subset (t^n)^*$ is defined by

$$\Delta_\epsilon = \{p \in (t^n)^* \mid \epsilon(i)\langle \pi^* p + h, X_i \rangle \geq 0 \quad \text{for any } i = 1, \ldots, N\}.$$

Then the associated polyhedral complex $\mathcal{C}(X(\nu))$ is defined to be a complex consisting of all *compact* faces of all polyhedra Δ_ϵ, where $\epsilon \in \Theta$. It should be remarked that, to define $\mathcal{C}(X(\nu))$, we need $h \in (t^N)^*$ such that $\iota^* h = \nu_1$. However, $\mathcal{C}(X(\nu))$ is determined by ν_1 up to parallel translation. So we use this notation.

For each $\epsilon \in \Theta$, we define a subspace V_ϵ of $\mathbf{H}^N$ as follows: $(z, w) \in V_\epsilon$ if and only if, for any $i = 1, \ldots, N$, $w_i = 0$ if $\epsilon(i) = 1$, and $z_i = 0$ if $\epsilon(i) = -1$. It is easy to see that if we set $M_\epsilon = \mu_{T^n}^{-1}(\Delta_\epsilon, 0, 0)$, then we have

$$M_\epsilon = \{V_\epsilon \cap \mu_{K,1}^{-1}(\nu_1)\}/K.$$

Since $V_\epsilon \cong \mathbf{C}^N$, M_ϵ is an ordinary toric manifold.

Let us recall the fundamental property of $X(\nu)$, which is proved in [10].

Lemma 3.1. (1) $\mu_{T^n}^{-1}((t^n)^*, 0, 0) = \bigcup_{\epsilon \in \Theta} M_\epsilon$.
(2) *Suppose that* $\Delta_\epsilon \cap F_i$ *is a face of* Δ_ϵ *with codimension one. Then the homology class represented by* $\mu_{T^n}^{-1}(\Delta_\epsilon \cap F_i, 0, 0)$ *is the Poincaré dual of* $\epsilon(i) c_1(L_{u_i})$ *in* M_ϵ.

Then we have the following fact, which was due to [5] in special cases and due to [2] for general toric hyperKähler manifolds.

Fact 3.2. *Let* $X(\nu)$ *be a toric hyperKähler manifold with* $\nu = (\nu_1, 0, 0)$ *and* $\mathcal{C} = \mathcal{C}(X(\nu))$ *the associated polyhedral complex. Then the following holds:*
(1) *For each* $\tau \in \mathcal{C}(X(\nu))$, $N_\tau = \mu_{T^n}^{-1}(\tau, 0, 0)$ *is a projective toric submanifold of* $X(\nu)$.
(2) $\bigcup_{\tau \in \mathcal{C}} N_\tau = \mu_{T^n}^{-1}(|\mathcal{C}|, 0, 0)$ *is a* T^n*-equivariant deformation retract of* $X(\nu)$.
(3) *The homeomorphism type of* $\bigcup_{\tau \in \mathcal{C}} N_\tau$ *is completely determined by the combinatorial structure of the associated polyhedral complex* $\mathcal{C}(X(\nu))$.

Definition. Due to Fact 3.2 we call the union of projective toric manifolds $\bigcup_{\tau \in \mathcal{C}} N_\tau$ *the core of the toric hyperKähler manifold* $X(\nu)$.

Example. Let us consider a toric hyperKähler manifold $X(\nu)$ in Section 2 again. Here we assume $\nu = (\nu_1, 0, 0)$. If we set $v_1 = \iota^* u_4 = \iota^* u_2$ and $v_2 = \iota^* u_5 = \iota^* u_3$, then k^* is devided into six chambers as in Figure 1. Suppose that $\nu_1 \in S_1$. If we define $\epsilon_1, \epsilon_2 \in \Theta$ by

$$\epsilon_1(i) = 1 \quad \text{for } i = 1, 2, 3, 4, 5, \qquad \epsilon_2(i) = \begin{cases} 1 & \text{for } i = 1, 2, 4, \\ -1 & \text{for } i = 3, 5, \end{cases}$$

then the associated polyhedral complex $\mathcal{C}(X(\nu))$ consists of all faces of Δ_{ϵ_1} and Δ_{ϵ_2} as in Figure 2, where we take an appropriate coordinate (a_1, a_2, a_3) in $(t^3)^*$ such that $F_i = \{(a_1, a_2, a_3)|a_i = 0\}$ for $i = 1, 2, 3$. We remark that the combinatorial structure of the associated polyhedral complex and the topology of the core depend on the chamber. However, the topology of $X(\nu)$ does not depend on it [10].

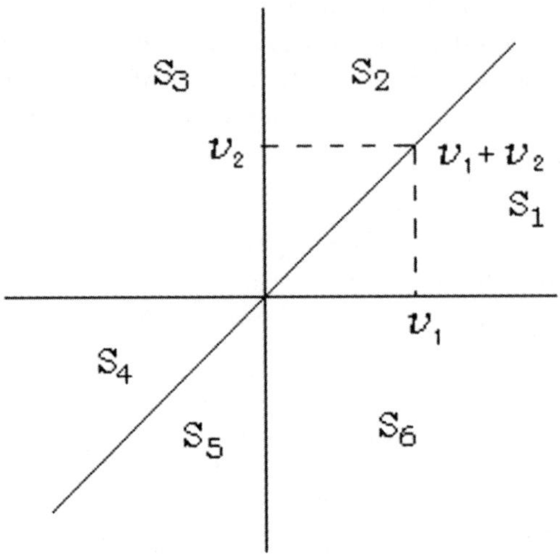

Figure 1.

Thus, to study the cohomology of $X(\nu)$, we have only to study its core $\bigcup_{\tau \in \mathcal{C}} N_\tau$. It is a union of projective toric manifolds, which intersect along toric submanifolds. The topology of projective toric manifolds N_τ is well-known [4]. However, since N_τ's intersect in a comlpicated way, it is not easy to study the topology of the core.

Let us recall the notion of star-collapsibility, which we learned from the earlier version of [2].

Definition . A polyhedral complex $\mathcal{C}$ is *star-collapsible* if there exists a filtration

$$\emptyset = \mathcal{C}_{r+1} \subset \mathcal{C}_r \subset \cdots \subset \mathcal{C}_1 = \mathcal{C}$$

by subcomplexes such that, for $i \leq r$, there exists a vertex $x_i \in \mathcal{C}_i$ and the following conditions are satisfied:

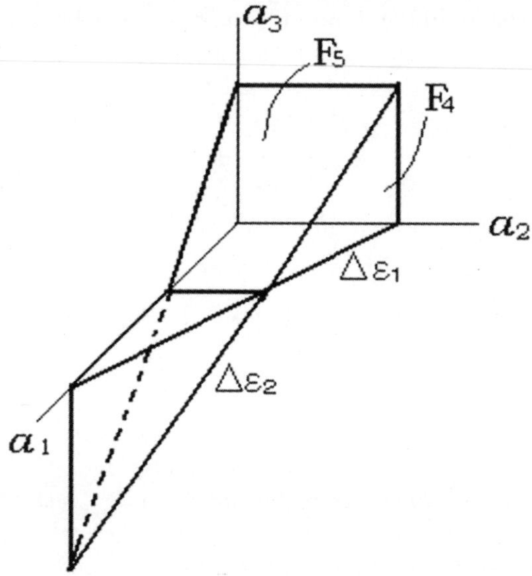

Figure 2.

1. There exists $\sigma_i \in \mathcal{C}_i$ uniquely such that $x_i \in \sigma_i$ and σ_i is a maximal element in $\mathcal{C}_i$.
2. $\mathcal{C}_i \setminus \mathcal{C}_{i+1} = \{\tau \in \mathcal{C}_i \mid x_i \in \tau, \ \ \tau \text{ is a face of } \sigma_i\}$.

Now we show the following lemma. The proof below was suggested by T. Gocho.

Lemma 3.3. *Let $X(\nu)$ be a toric hyperKähler manifold with $\nu = (\nu_1, 0, 0)$. Then the associated polyhedral complex $\mathcal{C}(X(\nu))$ is star-collapsible.*

Proof. Define the S^1-action on $\mathbf{H}^N$ by $(z, w)\beta = (z\beta, w\beta)$ for $\beta \in S^1$. This induces the S^1-action on $X(\nu)$. It is easy to see that this action preserves $\omega_{\nu,1}$, which is the Kähler form with respect to $I_{\nu,1}$. Note that the moment map for this action $\mu_{S^1} \colon X(\nu) \to \mathbf{R}$ is proper and T^n-invariant. If we perturb this function by a small and generic $\xi \in t^n$ as

$$f([z, w]) = \mu_{S^1}([z, w]) + \langle \mu_{T^n,1}([z, w]), \xi \rangle,$$

then f remains proper and the critical point set of f coincides with the fixed point set of T^n, which consists of finite points $\{p_1, \ldots, p_r\}$. We may also assume $f(p_1) > f(p_2) > \cdots > f(p_r)$.

Moreover the gradient flow of f is described by the action of 1-parameter subgroup of the complexification of $S^1 \times T^n$. Therefore the gradient flow preserves N_τ for every $\tau \in \mathcal{C}(X(\nu))$.

Note that $f|_{\mu_{T^n}^{-1}((t^n)^*,0,0)}$ desends to the function $\bar{f}$ on $(t^n)^*$. Since $\bar{f}$ is also proper and bounded below, it is easy to see that, for every $x_i = \mu_{T^n,1}(p_i)$, there exists a unique maximal $\sigma_i \in \mathcal{C}(X(\nu))$ such that $x_i \in \sigma_i$ and $\bar{f}|_{\sigma_i}$ has the maximum at x_i. Thus x_i's and σ_i's define a desired filtration on $\mathcal{C}(X(\nu))$. □

Now we discuss the relation between the topology of $X(\nu)$ and the combinatorial property of $\mathcal{C}(X(\nu))$.

Theorem 3.4. *Let $X(\nu)$ be a toric hyperKähler manifold with $\nu = (\nu_1, 0, 0)$ with the associated polyhedral complex $\mathcal{C} = \mathcal{C}(X(\nu))$. Let $\emptyset = \mathcal{C}_{r+1} \subset \mathcal{C}_r \subset \cdots \subset \mathcal{C}_1 = \mathcal{C}$, $x_i \in \mathcal{C}_i$ and $\sigma_i \in \mathcal{C}_i$ be a filtration, vertices and faces concerned with star-collapsibility, respectively. We set $N_i = \mu_{T^n}^{-1}(\sigma_i, 0, 0)$ for $i = 1, \ldots, r$. We denote the embedding of N_i into $X(\nu)$ by $\psi_i \colon N_i \to X(\nu)$. Then we have*

$$\ker \Phi = \bigcap_{i=1}^{r} \ker(\psi_i^* \circ \Phi).$$

Proof. Since $\ker \Phi \subset \bigcap_{i=1}^{r} \ker(\psi_i^* \circ \Phi)$ is trivial, we have only to show that $\ker \Phi \supset \bigcap_{i=1}^{r} \ker(\psi_i^* \circ \Phi)$. To prove this, it is sufficient to show that the map

$$\Psi = \bigoplus_{i=1}^{r} \psi_i^* \colon H^*(X(\nu); \mathbf{Z}) \to \bigoplus_{i=1}^{r} H^*(N_i; \mathbf{Z})$$

is injective.

We set $E_i = \mu_{T^n}^{-1}(|\mathcal{C}_i|, 0, 0)$. Since $|\mathcal{C}_i| = |\mathcal{C}_{i+1}| \cup \sigma_i$, we have $E_i = E_{i+1} \cup N_i$. Moreover we prove the following claim.

Claim. *The natural map $H^*(E_i; \mathbf{Z}) \to H^*(E_{i+1}; \mathbf{Z}) \oplus H^*(N_i; \mathbf{Z})$ is injective for $i = 1, \ldots, r$.*

Proof of Claim. Since N_i is a projective toric manifold, $H^{\mathrm{odd}}(N_i; \mathbf{Z}) = 0$. Moreover, since $N_i \setminus (E_{i+1} \cap N_i)$ is the biggest cell in N_i, we also have $H^{\mathrm{odd}}(E_{i+1} \cap N_i; \mathbf{Z}) = 0$. To show that $H^{\mathrm{odd}}(E_i; \mathbf{Z}) = 0$, we consider the cohomology exact sequence (This argument is due to Bielawski and Dancer):

$$\to H^{\mathrm{odd}}(E_i, E_{i+1}; \mathbf{Z}) \to H^{\mathrm{odd}}(E_i; \mathbf{Z}) \to H^{\mathrm{odd}}(E_{i+1}; \mathbf{Z}) \to .$$

Since $H^{\mathrm{odd}}(E_i, E_{i+1}; \mathbf{Z}) \cong H^{\mathrm{odd}}(N_i, N_i \cap E_{i+1}; \mathbf{Z}) \cong H^{\mathrm{odd}}(D, \partial D; \mathbf{Z}) \cong 0$, where D is the unit disk in $\mathbf{C}^{\dim \sigma_i}$, we see that $H^{\mathrm{odd}}(E_{i+1}; \mathbf{Z}) \cong 0$ implies $H^{\mathrm{odd}}(E_i; \mathbf{Z}) \cong 0$. Since $H^{\mathrm{odd}}(E_r; \mathbf{Z}) \cong 0$, by the inductive argument we have $H^{\mathrm{odd}}(E_i; \mathbf{Z}) = 0$.

Hence, by applying the standard Mayer-Vietoris argument to $E_i = E_{i+1} \cup N_i$, we can show the claim. □

By the above claim we can conclude that the map

$$H^*(X(\nu); \mathbf{Z}) \cong H^*(E_1; \mathbf{Z}) \to H^*(E_2; \mathbf{Z}) \oplus H^*(N_1; \mathbf{Z})$$

is injective. By using this argument repeatedly, we finish the proof of Theorem 3.4. □

§4. Chern classes

In this section we compute the total Chern class of a toric hyperKähler manifold as a simple application of Theorem 3.4.

Theorem 4.1. *Let $X(\nu)$ be a toric hyperKähler manifold. Let*

$$c(X(\nu)) = 1 + c_1(X(\nu)) + c_2(X(\nu)) + \cdots \in H^*(X(\nu); \mathbf{Z})$$

be the total Chern class of the holomorphic tangent bundle of $X(\nu)$ with respect to the complex structure $I_{\nu,1}$. Then we have

$$c(X(\nu)) = \Phi\left(\prod_{i=1}^{N}(1 - u_i^2)\right) \in H^*(X(\nu); \mathbf{Z}).$$

To prove Theorem 4.1, we need the following lemma, which is a simple generalization of the argument due to Bielawski and Dancer [2]. They showed it in the case $\epsilon_0 \in \Theta$ such that $\epsilon_0(i) = 1$ for all $i = 1, \ldots, N$.

Lemma 4.2. *Let $X(\nu)$ be a toric hyperKähler manifold with $\nu = (\nu_1, 0, 0)$. If M_ϵ is not empty, then its holomorphic cotangent bundle T^*M_ϵ is contained in $X(\nu)$ as an open subset.*

Proof. We first recall the notation in Section 3. Fix $\epsilon \in \Theta$. For $i = 1, \ldots, N$, we define $(q_i^\epsilon, p_i^\epsilon)$ by

$$(q_i^\epsilon, p_i^\epsilon) = \begin{cases} (z_i, w_i) & \text{if } \epsilon(i) = 1, \\ (w_i, -z_i) & \text{if } \epsilon(i) = -1. \end{cases}$$

Then $q^\epsilon = (q_1^\epsilon, \ldots, q_N^\epsilon)$ is a point in the vector space V_ϵ, and $p^\epsilon = (p_1^\epsilon, \ldots, p_N^\epsilon)$ is a point in the dual space V_ϵ^*. In other words, we identify the cotangent bundle T^*V_ϵ with $\mathbf{H}^N$ as above.

Let us recall that we have a holomorphic description of M_ϵ as follows:

$$M_\epsilon = U_\epsilon / K^{\mathbf{C}},$$

where $K^{\mathbf{C}}$ is the complexification of K, and U_ϵ is an open subset of V_ϵ. By the argument in [6], $q^\epsilon \in U_\epsilon$ if and only if the functional l_{q^ϵ} on k defined by

$$l_{q^\epsilon}(Y) = \langle \nu_1, Y \rangle + \frac{1}{4}\sum_{i=1}^{N} |q_i^\epsilon|^2 e^{-\epsilon(i)4\pi\langle u_i Y\rangle} \quad \text{for } Y \in k$$

has the minimum. Moreover, we have a holomorphic (with respect to the complex structure $I_{\nu,1}$) description of $X(\nu)$ as follows:

$$X(\nu) = W/K^{\mathbf{C}},$$

where W is a subset of $T^*V_\epsilon = \mathbf{H}^N$. Similarly, $(q^\epsilon, p^\epsilon) \in W$ if and only if $(\mu_{K,2} + \sqrt{-1}\mu_{K,3})(q^\epsilon, p^\epsilon) = 0$ and the functional $l_{q^\epsilon,p^\epsilon}$ on k defined by

$$l_{q^\epsilon,p^\epsilon}(Y) = \langle \nu_1, Y \rangle + \frac{1}{4}\sum_{i=1}^{N} |q_i^\epsilon|^2 e^{-\epsilon(i)4\pi\langle u_i Y\rangle} + \frac{1}{4}\sum_{i=1}^{N} |p_i^\epsilon|^2 e^{\epsilon(i)4\pi\langle u_i Y\rangle}$$

has the minimum.

Suppose that $q^\epsilon \in U_\epsilon \subset V_\epsilon$ and that $p^\epsilon \in V_\epsilon^*$ defines a cotangent vector of M_ϵ at $[q^\epsilon]$, that is,

$$\langle Y^*_{\text{at } q^\epsilon}, p^\epsilon \rangle = 0 \quad \text{for any } Y = \sum_{i=1}^{N} a_i X_i \in k,$$

where Y^* is a vector field on V_ϵ generated by Y. If we note

$$Y^*_{\text{at } q^\epsilon} = (2\pi\sqrt{-1}\epsilon(1)a_1 q_1^\epsilon, \ldots, 2\pi\sqrt{-1}\epsilon(N)a_N q_N^\epsilon),$$

then we have

$$\langle Y^*_{\text{at } q^\epsilon}, p^\epsilon \rangle = 2\pi\sqrt{-1}\sum_{i=1}^{N} a_i \epsilon(i) q_i^\epsilon p_i^\epsilon = 2\pi\sqrt{-1}\left\langle Y, \sum_{i=1}^{N} z_i w_i u_i \right\rangle.$$

Therefore, $p^\epsilon \in V_\epsilon^*$ defines a cotangent vector of M_ϵ at $[q^\epsilon]$ if and only if $(\mu_{K,2} + \sqrt{-1}\mu_{K,3})(q^\epsilon, p^\epsilon) = 0$. Moreover, if l_{q^ϵ} has the minimum, then it is easy to see that $l_{q^\epsilon,p^\epsilon}$ has also the minimum. Thus we have $(q^\epsilon, p^\epsilon) \in W$, which implies $T^*M_\epsilon \subset X(\nu)$. □

Proof of Theorem 4.1. We may assume $\nu = (\nu_1, 0, 0)$. Let $i_\epsilon : M_\epsilon \to X(\nu)$ be the embedding. By Lemma 4.2, we have

$$i_\epsilon^* TX(\nu) \cong TM_\epsilon \oplus T^* M_\epsilon.$$

By the same argument in [4] and Lemma 3.1, we have

$$c(TM_\epsilon) = \Phi_\epsilon \left(\prod_{i=1}^{N} (1 + \epsilon(i) u_i) \right), \quad c(T^* M_\epsilon) = \Phi_\epsilon \left(\prod_{i=1}^{N} (1 - \epsilon(i) u_i) \right),$$

where $\Phi_\epsilon : \mathbf{Z}[u_1, \ldots, u_N] \to H^*(M_\epsilon; \mathbf{Z})$ is a ring homomorphism defined by $\Phi_\epsilon(u_i) = c_1(i_\epsilon^* L_{u_i})$. Therefore we have

$$i_\epsilon^* c(X(\nu)) = c(TM_\epsilon) c(T^* M_\epsilon) = i_\epsilon^* \Phi \left(\prod_{i=1}^{N} (1 - u_i^2) \right).$$

On the other hand, by Theorem 2.2, there exsists $f \in \mathbf{Z}[u_1, \ldots, u_N]$ such that $\Phi(f) = c(X(\nu))$. Therefore we have

$$i_\epsilon^* \Phi \left(f - \prod_{i=1}^{N} (1 - u_i^2) \right) = 0 \quad \text{for any } \epsilon \in \Theta.$$

Recall now Lemma 3.4. Since any $\tau \in \mathcal{C}(X(\nu))$ is a face of Δ_ϵ for some $\epsilon \in \Theta$, we have

$$f - \prod_{i=1}^{N} (1 - u_i^2) \in \bigcap_{i=1}^{r} \ker(\psi_i^* \circ \Phi) = \ker \Phi.$$

This implies Theorem 4.1. □

References

[1] M. Audin, *The Topology of Torus Actions on Symplectic Manifolds*, Birkhäuser, Boston, 1991.

[2] R. Bielawski and A. Dancer, *The Geometry and Topology of Toric HyperKähler Manifolds*, Comm. Anal. Geom., to appear.

[3] T. Delzant, *Hamiltoniens périodiques et images convexes de l'application moment*, Bull. Soc. Math. France, **116** (1988), 315–339.

[4] W. Fulton, *Introduction to Toric Varieties*, Princeton Univ. Press, Princeton, 1993.

[5] R. Goto, *On toric hyper-Kähler manifolds given by the hyper-Kähler quotient method*, Infinite Analysis, World Scientific, Singapore, 1992, 317–338.

[6] V. Guillemin, *Moment Maps and Combinatorial Invariants of Hamiltonian T^N-spaces*, Birkhäuser, Boston, 1994.
[7] N. Hitchin, A. Karlhede, U. Lindström and M. Roček, *Hyperkähler metrics and supersymmetry*, Comm. Math. Phys., **108** (1987), 535–589.
[8] L. Jeffrey and F. Kirwan, *Localization for nonabelian group actions*, Topology, **34** (1995), 291–327.
[9] F. Kirwan, *Cohomology of Quotients in Symplectic and Algebraic Geometry*, Princeton Univ. Press, Princeton, 1984.
[10] H. Konno, *Cohomology Rings of Toric HyperKähler Manifolds*, Internat. J. Math., **11** (2000), 1001–1026.

Graduate School of Mathematical Sciences
University of Tokyo
3-8-1 Komaba, Meguro-ku
Tokyo, 153-8914
Japan
konno@ms.u-tokyo.ac.jp

Advanced Studies in Pure Mathematics 34, 2002
Minimal Surfaces, Geometric Analysis and Symplectic Geometry
pp. 185–199

Cyclic Hypersurfaces of Constant Curvature

Rafael López

Abstract.

We study hypersurfaces in Euclidean, hyperbolic or Lorentz-Minkowski space with the property that it is foliated by a one-parameter family of round spheres. We describe partially such hypersurfaces with some assumption on its curvature. In general, we shall consider the situation that the mean curvature or the Gaussian curvature is constant.

§1. Introduction

A cyclic hypersurface in $(n+1)$-dimensional Euclidean space $\mathbf{E}^{n+1}$ is a hypersurface defined by a smooth one-parameter family of round $(n-1)$-spheres. We say then that M is foliated by spheres. The first example of cyclic hypersurfaces is a hypersurface of revolution, that is, a hypersurface which is stable under a group of rotations that leave a straight-line pointwise fixed. It has been known that the only minimal cyclic surfaces in Euclidean 3-space $\mathbf{E}^3$ are the catenoid (which it is rotational [15]) and the examples discovered by Enneper and Riemann in the nineteenth century [2], [3], [19]. Riemann's surface is a (non-rotational) surface constructed by circles in parallel planes with the exception of a discrete set of straight-lines. Moreover, each of these surfaces is invariant by a family of translations. In higher dimensions, Jagy proved that a cyclic minimal hypersurface in $\mathbf{E}^{n+1}$, $n \geq 3$, must be rotational, that is, it is the n-dimensional catenoid [6]. In contrast with the minimal case, the only cyclic surfaces in $\mathbf{E}^3$ with nonzero constant mean curvature are surfaces of revolution [17]. This note is motivated by these examples and the possible extensions of these results for other space forms. We are interested in studying cyclic hypersurfaces under some assumptions on their curvatures. One of our goal in this paper is to exhibit the existence of a family of maximal spacelike surfaces in the Lorentz-Minkowski

---

2000 *Mathematics Subject Classification.* Primary 53A10; Secondary 53C42.

Research partially supported by DGICYT grant number PB97-0785.

3-space $\mathbf{L}^3$ that were announced in [9], [10], with similar properties as Riemann's examples in $\mathbf{E}^3$. These surfaces are foliated by circles in parallel planes with the exception of a discrete set of straight-lines and singularities. In this sense, we say that such surface is a 'Riemann type surface' in $\mathbf{L}^3$. See Figure 1 for an example.

We divide this paper into three parts:

1. Cyclic hypersurfaces of constant mean curvature in hyperbolic space.
2. Cyclic hypersurfaces of constant mean curvature in Lorentz-Minkowski space.
3. Cyclic surfaces of constant Gauss curvature in Euclidean space.

Fig. 1. A 'Riemann type surface' in $\mathbf{L}^3$.

§2. Cyclic hypersurfaces of constant mean curvature in hyperbolic space

In this section we study cyclic hypersurfaces in the $(n+1)$-dimensional hyperbolic space $\mathbf{H}^{n+1}$, for which the spheres that determine the hypersurfaces lie in parallel horospheres. Recall that horospheres are the umbilical hypersurfaces in $\mathbf{H}^{n+1}$ that are flat. Since there exist no intrinsic concept of parallelism in hyperbolic setting, we now give our precise definition.

Definition 2.1. A family of horospheres are called parallel if their asymptotic boundaries meet at exactly one point.

Since the asymptotic boundary of a horosphere is a point, two parallel horospheres meet at infinity in the same point. In the upper half-space model for $\mathbf{H}^{n+1}$, parallel horospheres can be viewed as horizontal parallel Euclidean hyperplanes or, after a rigid motion, Euclidean n-spheres tangent to the hyperplane $x_{n+1} = 0$ at the same point. On the other hand, note that in this model, $(n-1)$-spheres are Euclidean $(n-1)$-spheres.

Theorem 2.2 ([7], [10]). *Let M be a hypersurface in $\mathbf{H}^{n+1}$ of constant mean curvature which is foliated by spheres in parallel horospheres. Then M is a hypersurface of revolution.*

Proof. Consider the upper half-space model for $\mathbf{H}^{n+1}$, that is, $\mathbf{R}^{n+1}_{+} = \{(x_1, \ldots, x_{n+1}) \in \mathbf{R}^{n+1}; x_{n+1} > 0\}$ endowed with the metric

$$\langle \, , \, \rangle = \frac{dx_1^2 + \cdots + dx_{n+1}^2}{x_{n+1}^2}.$$

After a rigid motion in the ambient space, we may assume that the horospheres are Euclidean hyperplanes in $\mathbf{R}^{n+1}_{+}$ parallel to the hyperplane $x_{n+1} = 0$. We pick a piece M' of M bounded by two spheres $S_1 \cup S_2$. The proof consists of two parts.

The first part is done by a standard application of the Alexandrov reflection method [1]. We consider reflections across a family of vertical parallel geodesic hyperplanes (in Euclidean sense). These hyperplanes are also geodesic hyperplanes in $\mathbf{H}^{n+1}$. Reflections across vertical hyperplanes are isometries in $\mathbf{H}^{n+1}$, so the mean curvature remains unchanged. Consider a vertical hyperplane P disjoint from M' and move P parallel to itself (for example, to the right) until it touches M' at a first point. One continues to move P a little more to the right and considers the symmetry through P of the part of M' on the left-side of P. Now continue moving P to the right and reflecting the left-side of M' until this part touches the part of M' on the right-side of P. The strong maximum principle implies reflection symmetry if they contact and the Alexandrov reflection process yields that P is a hyperplane of symmetry of M'. Thus, for each vertical hyperplane P, some parallel translate of P is a hyperplane of symmetry of M' and M' inherits the symmetries of its boundary $S_1 \cup S_2$. So, the Euclidean centers of the spheres that define M' lies in a 2-plane. Without loss of generality, we may suppose that the curve of centers is parametrized by $(c(t), 0, \ldots, t)$. It then follows that M' is defined as the level hypersurface of the function

$$f(x_1, \ldots, x_n, t) = (x_1 - c(t))^2 + \sum_{i=2}^{n} x_i^2 - r(t)^2,$$

where $r(t) > 0$ denotes the Euclidean radius of each sphere $M' \cap \{x_{n+1} = t\}$.

The second part of the proof is done by computing of the mean curvature of M' in terms of the function f. For this, let $N = -\nabla f / |\nabla f|$ be a unit normal vector field of the immersion $M' \to \mathbf{E}_+^{n+1}$. Then the following formula is well-known:

$$n\, H_{\mathrm{e}} |\nabla f|^3 = \Delta f\, |\nabla f|^2 - \operatorname{Hess} f(\nabla f, \nabla f), \tag{1}$$

where H_{e} denotes the mean curvature of $M \subset \mathbf{E}_+^{n+1}$, and Δ and Hess are the Laplacian and Hessian operators in $\mathbf{E}^{n+1}$, respectively. Choose $x_{n+1}N$ as the Gauss map of $M' \subset \mathbf{H}^{n+1}$. Then its mean curvature H is related with H_{e} by the formula $H = x_{n+1}H_{\mathrm{e}} + N_{n+1}$, where $N = (N_1, \ldots, N_{n+1})$. Thus (1) yields

$$nH|\nabla f|^3 = nN_{n+1}|\nabla f|^3 + x_{n+1}\left(\Delta f |\nabla f|^2 - \operatorname{Hess} f(\nabla f, \nabla f)\right). \tag{2}$$

If the function $c(t)$ is constant, the curve of centers is a straight-line orthogonal to each hyperplanes of the foliation. Consequently, the spheres that define M' are coaxial and hence M' is a hypersurface of revolution.

Assume, on the contrary, that M' is not a hypersurface of revolution, that is, $c' \neq 0$. It is computed that

$$\nabla f = 2(x_1 - c, x_2, \ldots, x_n, -r\lambda), \quad |\nabla f|^2 = 4r^2(1+\lambda)^2,$$
$$\Delta f = 2\left(n - r'\lambda - r\frac{\partial \lambda}{\partial r}\right),$$
$$\operatorname{Hess} f(\nabla f, \nabla f) = 8r^2\left\{1 + 2\lambda(\lambda - r') - \lambda^2\left(r'\lambda + r\frac{\partial\lambda}{\partial t}\right)\right\},$$

where $\lambda = \lambda(x_1, t) = ((x_1 - c)c' + rr')/r$.

On the other hand,

$$\frac{\partial \lambda}{\partial t} = \frac{1}{c'\lambda}\{(\lambda - r')(c''r - c'r') - (c')^3 + c'rr''\}.$$

Let fix a level $x_{n+1} = t_0$. We introduce a new variable $\lambda = \lambda(x_1, t_0)$ instead of x_1. Then (1) and (2) are written respectively as

$$nrH(1+\lambda^2)^{3/2} = a_0 + a_1\lambda + a_2\lambda^2, \tag{3}$$

$$nrH(1+\lambda^2)^{3/2} = nr\lambda(1+\lambda^2) + t_0(a_0 + a_1\lambda + a_2\lambda^2), \tag{4}$$

where the coefficients a_i are independent of λ. We take the square of both sides of the equation (4) and compare terms by terms. The term of the highest degree corresponds to λ^6. Then $n^2r^2H^2 = n^2r^2$ and this yields $H^2 = 1$. Since the square of the left-hand side of (4) is a polynomial with only terms of even degree in λ, the coefficients of λ^5 and λ^3 vanish on the right-hand side. This yields $t_0a_2 = 0$ and $t_0a_0 = 0$, respectively. However the constant term on the left-hand side of (4) is $n^2r^2H^2 = n^2r^2 \neq 0$, obtaining a contradiction. □

In this context, we recall a theorem of Hsiang [5], which is proved by using the Alexandrov reflection principle, stating that a complete embedded hypersurface $M \subset \mathbf{H}^{n+1}$ that remains within a uniform distance from a geodesic is a hypersurface of revolution.

Remark 1. The same reasoning can be carried over to the case of Euclidean space $\mathbf{E}^{n+1}$. Indeed, squaring (3), the coefficient of λ^6 on the rights-hand side is 0. As a consequence, we obtain that *'the only hypersurfaces in $\mathbf{E}^{n+1}$ with nonzero constant mean curvature which are foliated by $(n-1)$-spheres in parallel hyperplanes are the hypersurfaces of revolution'.*

§3. Cyclic hypersurfaces of constant mean curvature in Lorentz-Minkowski space

Let $\mathbf{L}^{n+1}$ be the $(n+1)$-dimensional Lorentz-Minkowski space, that is, $\mathbf{R}^{n+1}$ equipped with the metric $ds^2 = dx_1^2 + \ldots + dx_n^2 - dx_{n+1}^2$. We study cyclic hypersurfaces of constant mean curvature in $\mathbf{L}^{n+1}$. First, we prove the Lorentzian counterpart of the previous section. Then we investigate the 3-dimensional case, that is, constant mean curvature surfaces of $\mathbf{L}^3$ foliated by circles.

3.1. Cyclic hypersurfaces of constant mean curvature in $\mathbf{L}^{n+1}$

By a similar reasoning as in Theorem 2.2, we obtain the next result which is analogous to what occurs in Euclidean space $\mathbf{E}^{n+1}$ (see Introduction and Remark 1).

Theorem 3.1 ([10]). *Let M^n be a spacelike hypersurface in $\mathbf{L}^{n+1}$ of constant mean curvature H which is foliated by $(n-1)$-spheres in parallel spacelike hyperplanes. Then the following hold.*

1. *If $H \neq 0$, then M is a hypersurface of revolution.*
2. *If $H = 0$ and*
   (a) *$n \geq 3$, then M is a hypersurface of revolution.*
   (b) *$n = 2$, then M is a surface of revolution or is a 'Riemann type surface'.*

Proof. After a rigid motion of $\mathbf{L}^{n+1}$, we may assume that the parallel spacelike hyperplanes are parallel to $x_{n+1} = 0$ (in this case, 'spheres' are 'Euclidean spheres'). The proof is similar to that for Theorem 2.2 and so we only give an outline. Note that Alexandrov reflection method works as in $\mathbf{E}^{n+1}$ and $\mathbf{H}^{n+1}$: indeed, a spacelike hypersurface in $\mathbf{L}^{n+1}$ of constant mean curvature locally satisfies an elliptic partial differential equation for which we can use the standard maximum principle. We compute H through the identity:

$$nH|\nabla f|^3 = \langle \nabla f, \nabla f\rangle \Delta f - \operatorname{Hess} f(\nabla f, \nabla f), \tag{5}$$

where in this case

$$\nabla f = (f_1, \ldots, f_n, -f_{n+1}), \quad \Delta f = \sum_{i=1}^{n} f_{i,i} - f_{n+1,n+1},$$

$$\operatorname{Hess} f(\nabla f, \nabla f) = \sum_{i,j} f_i f_j f_{i,j},$$

$$f = f(x_1, \ldots, x_{n+1}), \quad f_i = \frac{\partial f}{\partial x_i}, \quad f_{i,j} = \frac{\partial^2 f}{\partial x_i \partial x_j}.$$

With the same variable λ defined in Section 2, the identity (5) reads as

$$nrH(-1+\lambda^2)^{3/2} = g_0 + g_1\lambda + g_2\lambda^2,$$

where the coefficients g_i do not depend on λ. We follow the same argument by squaring the above equation. Special attention should be paid to the case $H = 0$ and $n = 2$. In this situation, $g_1 = g_0 = 0$, which yields the next two ordinary differential equations:

$$\frac{rc''}{c'} - 2r' = 0,$$

$$1 + r'^2 - c'^2 + rr'' - \frac{rr'c''}{c'} = 0.$$

A first integration of both equations can do as in [16, p. 87]. So, the first equation leads $c' = ar^2$, for a constant a. This yields in the second equation $1 - r'^2 + rr'' - a^2r^4 = 0$. Consider $x = r^2$ and $y = (r^2)'$ as the new dependent and independent variables. Thus

$$\frac{dt}{dr} = \frac{1}{\sqrt{a^2r^4 + 2br^2 + 1}}.$$

Then M is parametrized by $\mathbf{X}_{a,b}(u,\theta) = (x(u,\theta), y(u,\theta), z(u,\theta))$, where

$$\begin{aligned} x(u,\theta) &= a\int^u \frac{u^2}{\sqrt{a^2u^4 + 2bu^2 + 1}}du + u\cos\theta, \\ y(u,\theta) &= u\sin\theta, \\ z(u,\theta) &= \int^u \frac{du}{\sqrt{a^2u^4 + 2bu^2 + 1}}, \end{aligned}$$

and $a, b \in \mathbf{R}$. □

The integrals that appear in this parametrization are of elliptic type (as it occurs with Riemann's examples in $\mathbf{E}^3$). We illustrate Theorem 3.1 by presenting two examples.

Example 1. Let $a = 0$. In this case, $c' = 0$ and the surface is rotational. This surface is the Lorentzian catenoid:

$$\mathbf{X}_{0,b}(u,\theta) = \left(u\cos\theta, u\sin\theta, \frac{1}{\sqrt{2b}}\operatorname{arc\,sinh}(\sqrt{2b}u)\right),$$

which is generated by the rotation of the curve $((1/\sqrt{2b})\sinh(\sqrt{2b}u), 0, u)$ with respect to the x_3-axis. The Lorentzian catenoid is the only maximal spacelike surface of revolution in $\mathbf{L}^3$ with respect to a timelike rotation axis.

Example 2. Let $a = b = 1$. The integrals that define M can be explicitly calculated. Then M is given by

$$\mathbf{X}_{1,1}(u,\theta) = (u - \arctan u + u\cos\theta, u\sin\theta, \arctan u).$$

This surface has a singularity of cone type at the origin. Moreover it is asymptotic to the planes $x_3 = \pm\pi/2$ and at these heights, M has two flat ends. The circles that define M converge to straight-lines as $u \to \pm\infty$. Thus, we can reflect M across them to obtain a simply periodic embedded maximal surface invariant by a family of translations. Figure 1 in Introduction represents precisely this surface. Up to homotheties in $\mathbf{L}^3$, the immersions $\mathbf{X}_{a,b}$ define a one-parameter family of maximal spacelike surfaces in $\mathbf{L}^3$ that, in a sense, correspond with Riemann's examples in $\mathbf{E}^3$.

3.2. Cyclic maximal surfaces in $\mathbf{L}^3$

In this and the next subsection we focus on cyclic surfaces with constant mean curvature in $\mathbf{L}^3$. Recall that a surface in $\mathbf{L}^3$ is called nondegenerate if the induced metric on it is nondegenerate. In $\mathbf{L}^3$, we have two possibilities: the induced metric is Riemannian and the surface is called spacelike; or the induced metric is Lorentzian and the surface is called timelike. In Theorem 3.1 we studied the case that the spheres that form the hypersurface are contained in parallel spacelike hyperplanes. We want to consider a more general situation on the hyperplanes that determine the foliation. First, we give the following definition.

Definition 3.2. A circle in $\mathbf{L}^3$ is the orbit of a point p under the action of a group of rotations in $\mathbf{L}^3$.

There exist three families of rotations in $\mathbf{L}^3$ according to the causal character of the line L that define each family (see for example [4]). For an easy description of the circles obtained in each case, let $B = (e_1, e_2, e_3)$ be the standard basis in $\mathbf{L}^3$. Then the following hold:

1. (timelike axis) If $L = \text{span}(e_3)$, then the circles are Euclidean circles in horizontal planes.
2. (spacelike axis) If $L = \text{span}(e_1)$, then the circles are hyperbolas in vertical planes.
3. (lightlike axis) If $L = \text{span}(e_2 + e_3)$, then the circles are parabolas in planes parallel to $x_2 - x_3 = 0$.

Surfaces of revolution in $\mathbf{L}^3$ of constant mean curvature have been studied in [4], [8], [20]. In the Lorentzian case, a surface with $H = 0$ everywhere is called maximal. Now we are in a position to give the following two results for (spacelike or timelike) surfaces (see [9]):

Theorem 3.3. *Let M be a nondegenerate maximal cyclic surface in $\mathbf{L}^3$. Then the planes containing pieces of circles must be parallel.*

Theorem 3.4. *Let M be a nondegenerate maximal surface in $\mathbf{L}^3$ foliated by pieces of circles in parallel planes. Then M is a surface of revolution or it is contained in a 'Riemann type surface'.*

Proof. [Sketch] For simplicity of the proof of Theorem 3.3, we consider the case where the planes containing the circles are spacelike. The proof is done by contradiction. Assume that these planes are not parallel. Let $\Gamma(u)$ be an orthogonal curve to each u-plane of the foliation. Since Γ is not a straight-line, we can consider the Frenet frame of Γ. Remark that the unit tangent vector field $\mathbf{t}(u)$ to Γ has a timelike causal character. Then $\mathbf{t}'$ is a spacelike vector field. Let $\mathbf{n}(u)$ be the unit spacelike vector field such that $\mathbf{t}' = \kappa\mathbf{n}$, for some function $\kappa \neq 0$. Put

$\mathbf{b} = \mathbf{t} \wedge \mathbf{n}$. Then Frenet basis associated to Γ is given by $\{\mathbf{t}, \mathbf{n}, \mathbf{b}\}$. Hence M can be parametrized by

$$\mathbf{X}(u,v) = \mathbf{c}(u) + r(u)(\cos v\, \mathbf{n}(u) + \sin v\, \mathbf{b}(u)) \quad r > 0, \mathbf{c} \in \mathbf{L}^3.$$

Let us compute the mean curvature H of $\mathbf{X}$ by the classical local theory (see [18]). Let $I = (E, F, G)$ and $II = (e, f, g)$ be the coefficients of the first and the second fundamental forms respectively, and set $W = EG - F^2$ (W is positive if M is spacelike and negative if M is timelike). Then the mean curvature H is given by

$$H = \frac{eG - 2fF + gE}{2W}. \tag{6}$$

Put $\mathbf{c}' = \alpha \mathbf{t} + \beta \mathbf{n} + \gamma \mathbf{b}$, where α, β, γ are smooth functions on u. Let us use the corresponding Frenet equations of Γ:

$$\begin{aligned} \mathbf{t}' &= \kappa \mathbf{n}, \\ \mathbf{n}' &= \kappa \mathbf{t} + \sigma \mathbf{b}, \\ \mathbf{b}' &= -\sigma \mathbf{n}. \end{aligned}$$

Remark that these equations are slightly different from the Euclidean case. It follows from (6) that $H = 0$ is written as

$$\sum_{n=0}^{3} A_n(u) \cos nv + \sum_{n=1}^{3} B_n(u) \sin nv = 0$$

for some functions A_n and B_n. This is a linear combination of the independent functions $\sin nv$ and $\cos nv$. Thus $A_n = B_n = 0$ for all n. A hard work to obtain explicit expressions of the coefficients A_n and B_n together with the fact that $W, r \neq 0$ gives a contradiction. Therefore $\kappa = 0$ and Γ is a straight-line. When the circles are contained in timelike or lightlike planes, the reasoning is analogous, with the observation that in each case the Frenet frame of Γ changes as well as the corresponding Frenet equations. A more explicit example of the reasoning of this kind can be seen in Theorem 3.5 below.

The proof of Theorem 3.4 is easier. Now, after a rigid motion in $\mathbf{L}^3$ we may assume that the circles of M are Euclidean circles, hyperbolas or parabolas, depending on the causal character of the planes containing the circles. For example, in the case where the planes of the foliation are spacelike, we assume without loss of generality that the surface is given by

$$\mathbf{X}(u,v) = (a(u) + r(u) \cos v, b(u) + r(u) \sin v, u),$$

where a and b are smooth functions on u. If we compute the mean curvature, then the similar reasoning to the proof of Theorem 3.1 for $n = 2$ applies. □

Remark 2. Maximal spacelike surfaces in $\mathbf{L}^3$ can be described in terms of complex data. More exactly, there exists a Weierstrass representation as in the case of minimal surfaces in $\mathbf{E}^3$. Let M be a Riemann surface and $X : M \to \mathbf{E}^3$ a conformal minimal immersion. If $(M, (\phi_1, \phi_2, \phi_3))$ is the corresponding Weierstrass representation, then it is easy to prove that $(M, (i\phi_1, i\phi_2, \phi_3))$ defines a maximal spacelike immersion of M in $\mathbf{L}^3$. This process allows us to obtain a correspondence between minimal surfaces in $\mathbf{E}^3$ and maximal spacelike surfaces in $\mathbf{L}^3$. Therefore it is possible to use the complex analysis machinery in the study of maximal spacelike surfaces and, in particular, of cyclic surfaces. This point of view is developed in [9].

3.3. Cyclic surfaces of nonzero constant mean curvature in $\mathbf{L}^3$

The case $H \neq 0$ in $\mathbf{L}^3$ is different from the maximal one, as it is the case in the Euclidean ambient (see Introduction and [17]).

Theorem 3.5 ([11], [12]). *Let M be a nondegenerate cyclic surface in $\mathbf{L}^3$ with nonzero constant mean curvature. Then either the planes containing the circles must be parallel or M is a subset of a pseudohyperbolic surface or a pseudosphere.*

Comparing with Theorem 3.3, let us first observe that possibly the planes are not parallel. But in this case, the surface is contained in a surface of revolution. This phenomenon also occurs in $\mathbf{E}^3$: the intersection between any smooth one-parameter family of (not necessarily parallel) planes with a sphere produces circles. In the Lorentzian space, the role of spheres is played by the pseudohyperbolic surfaces $\mathbf{H}^{2,1}(r)$ and the pseudospheres $\mathbf{S}^{2,1}(r)$:

$$\mathbf{H}^{2,1}(r) = \{p \in \mathbf{L}^3; \langle p, p\rangle = -r^2\},$$
$$\mathbf{S}^{2,1}(r) = \{p \in \mathbf{L}^3; \langle p, p\rangle = r^2\}.$$

The surfaces $\mathbf{H}^{2,1}(r)$ and $\mathbf{S}^{2,1}(r)$ are spacelike and timelike, respectively. Moreover, both surfaces have nonzero constant mean curvature $|H| = 1/r$. Theorem 3.5 is a revised and corrected version of [11, Th. 1] and examples therein: although the examples exhibited in [11] are spacelike surfaces with nonzero constant mean curvature and foliated by pieces of circles in non-parallel planes, these surfaces are subsets of $\mathbf{S}^{2,1}(r)$ or $\mathbf{H}^{2,1}(r)$.

Proof. The case that the planes of the foliation are spacelike is studied in [11]. It remains the cases that they are timelike or lightlike. In order to simplify the presentation, we explicitly discuss the case that the planes are lightlike. By contradiction, we assume that the planes are not parallel and that, after a homothety in $\mathbf{L}^3$, the mean curvature of M is $H = 1/2$. In each u-plane of the foliation that defines M, let $\mathbf{e}_1(u)$ and $\mathbf{e}_2(u)$ be vector fields such $\langle \mathbf{e}_1, \mathbf{e}_1 \rangle = 1$ and $\langle \mathbf{e}_1, \mathbf{e}_2 \rangle = \langle \mathbf{e}_2, \mathbf{e}_2 \rangle = 0$. Then M is parametrized as

$$\mathbf{X}(u,v) = \mathbf{c}(u) + v\mathbf{e}_1(u) + r(u)v^2\mathbf{e}_2(u), \quad r \neq 0.$$

Denote $\mathbf{n} = \mathbf{e}_1$ and $\mathbf{t} = \mathbf{e}_2$, and use null coordinates: for each u, let $\mathbf{b}(u)$ be the unique lightlike vector orthogonal to $\mathbf{n}(u)$ such that

$$\langle \mathbf{t}, \mathbf{b} \rangle = 1, \quad \text{determinant}(\mathbf{t}, \mathbf{n}, \mathbf{b}) = 1.$$

With a change on the variables u and v, we assume that $\mathbf{t}' = \kappa\mathbf{n}$ for some function κ (see discussion in [9]). Remark that $\kappa \neq 0$ because the planes are not parallel. The Frenet equations are

$$\begin{aligned} \mathbf{t}' &= \kappa\mathbf{n}, \\ \mathbf{n}' &= \sigma\mathbf{t} - \kappa\mathbf{b}, \\ \mathbf{b}' &= -\sigma\mathbf{n}. \end{aligned}$$

In the above notation, the surface is parametrized as

$$\mathbf{X}(u,v) = \mathbf{c} + v\mathbf{n} + rv^2\mathbf{t}.$$

Put $\mathbf{c}' = \alpha\mathbf{t} + \beta\mathbf{n} + \gamma\mathbf{b}$, for smooth functions α, β, γ. Squaring the identity (6), we obtain

$$\sum_{n=0}^{9} A_n(u)v^n = 0.$$

This is a polynomial equation on the variable v and thus the coefficients A_n vanish everywhere. The coefficient A_9 is given by $A_9 = 8\kappa^3(2\gamma r^2 - r')^3$. Then $r' = 2\gamma r^2$. Hence $A_7 = A_8 = 0$ and

$$A_6 = 8\kappa^3(2\beta r - 8\kappa r^2 - \sigma)(-2\beta r + \sigma)^2.$$

We have two possibilities:

1. $\sigma = 2\beta r$. Then $A_4 = -81\alpha^2\kappa^4 r^2$. In particular, $\alpha = 0$ and this implies $W = 0$, a contradiction because M is a nondegenerate surface.

2. $\sigma = 2\beta r - 8\kappa r^2$. The computation of A_4 leads

$$A_4 = 384\kappa^5 r^4(-\alpha + 8\gamma r^2).$$

Then $\alpha = 8\gamma r^2 = 4r'$. By using the Frenet equations, we obtain

$$\mathbf{c}' = 4r'\mathbf{t} + \beta\mathbf{n} + \gamma b = -\left(\frac{\mathbf{b}}{2r} - 4r\mathbf{t}\right)'.$$

Therefore there exists a point $\mathbf{c}_0 \in \mathbf{L}^3$ such that $\mathbf{c} = \mathbf{c}_0 - \mathbf{b}/(2r) + 4r\mathbf{t}$ and the parametrization of M is

$$\mathbf{X}(u,v) = \mathbf{c}_0 + r(4+v^2)\mathbf{t} + v\mathbf{n} - \frac{1}{2r}\mathbf{b}.$$

Thus

$$\langle \mathbf{X}(u,v) - \mathbf{c}_0, \mathbf{X}(u,v) - \mathbf{c}_0 \rangle = -4,$$

and M is contained in the pseudohyperbolic surface $\mathbf{H}^{2,1}(2)$.

□

Let us study the case where the planes containing the circles are parallel. As in Theorem 3.4, an easy reasoning leads to

Theorem 3.6. *Let M be a nondegenerate surface in $\mathbf{L}^3$ with nonzero constant mean curvature which are foliated by pieces of circles in parallel planes. Then M is a surface of revolution.*

From Theorems 3.5 and 3.6, we have

Corollary 3.7 ([11], [12]). *All cyclic nondegenerate surfaces in $\mathbf{L}^3$ with nonzero constant mean curvature are surfaces of revolution.*

This result claims that there exist no 'Riemann type surfaces' in $\mathbf{L}^3$ with nonzero constant mean curvature.

§4. Cyclic surfaces of constant Gauss curvature in Euclidean space

We close this paper with a study of cyclic surfaces in $\mathbf{E}^3$ with constant Gaussian curvature. We have the following two results:

Theorem 4.1. *Let M be a surface in $\mathbf{E}^3$ with constant Gauss curvature which is foliated by pieces of circles. Then M is contained in a sphere or, in the non-spherical case, the planes containing the circles of the foliation must be parallel.*

Theorem 4.2. *Let M be a surface in $\mathbf{E}^3$ with constant Gauss curvature K which is foliated by pieces of circles in parallel planes.*

1. *If $K \neq 0$, then M is a surface of revolution.*
2. *If $K = 0$, then the surface is not necessarily rotational. However, the curve of centers is a straight-line and the radius of the circles is given by a linear function on the parameter of the foliation.*

As a consequence of these theorems, we have:

Corollary 4.3 ([13]). *All cyclic surfaces in $\mathbf{E}^3$ with nonzero constant Gauss curvature are surfaces of revolution.*

In a sense, this corollary is the analogue of Nitsche's theorem for nonzero constant mean curvature cyclic surfaces in $\mathbf{E}^3$.

Proof. The proof of Theorem 4.1 is similar to Theorem 3.3. By contradiction, assume that the u-planes containing the circles are not parallel. Consider a curve $\Gamma(u)$ orthogonal to each u-plane. Then M can be parametrized in the form

$$\mathbf{X}(u,v) = \mathbf{c}(u) + r(u)(\cos v\, \mathbf{n}(u) + \sin v\, \mathbf{b}(u)),$$

where $(\mathbf{t}, \mathbf{n}, \mathbf{b})$ denotes the Frenet frame of Γ. The formula for the Gaussian curvature in local coordinates with respect to $\mathbf{X}$ is:

$$K = \frac{eg - f^2}{EG - F^2}.$$

By using the Frenet equations as in Theorem 3.3, the above equation implies that

$$\sum_{n=0}^{4} A_n(u) \cos nv + \sum_{n=1}^{4} B_n(u) \sin nv = 0.$$

This is a linear combination of the independent functions $\cos nv$ and $\sin nv$. Thus $A_n = B_n = 0$ for all n. A delicate study with the coefficients A_n, B_n concludes that M is contained in a sphere in the case that $K > 0$ or a contradiction. Theorem 4.2 is proved by considering a more explicit parametrization of the surface. After a rigid motion in $\mathbf{E}^3$, we may assume that the planes containing the circles are parallel to the plane $x_3 = 0$. Then the parametrization of M is in the form

$$\mathbf{X}(u,v) = (a(u) + r(u)\cos v, b(u) + r(u)\sin v, u),$$

where $a, b, r > 0$ are smooth functions on u. Then we compute the Gaussian curvature K. If $K \neq 0$, we conclude that $a' = b' = 0$, that is, the curve of centers of the circles is a vertical straight-line orthogonal to each u-plane of the foliation. Thus M is a surface of revolution. In the case $K = 0$, we obtain $a'' = b'' = r'' = 0$. □

Remark 3. Recently the present author has extended Theorems 4.1 and 4.2 to the case of the Lorentz-Minkowski space $\mathbf{L}^3$ [14]: a non-degenerate cyclic surface in $\mathbf{L}^3$ with nonzero constant Gauss curvature is a surface of revolution. The result is divided into two parts. First, it is proved that the planes of the foliation are parallel and secondly, we prove that the surface is rotational. The proof follows the same steps as in Theorems 4.1 and 4.2, but needs to take care of extra complication that there are three cases to distinguish according to the causal character of the planes that define the surface (see Theorems 3.3 and 3.5.)

References

[1] A. D. Aleksandrov, *Uniqueness theorems for surfaces in the large, V*, Amer. Math. Society Transl., **21** (1962), 412–416.

[2] A. Enneper, *Ueber die cyclischen Flächen*, Nach. Königl. Ges. d. Wissensch. Göttingen, Math. Phys. Kl., (1866), 243–249.

[3] A. Enneper, *Die cyklischen Flächen*, Z. Math. Phys., **14** (1869), 393–421.

[4] J. I. Hano and K. Nomizu, *Surfaces of revolution with constant mean curvature in Lorentz-Minkowski space*, Tohoku Math. J., **36** (1984), 427–435.

[5] W. Y. Hsiang, *On generalization of theorems of A. D. Alexandrov and C. Delaunay on hypersurfaces of constant mean curvature*, Duke Math. J., **49** (1982), 485–496.

[6] W. Jagy, *Minimal hypersurfaces foliated by spheres*, Michigan Math. J., **38** (1991), 255–270.

[7] W. Jagy, *Sphere-foliated constant mean curvature submanifolds*, Rocky Mountain J. Math., **28** (1998), 983–1015.

[8] O. Kobayashi, *Maximal surfaces in the 3-dimensional Minkowski space L^3*, Tokyo J. Math., **6** (1983), 297–303.

[9] F. J. López, R. López and R. Souam, *Maximal surfaces of Riemann-type in the Lorentz-Minkowski space* $\mathbf{L}^3$, Michigan Math. J., **47** (2000), 469–497.

[10] R. López, *Constant mean curvature hypersurfaces foliated by spheres*, Diff. Geom. Appl., **11** (1999), 245–256.

[11] R. López, *Constant mean curvature surfaces foliated by circles in Lorentz-Minkowski space*, Geom. Dedicata, **76** (1999), 81–95.

[12] R. López, *Timelike surfaces in Lorentz 3-space with constant mean curvature*, Tohoku Math. J., **52** (2000), 515–532.

[13] R. López, *Cyclic surfaces of constant Gauss curvature*, Houston J. Math., **27** (2001), to appear.

[14] R. López, *Surfaces of constant Gauss curvature in Lorentz-Minkowski 3-space*, Rocky Mount. J. Math., to appear.

[15] J. B. Meusnier, *Mémoire sur la courbure des surfaces*, Mém. des Savants étrangers, **10** (1785), 504.

[16] J. C. C. Nitsche, Lectures on Minimal Surfaces, Cambridge Univ. Press, Cambridge, 1989.

[17] J. C. C. Nitsche, *Cyclic surfaces of constant mean curvature*, Nachr. Akad. Wiss. Göttingen Math.-Phys. Kl. II, **1** (1989), 1–5.

[18] B. O'Neill, Semi-Riemannian Geometry with Applications to Relativity, Academic Press, New York, 1983.

[19] B. Riemann, *Über die Flächen vom Kleinsten Inhalt be gegebener Begrenzung*, Abh. Königl. Ges. d. Wissensch. Göttingen, Mathema. Cl., **13** (1868), 329–333.

[20] I. Van de Woestijne, Minimal surfaces of the 3-dimensional Minkowski space, Geometry and Topology of Submanifolds, II, Word Scientific, Singapore, 1990, 344–369.

Departamento de Geometría y Topología
Universidad de Granada
18071 Granada
Spain
`rcamino@goliat.ugr.es`

Advanced Studies in Pure Mathematics 34, 2002
Minimal Surfaces, Geometric Analysis and Symplectic Geometry
pp. 201–217

A Generalized Height Estimate for H-graphs, Serrin's Corner Lemma, and Applications to a Conjecture of Rosenberg

John McCuan

Abstract.

In this note we give a generalized form of the well-known height estimate for constant mean curvature graphs due to J. Serrin. An application is also given that effectively relates the global rate of convergence of a family of constant mean curvature surfaces recently considered by A. Ros and H. Rosenberg to their convergence behavior on the boundary. Some information concerning boundary behavior is also obtained by applying reflection techniques and corner comparison in particular. This latter analysis allows one to construct a counterexample to one part of a conjecture of Rosenberg.

§ Introduction

Let Ω be a bounded domain in $\mathbf{R}^2$ and $u \in C^2(\Omega) \cap C^1(\bar{\Omega})$ satisfy the equation of constant mean curvature (CMC)

$$-\operatorname{div} \mathbf{T}u = 2H \quad (= \text{constant}) \quad \text{on } \Omega, \tag{0}$$

where $\mathbf{T}u = \mathrm{D}u/\sqrt{1+|\mathrm{D}u|^2}$ and without loss of generality $H > 0$. If $u_{|\partial\Omega} \equiv 0$, then one has the well-known height estimate given by Serrin [S1]

$$\max_{x\in\Omega} u \leq \frac{1}{H}. \tag{1}$$

This bound can be obtained by noting that on $\mathcal{G} \equiv \operatorname{graph} u$, the function

$$\phi \equiv X_3 + \frac{1}{H} N_3 \tag{2}$$

2000 *Mathematics Subject Classification.* Primary 53C21; Secondary 58J70.

Key words and prases. constant mean curvature, height estimates, reflection.

Research at MSRI supported in part by NSF grant DMS-9022140.

is subharmonic, where $X_3 = u$ and N_3 is the third component of the (downward pointing) unit normal to $\mathcal{G}$. More precisely, one has for the intrinsic Laplacian of ϕ

$$\Delta_{\mathcal{G}}\phi = -\frac{2}{H}(H^2 - K)N_3 \geq 0,$$

where K is the Gaussian curvature of $\mathcal{G}$.

In the context of parametric CMC surfaces, the reflection technique of Alexandrov [A] was recently extended in [M] by noting that on a CMC graph $\mathcal{S}$ over a domain $\Omega \subset S^2$, the function

$$\Phi \equiv \left\| X + \frac{1}{H}N \right\|^2 \tag{3}$$

is subharmonic, where $X \in \mathbf{R}^3$ denotes a point on $\mathcal{S}$ and N is the unit normal to $\mathcal{S}$ at X. [1] In fact,

$$\Delta_{\mathcal{S}}\Phi = -\frac{4}{H}(H^2 - K)X \cdot N \geq 0.$$

While reading a recent paper of Ros and Rosenberg [R] in which the height estimate (1) is used extensively in conjunction with the Alexandrov technique, we guessed that (3) should be an appropriate generalization of (2) by which a "height" estimate can be obtained. We give such an estimate (6) for spherical graphs that yields (1) as a limiting special case.

The paper of Ros and Rosenberg was partially motivated by a conjecture of Rosenberg which we now describe. Let Γ be a convex planar curve that bounds a domain $\Omega \subset \{x_3 = 0\}$. For each volume $V > 0$, let M_V be an embedded CMC surface with boundary Γ such that $M_V \cup \Omega$ encloses a volume V. It is clear at least for small V, that M_V exists and is a graph which, without loss of generality, is above the plane $\{x_3 = 0\}$. As the volume increases, the mean curvature will increase until the surface ceases to be a graph, and sometime thereafter a critical volume is reached for which the mean curvature starts to decrease. Let us assume that the surfaces M_V continue to enclose a volume V and that the normal $(0, 0, 1)$ on Ω always points into V.

Conjecture 1 (Rosenberg). *If Γ is not a circle, then there is a second critical volume V_c for which the following hold:*

[1]Note that $\mathcal{S}$ is a graph over the *sphere* S^2 centered at $0 \in \mathbf{R}^3$ instead of the plane. The condition that $\mathcal{S}$ is a spherical graph can be expressed by requiring $X \cdot N < 0$ in the interior of $\mathcal{S}$.

(1) *For $V \leq V_c$, M_V lies in $\{x_3 \geq 0\}$.*
(2) *M_{V_c} is tangent to the plane $\{x_3 = 0\}$.*
(3) *The point of tangency occurs at a point of minimum curvature of $\partial M_{V_c} = \Gamma$.*
(4) *For $V > V_c$, M_V intersects $\{x_3 < 0\}$.*

If one assumes, by way of contradiction, that M_V lies in $\{x_3 \geq 0\}$ for all V, then the result of Ros and Rosenberg applies to show that the family $\{M_V\}$, suitably scaled, converges as $V \to \infty$ to a round sphere. We use a nominally more general form (5) of the height estimate, with some additional assumptions motivated by the conjecture, to describe the *rate* of convergence. If the additional assumptions can be justified, then the result presented here should be of use in proving the first, second, and fourth assertions of the conjecture. In an essentially independent part of the paper, we show that for certain domains the third assertion cannot hold.

§ The estimate

Let $\mathcal{S}$ be a spherical graph as described above.[1] Since $H\Phi - 1/H$ is subharmonic, one has immediately from the maximum principle

$$\|X\|^2 H + 2X \cdot N \leq \max_{Y \in \partial\mathcal{S}} (\|Y\|^2 H + 2Y \cdot N). \tag{4}$$

Let the expression appearing on the right in this inequality be denoted by M. Using the fact that $X \cdot N \geq -\|X\|$, one has

$$H\|X\|^2 - 2\|X\| - M \leq 0.$$

This inequality is quadratic in $\|X\|$, and we easily obtain from it

$$\|X\| \leq \frac{1}{H}(1 + \sqrt{1 + HM}). \tag{5}$$

Note that $M \geq \max(\|Y\|^2 H - 2\|Y\|) = \max(\|Y\|\sqrt{H} - 1/\sqrt{H})^2 - 1/H \geq -1/H$. Thus, the expression on the right in (5) is real (and positive).

In order to complete our comparison to CMC graphs over the plane that satisfy the Dirichlet condition $u_{|\partial\Omega} \equiv 0$, let us assume that $\|X\|_{|\partial\mathcal{S}} \equiv \rho$ =constant. In this case, $M \leq \rho^2 H$ and (5) yields

$$\|X\| \leq \frac{1}{H} + \sqrt{\frac{1}{H^2} + \rho^2}. \tag{6}$$

Just as equality in (1) holds for a hemisphere over a disc, we get equality in (6) when $\mathcal{S}$ is a portion of a sphere that meets the sphere of radius ρ

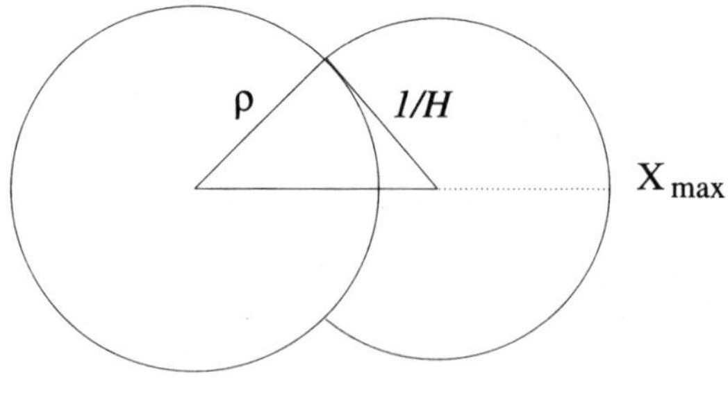

Fig. 1. Extremal spherical H-graph.

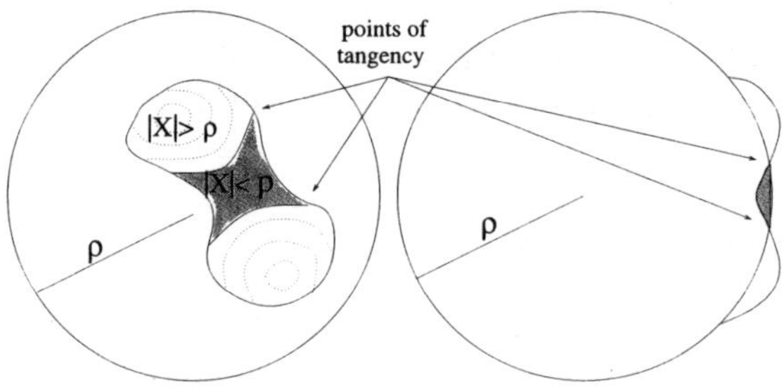

Fig. 2. Possible CMC graph over S_ρ with $H = 1/\rho$.

orthogonally along its boundary (See Fig. 1). Notice that (1) is obtained from (6) by subtracting ρ from each side and letting ρ tend to $+\infty$.

Before considering more general boundary conditions, for which the form (5) is useful, we note the curious fact that, by the maximum principle, the function u in (1) must satisfy $u > 0$ in $\Omega \subset \mathbf{R}^2$ and $\partial u/\partial n > 0$ on $\partial\Omega$, while one expects in general the spherical graph $\mathcal{S}$ in (6) to have interior points with $\|X\| < \rho$ and/or boundary points where $\mathcal{S}$ is tangent to the sphere of radius ρ (See Fig. 2). Notice that tangency of a spherical graph to the sphere of radius $\|X_{\min}\| < \rho$ does not in general contradict the maximum principle (e.g., if $\rho = 1/H$), nor does a tangency to the sphere of radius ρ on $\partial\mathcal{S}$ (i.e., $Y \cdot N = -\|Y\|$ for some $Y \in \partial\mathcal{S}$) imply that $\|X\| \leq \rho$. We use similar conditions below, however, to get some useful information.

§ An application

Some interesting estimates follow from (5) when the spherical graph $\mathcal{S}$ very nearly "encompasses" a sphere of radius $1/H$ (See Fig. 3). To

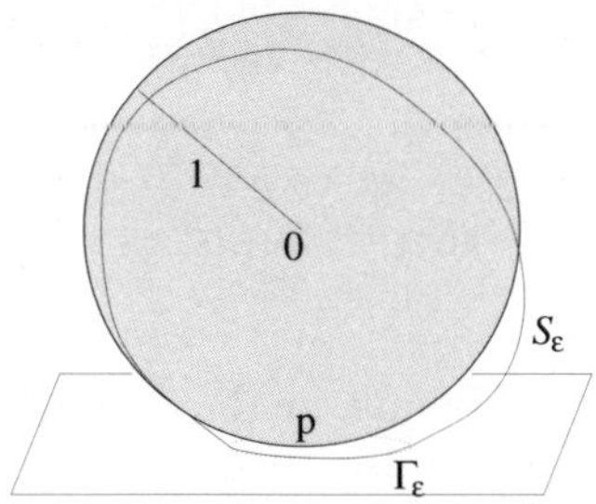

Fig. 3. A Rosenberg Bubble.

simplify notation, we specialize in this section to the case $H = 1$. To be even more specific, let $\mathcal{S}_\epsilon = \mathcal{S}_\epsilon(V)$ be the appropriate ϵ-homothety of M_V such that $\mathcal{S}_\epsilon$ has mean curvature 1. Recall that $\{M_V\}$ is the family of surfaces from Conjecture 1, and we are making the additional assumption that each surface lies in the half space $\{x_3 \geq 0\}$. We have then (as $V \to \infty$) that $\epsilon \to 0$ and the boundaries of the scaled surfaces $\partial\mathcal{S}_\epsilon = \Gamma_\epsilon$ converge as sets to a single point $p \in \mathbf{R}^3$.

Ros and Rosenberg show, moreover, that this particular family of surfaces, converges smoothly to a unit sphere on any compact set in $\mathbf{R}^3$ that does not contain p. Next, we slide our coordinates up one unit so that $p = (0, 0, -1)$ and the boundaries Γ_ϵ lie in $\{x_3 = -1\}$. The convergence of the surfaces is then to the unit sphere $S(0)$ centered at the origin, and we may assume (just by scaling the original curve) that $\max_{\Gamma_\epsilon} \|p - X\| \leq \epsilon$.

We would like to say something about the *rate* of convergence of $\mathcal{S}_\epsilon$ to $S(0)$. In order to do this, we will make an assumption about the behavior of $X \cdot N$ on Γ_ϵ. In fact, Ros and Rosenberg show that

$$\max_{\Gamma_\epsilon} |-1 - X \cdot N| \to 0.$$

We will need to assume more detailed information. As motivation, let $S(t)$ denote the unit sphere with center $(0, 0, t)$, and consider the family of spheres $S(-1 + \sqrt{1 - \epsilon^2})$ which translate up to the unit sphere $S(0)$ and intersect the plane $x_3 = -1$ in a circle of radius ϵ. On the boundary of one of these spheres, explicit calculation yields $X \cdot N = -(\sqrt{1 - \epsilon^2} + \epsilon^2) = -(1 + \epsilon^2/2 - \epsilon^4/8 - \cdots)$. In the result below, we require the boundary behavior to be bounded by that of these translated spheres up to second order.

Finally, before stating the main result we mention that in the case considered in [R] the method of [M] applies to show that $\mathcal{S}_\epsilon$ is a graph over $S(0)$ when ϵ is sufficiently small.

Theorem 1. *If $\mathcal{S}_\epsilon$ is a family of spherical graphs over $S(0)$ of CMC $H = 1$, and*

(1) $\max_{\partial\mathcal{S}_\epsilon} \|X - p\| \le \epsilon$, *where* $p = (0, 0, -1)$, *while*
(2) $\max_{\partial\mathcal{S}_\epsilon} X \cdot N \le -(1 + \epsilon^2/2 - \alpha(\epsilon))$, *where* $0 < \alpha(\epsilon) = \circ(\epsilon^2)$,

then $\max_{\mathcal{S}_\epsilon} \|X\| \le 1 + \sqrt{2\alpha}$.

Proof. From the first bound in the statement of the theorem, it follows that $\|Y\| \le \sqrt{1 + \epsilon^2}$ for each $Y \in \partial\mathcal{S}_\epsilon$. Using this and the second bound to estimate $M = \max_{Y\in\Gamma_\epsilon}(\|Y\|^2 + 2Y\cdot N)$, we get $M \le -1 + 2\alpha(\epsilon)$. Thus, (5) gives the conclusion of the theorem. □

§ Generalities

It should be remarked that Serrin's result was actually stated in [S1] as follows.

Theorem S. *If $u \in C^2(\Omega) \cap C^0(\bar{\Omega})$ is a solution of (0) on the bounded plane domain Ω, then*

$$\min_{y\in\partial\Omega} u(y) < u \le \max_{y\in\partial\Omega} u(y) + \frac{1}{H} \quad \text{in } \Omega \tag{7}$$

with the equality being attained only if graph u is a hemisphere.

We now discuss how this theorem can be generalized to spherical graphs. The proofs of this section are straightforward applications of the reasoning above and that of [S1], and we omit them. Our main result along these lines is the following.

Theorem 2. *If X parameterizes a spherical H-graph $\mathcal{S}$ over a domain $\Omega \subset S^2$ and $\mathcal{S}$ extends continuously to $\bar{\Omega}$, then for each $X \in \operatorname{int}\mathcal{S}$*

$$\|X\| \le \frac{1}{H} + \sqrt{\frac{1}{H^2} + \max_{Y\in\partial\mathcal{S}} \|Y\|^2}. \tag{8}$$

Moreover, the equality is obtained only if $\mathcal{S}$ is a portion of a sphere.

By a limiting argument, (8) can be used to obtain the upper bound in (7). The equality condition, however, does not survive the limiting procedure.

A more serious omission is the lower bound, and one sees immediately that a spherical graph $\mathcal{S}$ can come arbitrarily close to the origin, while $\partial\mathcal{S}$ lies on $S(0)$; consider spheres of radius $1/H$ that intersect $S(0)$ at its equator and have $H \searrow 0$. Nevertheless, the following result shows that Fig. 2 should not be taken too literally.

Theorem 3. *If $\mathcal{S}$ is a spherical H-graph over a domain in a* hemisphere *of* S^2 *and* $\min_{Y\in\partial\mathcal{S}}|Y| \geq 1/H$, *then*

$$\min_{Y\in\partial\mathcal{S}} \|Y\| \leq \|X\|$$

for every $X \in \operatorname{int}\mathcal{S}$ *with equality only if* $\mathcal{S}$ *is a portion of a sphere.*

Finally, Serrin noted that for a *minimal* graph u

$$u(x) \leq \max_{y\in\partial\Omega} u(y), \tag{9}$$

while the upper bound in (7) tends to $+\infty$ as $H \to 0$. Thus, he gives a complementary result. [2]

Theorem S′. *If u is described by* Theorem S, *and Ω is contained in a disc of radius $a \leq 1/H$, then*

$$u(x) \leq \max_{y\in\partial\Omega} u(y) + \frac{1}{H} - \sqrt{\frac{1}{H^2} - a^2} \tag{10}$$

with equality only if graph u is a hemisphere.

Notice in particular that, for a fixed domain Ω, (9) follows from (10) by letting H tend to 0.

For a spherical graph, the disc with boundary on the equator of $S(0)$ displays the futility of trying to bound $\|X\|$ away from 0. For an outer bound we have the following.

Theorem 2′. *Let $\mathcal{S}$ be a spherical H-graph described by* Theorem 2 *and let B be the expression on the right in* (8). *Let* $D = \{\xi \in S^2 : 0 \leq \phi < \arctan(H^2B^2 - 1)^{-1/2}\}$, *where* $\phi = \phi(\xi)$ *is the angle that* ξ *makes with the positive* x_3*-axis. If* $\Omega \subset D$, *then*

$$\|X\| \leq \max\left\{ \max_{Y\in\partial\mathcal{S}} \|Y\|, \right.$$
$$\left. \max_{Y\in\partial\mathcal{S}} \|Y\| \sqrt{1 - \frac{1}{H^2B^2}} + \frac{1}{H} - \sqrt{\frac{1}{H^2} - \frac{\max_{Y\in\partial\mathcal{S}}\|Y\|^2}{H^2B^2}} \right\}.$$

Note that $HB \to 2$ as $H \to 0$. In particular, as a set D converges to $D_0 \equiv \{\xi \in S^2 : 0 \leq \phi < \pi/6\}$. Thus, if Ω is compactly contained in D_0, then the theorem can be used to conclude that $\|X\| \leq \max_{Y\in\partial\mathcal{S}} \|Y\|$ when $\mathcal{S}$ is a minimal surface. While the bound in Theorem 2′ looks complicated, it is completely elementary and can be read off from Fig. 4 and a straightforward application of Serrin's argument.

---

[2]The lower bound, of course, is no problem.

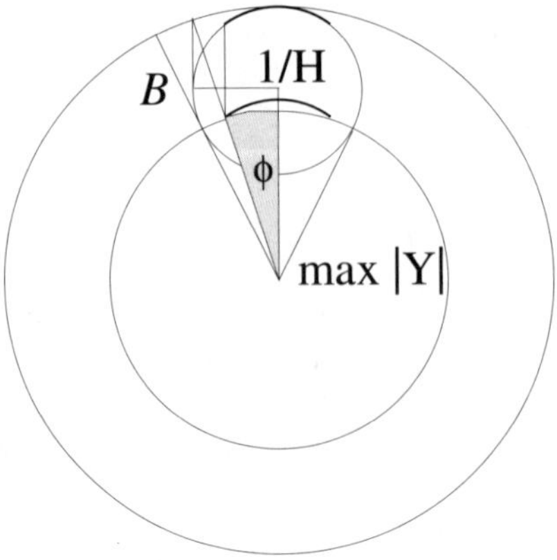

Fig. 4. Diagram for proof of Theorem 2′.

§ Serrin's corner lemma

For this section we assume Γ is a non-circular convex curve and that $\{M_V\}$ is a family of surfaces as in Conjecture 1. We assume furthermore that the family $\{M_V\}$ satisfies conclusions (1) and (2) of Conjecture 1.

Now, let $M = M_{V_c}$ be the bubble of critical volume which is tangent to the plane Π containing $\Gamma = \partial M$. We apply the Alexandrov reflection procedure (see, e.g., [W]) to M with planes that are orthogonal to Π. Recall that as the reflection plane P begins to cut M, the small cap M^- that is cut off reflects across P into V. We can continue to push P parallel to itself farther through M until the procedure terminates (with the reflection $\hat{M}$ of M^- about to exit V). At this terminal point, there are several possibilities.

It may be that the reflected surface $\hat{M}$ is tangent to M at a point interior to $\hat{M}$. In this case, it follows that the terminal plane P is a symmetry plane for M. Another possibility is that $\hat{M}$ is tangent to M at a point on ∂M^- but away from Π. In this case, P is again a plane of symmetry (which follows from the Hopf boundary point lemma). In both of these first two cases, the curve Γ, in particular, must possess a symmetry, and we say that this is a symmetry *in the direction of* P.

Of course, it may be the case that Γ does not have a symmetry in a certain direction. In this case, the touching described above cannot occur. One alternative is that $\hat{M}$ touches M along Γ but away from P. This touching may be tangential (in which case there must be a symmetry) or non-tangential (which case we ignore).

The last possibility is that $\hat{M}$ is tangent to M at a point p in $\Pi \cap P$. Each of the two surfaces may be expressed locally as a graph over a corner domain in the common tangent plane at p. It is customary at this point to apply a version of the Hopf boundary point lemma at a corner due in various forms to Serrin [S2] and Gidas, Ni, and Nirenberg [G]. Let u be the nonnegative difference of the two graphs over the corner domain K. A simple version of the corner lemma is as follows.

Lemma S. *If u satisfies*

(1) $u \geq 0$ *in* K,
(2) $u(p) = 0$ (*where* p *is the corner point*),
(3) $Du(p) = 0$, *and*
(4) $D_{\eta\eta}u(p) = 0$ *for every non-tangential direction* η *pointing into* K *from* p,

then $u \equiv 0$ *in* K.

If the hypotheses (1)–(4) hold, then the conclusion yields that M, and hence Γ, has a symmetry. Furthermore, note that if the $\Pi \cap P$ touching occurs, then conditions (1)–(3) automatically hold. Also, recall that the function to which we apply Lemma S is a difference $u = f(x,y) - \hat{f}(x,y)$, and one of the functions, $\hat{f}$, is obtained from the other by reflection so that in the appropriate coordinates $\hat{f}(x,y) = f(-x,y)$. It follows that $u_{xx}(p) = u_{yy}(p) = 0$ and $u_{xy}(p) = 2f_{xy}(p)$.

On the other hand, if $\eta = (\eta_1, \eta_2)$ is a non-tangential direction, then since $u \geq 0$ and $Du(p) = 0$, one has

$$0 \leq D_{\eta\eta}u(p) = 4f_{xy}(p)\eta_1\eta_2.$$

We conclude that either (1) there is a symmetry *or* (2) $f_{xy}(p) > 0$.

Notice that the x-direction in this discussion is the direction of reflection and is tangent to $\Gamma = \partial M$ at p. Let us assume that M is a graph over Ω locally near p. Let $\phi = \phi(x,n)$ represent the graph where x is measured in the tangent direction to Γ as before and n is measured in the inward normal direction to Ω. An elementary computation shows that

$$f_{xy}(p) = -\frac{1}{1+\phi_n^2}(\phi_n)_x,$$

where ϕ_n is the usual inward normal derivative of ϕ. This means that *either there is a symmetry in the x direction or the angle between M and Π measured within V is strictly increasing along Γ in the direction of reflection at p.*

Similarly, and more interesting for what we say below, the same conclusion holds when M is locally a graph over $\mathbf{R}^2 \backslash \Omega$ near p.

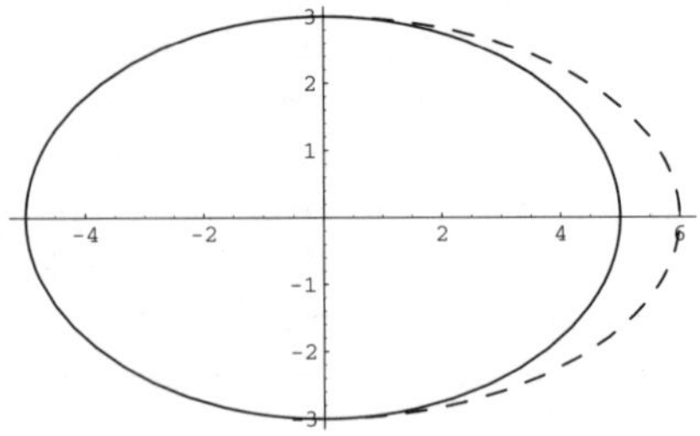

Fig. 5. Modified ellipse.

§ Counterexample

Consider what happens if conclusions (1) and (2) hold when $\Gamma = \mathcal{E}$ is an ellipse. In this case, there are only two symmetries of Γ and hence of $M = M_{V_c}$. Consequently, in every other direction of reflection, the reflection procedure described above terminates due to a boundary touching. Furthermore, due to the monotonicity properties possessed by the curvature of an ellipse as the boundary is traversed, one can check (as we do in the next section for all curves with two symmetries and four vertices) that the procedure must end in a $\Pi \cap P$ touching without symmetry. One concludes that the angle between M and Π is increasing with decreasing curvature along the boundary of the ellipse and, in accord with the conjecture, the maximum angle which occurs at the tangency must be at the minimum point of curvature. We have shown that conclusions (1) and (2) of Conjecture 1 imply (3) when Γ is an ellipse.

It is possible, however, to construct a smooth convex curve Γ with the following properties; see Fig. 5.

(1) $\Gamma \cap \{x_1 \leq 0\}$ is half of an ellipse $\mathcal{E}$ given by $x_1^2/a^2 + x_2^2/b^2 = 1$ with $b < a$.

(2) $\Gamma \cap \{x_1 \geq 0\}$ is a curve that is symmetric with respect to the x_1-axis and has monotone curvature which increases with x_1 to a vertex $(a', 0)$ with $a' > a$.

(3) The domain Ω bounded by Γ contains the half of $\mathcal{E}$ with $x_1 > 0$.

It is not at all difficult to believe that such a modification of an ellipse exists (though Fig. 5 actually shows two ellipses and hence the resulting modification is not smooth). We give an explicit construction in the next section. For now, observe that, given such a curve, the points of minimum curvature still occur at $(0, \pm b)$ and have the same value b/a^2 as $\mathcal{E}$. Reflection in the positive x_1 direction, however, does not result in

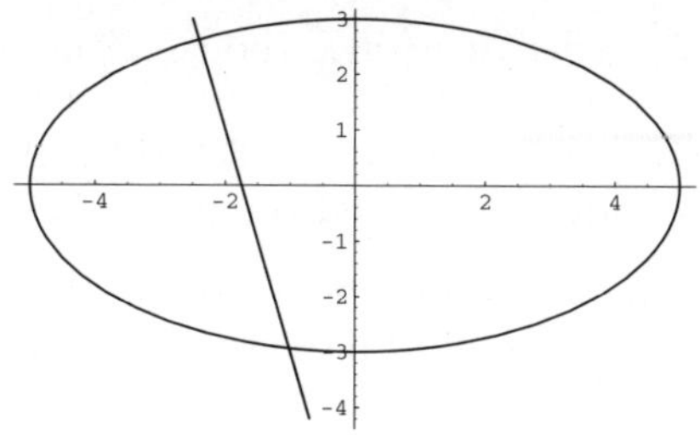

Fig. 6. The line l normal to $\mathcal{E}$ at p.

a symmetry, but in a $\Pi \cap P$ corner touching without symmetry. Thus, the angle between M and Π is increasing along Γ at $(0, \pm b)$; it cannot be a maximum there as required by Conjecture 1 part (3).

§ Details

We first show that any curve $\mathcal{E}$ with the symmetries of an ellipse and four vertices must reflect strictly inside the domain it bounds until a tangency occurs on the plane of reflection P.

Suppose that $\mathcal{E}$ is symmetric with respect to the x_1 and x_2-axes with minimum curvature at $(0, \pm b)$ and maximum curvature at $(\pm a, 0)$. of $\mathcal{E}$ in the j-th quadrant. Let l be a line of negative slope whose normal points into the first quadrant. We move l parallel to itself in the direction of its normal through the domain Ω bounded by $\mathcal{E}$. (The line l may be thought of as $\Pi \cap P$ in our previous discussion.) There are two points of $\mathcal{E}$ where the normal to $\mathcal{E}$ is parallel to l. There is one, p_2, in the second quadrant and another, p_4, in the fourth quadrant.

Lemma 1. *The portion $\mathcal{E}^-$ of $\mathcal{E}$ through which l has passed when l contains $p = p_2$ is the graph of a function $u(x)$ over l.*

Proof. Since there are only two points where l can be normal to $\mathcal{E}$, the statement is equivalent to saying p_2 is reached first—before p_4. When the slope of l is very close to $-\infty$, this is clear because the radius of curvature near $(0, b)$ is greater than b. We conclude by continuity that the only way for the assertion to fail is if, for some slope, p_2 and p_4 are reached at the same time.

In this case, it follows from symmetry that the circle centered at the origin and passing through p_2 and p_4 is tangent to $\mathcal{E}$ at both of these points. A comparison of the curvature of $\mathcal{E}$ with the curvature of this circle yields a contradiction. If, for example, the curvature of $\mathcal{E}$ at p_2 is less than that of the circle, then since the curvature of $\mathcal{E}$ decreases up to $(0, b)$, one sees that the slope of $\mathcal{E}$ at $(0, b)$ must be less than that of the circle when it crosses the x_2-axis (which is zero). This contradicts the fact that $\mathcal{E}$ is smooth at $(0, b)$ and completes the proof. □

We position l as indicated in Fig. 6, passing through the point $p = p_2$. Denote by q the other point of intersection of l with $\mathcal{E}$. We need to prove the following.

Theorem 4. *The reflection $\hat{\mathcal{E}}$ of $\mathcal{E}^-$ lies strictly inside the domain Ω except at the endpoints p and q.*

Let us slightly abuse notation by letting l be the first axis of a plane coordinate system with real coordinates p and q corresponding to the points p and q. The entire curve $\mathcal{E}$ is then the union of the graphs of two smooth functions u and v defined on some interval $p \leq x \leq q'$, where $(q', u(q')) = (q', v(q')) = p_4$. In harmony with the notation of Lemma 1, we require that $\mathcal{E}^- \subset \text{graph}(u)$. In order to prove the theorem, it is enough to show that $\hat{u}(x) = -u(x) < v(x)$ for each x between p and q.

Let $\kappa_u(x)$ be the curvature of graph (u) at $(x, u(x))$ and $\kappa_v(x)$ that of graph (v) at $(x, v(x))$. Due to the simple nature of the curve $\mathcal{E}$ the curvatures satisfy some simple relations.

Lemma 2. *Let $x_0 = (p + q')/2$. Then the following hold:*

(1) $\kappa_u(p) = \kappa_v(p)$.
(2) $\kappa_u(x) > \kappa_v(x)$ *for* $p < x < x_0$.
(3) $\kappa_u(x_0) = \kappa_v(x_0)$.
(4) $\kappa_u(x) < \kappa_v(x)$ *for* $x_0 < x < q'$.

Proof. For x near p, it follows from the monotonicity of curvature that $\kappa_u(x) > \kappa_v(x)$. Since κ_u is increasing and κ_v decreasing initially, this inequality will be maintained at least until $(x, u(x))$ or $(x, v(x))$ is a vertex—and for some time thereafter. There are three cases determined by whether a vertex of minimum curvature is reached first, one of maximum curvature, or both at the same time. In all cases, one sees that the assertions of the lemma hold. □

The key points from Lemma 2 are (1) κ_u is initially greater than κ_v, and (2) κ_u may become smaller than κ_v, but once the curvatures are equal at x_0, κ_u will always be less from there on.

In the event that $q \leq x_0$, we can easily finish the proof of Theorem 4 since the reflection $\hat{\mathcal{E}}$ of $\mathcal{E}^-$ is given by the graph of $\hat{u}(x) = -u(x)$ and the curvature $\kappa_{\hat{u}} = \kappa_u \geq \kappa_v$. The two graphs are tangent at p and that of $\hat{u}$ has greater curvature at each point. Hence, $\hat{u}(x) \leq v(x)$ for $p \leq x \leq q$ with equality only at p as required.

In the other case ($x_0 < q$) we proceed by contradiction and use Lemma 3 below. If Theorem 4 is to fail, then there must be a first value x_1 with $p < x_1 < q$ for which $\hat{u}(x_1) = v(x_1)$. Moreover, we have that $x_0 < x_1 < q$ and $\hat{u}'(x_1) \geq v'(x_1)$. This is enough.

Lemma 3. *If $\hat{u}(x_1) = v(x_1)$ and $\hat{u}'(x_1) \geq v'(x_1)$, but the curvature of graph $(\hat{u})$ is less than the curvature of graph (v) on some interval $x_1 < x \leq q$, then $\hat{u}(x) > v(x)$ for $x_1 < x \leq q$.*

Since we know that $\hat{u}(q) < v(q)$, we have a contradiction—and a theorem. □

We now turn our attention to constructing a modification Γ of an ellipse as described above.

By symmetry it is enough to construct the portion of Γ in the fourth quadrant. We recall furthermore that a curve can be constructed from its curvature. In our case, if θ denotes the angle made by the tangent to the curve with the positive x_1-axis and we specify a positive function $\kappa(\theta)$ on $[0, \pi/2]$, then the formula

$$X(\theta) = (0, -b) + \int_0^\theta \frac{1}{\kappa(\vartheta)} (\cos\vartheta, \sin\vartheta) d\vartheta$$

defines a convex curve that extends $\mathcal{E}$ from the third quadrant into the fourth; see Fig. 7. Various other conditions must by satisfied by κ as described below. In order to introduce these conditions, we consider the curvature κ_0 and parameterization X_0 of the corresponding portion of the ellipse $\mathcal{E}$.

In order for Γ to be a smooth extension of $\mathcal{E}$ at $(0, -b)$, we need

(1a) $$\kappa(0) = \kappa_0(0) \quad \text{and} \quad \kappa'(0) = 0.$$

The condition on monotone curvature is

(1b) $$\kappa'(\theta) > 0 \quad \text{for} \quad 0 < \theta < \frac{\pi}{2}.$$

In order for Γ to close smoothly with it's reflection at a point $(a', 0)$, we need

(2a) $$x_2(\pi/2) = -b + \int_0^{\pi/2} \frac{1}{\kappa(\vartheta)} \sin\vartheta \, d\vartheta = 0,$$

and

(2b) $$\kappa'\left(\frac{\pi}{2}\right) = 0.$$

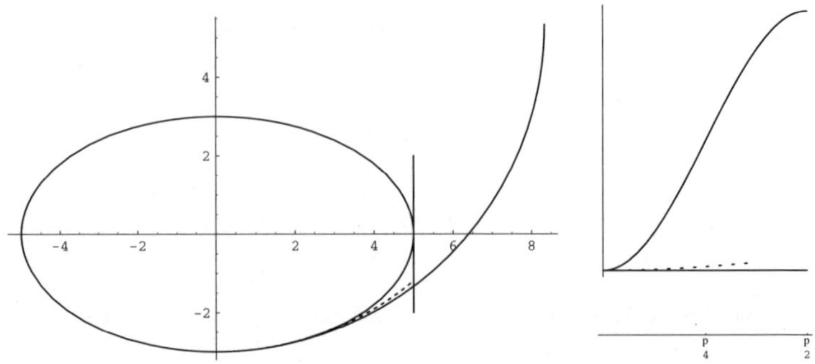

Fig. 7. Shown on the left are $\mathcal{E}$ and a quarter of the osculating circle at $(0, -b)$. On the right are the corresponding curvature functions. A possible modification is shown (dashed) which stays close to the circle for $0 \le \theta \le \theta_c(a)$.

In order to describe a sufficient condition that $\mathcal{E}$ be enclosed by Γ, we note that given a positive function κ, the function

$$\xi(\theta) = \int_0^\theta \frac{1}{\kappa(\vartheta)} \cos\vartheta \, d\vartheta$$

has an inverse. In particular, this is the case for κ_0, and we denote the corresponding inverse, defined on the interval $[0, a]$, by θ_0. Let the inverse of the function ξ defined by κ be denoted by θ, then we will require

$$\kappa(\theta(x_1)) < \kappa_0(\theta_0(x_1)) \quad \text{for} \quad 0 < x_1 < a. \tag{3}$$

Thus, we are done, if we can find a positive function κ defined on $[0, \pi/2]$ such that (1a), (1b), (2a), (2b), and (3) are satisfied.

We begin by considering the circle shown in Fig. 7. It has parameterization

$$X_c(\theta) = (0, -b) + \int_0^\theta \frac{1}{\kappa_0(0)} (\cos\vartheta, \sin\vartheta) \, d\vartheta.$$

This curve does have the virtue that it satisfies (3), i.e., it encloses the part of $\mathcal{E}$ in the fourth quadrant. We have already considered the inverse θ of the x_1-coordinate ξ. Similarly, the x_2-coordinate

$$\eta(\theta) = -b + \int_0^\theta \frac{1}{\kappa(\vartheta)} \sin\vartheta \, d\vartheta$$

has an inverse, which we denote by $\bar{\theta}$. We denote quantities corresponding to the circle by a subscript c. Note that $\theta_c(a) < \bar{\theta}_c(0)$. Let $\theta_1 = (\theta_c(a) + \bar{\theta}_c(0))/2 > \theta_c(a)$.

In order to make sure that (3) is satisfied, we will require that κ stay close to κ_c for $0 \leq \theta \leq \theta_1$.

To be precise, we choose a function κ of the form

$$\kappa(\theta) = \kappa_0(0) + \epsilon A(\theta) + hB(\alpha(\theta - \beta)).$$

In this definition $\epsilon < 1$, h, α, and β are positive numbers to be chosen. $A(\theta)$ is a fixed smooth increasing function that satisfies $A(\theta) \leq (\kappa_0(\theta) - \kappa_0(0))/2$ for $0 \leq \theta \leq \theta_1$ and $A'(\theta) \equiv A'(\theta_1)$ for $\theta_1 \leq \theta \leq \pi/2$. $B(\mu)$ is a bump function to be chosen with the standard properties ($B \geq 0$, $\operatorname{supp} B = [-1, 1]$, symmetric, one inflection on $[0, 1)$).

If we stipulate that $\beta - 1/\alpha = \theta_1$, then $\kappa(\theta) = \kappa_0(0) + \epsilon A(\theta)$ for $0 \leq \theta \leq \theta_1$ and, for ϵ small enough, conditions (1a) and (3) will be satisfied. Condition (1a) follows immediately from the properties of A and κ_0. To see that (3) holds, note first that the restriction of ξ to the interval $[0, \theta_1]$ has a well-defined inverse θ. Secondly, $\xi(\theta)$ converges uniformly to $\xi_c(\theta)$ on $[0, \theta_1]$ as $\epsilon \to 0$. In particular, $\xi_c(\theta_1) > a$ so that for ϵ small $\theta(x_1)$ is well-defined on $[0, a]$. Thirdly, $\kappa_0(\theta) \geq \kappa(\theta) = \kappa_0(0) + \epsilon A(\theta)$ for $0 \leq \theta \leq \theta_1$. In fact, $\kappa_0(\theta) - \kappa(\theta) \geq (\kappa_0(\theta) - \kappa_0(0))(1 - \epsilon/2) = \theta\kappa_0'(\theta_*)(1 - \epsilon/2) \geq 0$ for some $\theta_* \in (0, \theta)$, and equality holds only for $\theta = 0$. It follows that whenever $\xi(\vartheta) = \xi_0(\varphi)$, one must have $\vartheta < \varphi$, i.e., $\theta(x_1) < \theta_0(x_1)$. Finally then, we see that $\kappa(\theta(x_1)) < \kappa_0(\theta(x_1)) < \kappa_0(\theta_0(x_1))$. Let $\epsilon_1 > 0$ be such that (1a) and (3) hold whenever $\epsilon < \epsilon_1$.

Let μ_1 be the unique value in $[0, 1)$ such that $B''(\mu_1) = 0$. We then have $B'(\mu_1) < B'(\mu) < 0$ for $0 < \mu < \mu_1$. For any $\mu \in (0, \mu_1]$, we can define $\alpha = \alpha(\mu)$ and $\beta = \beta(\mu)$ by the equations

$$\frac{\mu}{\alpha} + \beta = \frac{\pi}{2} \quad \text{and} \quad \beta - \frac{1}{\alpha} = \theta_1.$$

Consider

$$\kappa = \kappa_1(\theta) = \kappa_0(0) + \epsilon A(\theta) + hB(\alpha(\theta - \beta)),$$

as before except that we have substituted the values $\alpha(\mu)$ and $\beta(\mu)$. Consider conditions (1b) and (2b). If $0 \leq \theta \leq \theta_1$, then $\kappa_1' = \epsilon A'(\theta) \geq 0$ with equality only at $\theta = 0$. For $\theta_1 \leq \theta \leq \pi/2$,

$$\kappa_1'(\theta) = \epsilon A'(\theta_1) + h\alpha B'(\alpha(\theta - \beta)) > \epsilon A'(\theta_1) + h\alpha B'\left(\alpha\left(\frac{\pi}{2} - \beta\right)\right).$$

Thus, there is a value

$$h = h(\epsilon, \mu) = -\frac{\epsilon A'(\theta_1)}{\alpha B'(\alpha(\pi/2 - \beta))}$$

for which the last expression is zero and, hence, (1b) and (2b) are satisfied.

The only remaining condition is (2a). Notice that for fixed μ we have $h(\epsilon, \mu) \to 0$ as $\epsilon \to 0$. It follows that $\kappa_1(\theta)$ converges uniformly to $\kappa_0(0)$ as $\epsilon \to 0$. Hence,

$$\eta_1\left(\frac{\pi}{2}\right) = -b + \int_0^{\pi/2} \frac{1}{\kappa_1(\vartheta)} \sin\vartheta \, d\vartheta \to \eta_c\left(\frac{\pi}{2}\right) > 0.$$

Consequently, there is a pair $(\epsilon_2, \mu_2) \in (0, \epsilon_1) \times (0, \mu_1)$ for which (1a), (1b), (2b), (3), and $\eta_1(\pi/2) > 0$ are satisfied.

On the other hand, the same reasoning concerning convergence of κ_1 to κ_0 works for $0 \le \theta \le \theta_1$ to show that for some $\epsilon = \epsilon_3$ small enough one has $\eta_1(\theta_1) < 0$. But when $\epsilon = \epsilon_3$ is fixed and μ is small, then α is small, β is close to $\pi/2$, and h gets monotonically *large*. More precisely, $\kappa_1(\theta)$ is a decreasing family of functions in μ, and we can make $\kappa_1(\theta)$ uniformly large on any interval $[\theta_1 + \delta, \pi/2]$. If we choose δ small enough so that $\eta_1(\theta_1 + \delta) < 0$, then it follows that

$$\eta_1\left(\frac{\pi}{2}\right) = \eta_1(\theta_1 + \delta) + \int_{\theta_1+\delta}^{\pi/2} \frac{1}{\kappa_1(\vartheta)} \sin\vartheta \, d\vartheta \to \eta_1(\theta_1) < 0$$

as $\mu \to 0$. Thus, for some (ϵ_3, μ_3), we have (1a), (1b), (2b), (3), and $\eta_1(\pi/2) < 0$.

By continuity, there is a point (ϵ_*, μ_*) on the line segment between (ϵ_2, μ_2) and (ϵ_3, μ_3) for which all the conditions (1a)-(3) hold. □

Acknowledgment. I would like to express my thanks to David Hoffman for several comments that improved the presentation above.

References

[A] A. D. Alexandrov, *Uniqueness theorems for surfaces in the large V*, Vestnik Leningrad Univ. Math., **13** (1958), 5–8; English translation in Amer. Math. Soc. Trans. Ser. 2, **21**, 412–416.

[G] B. Gidas, W-M. Ni and L. Nirenberg, *Symmetry and related properties via the maximum principle*, Comm. Math. Phys., **68** (1979), 209–243.

[M] J. McCuan, *Symmetry via spherical reflection*, J. Geom. Anal., **10(3)** (2000), 545–564.

[R] A. Ros and H. Rosenberg, *Constant mean curvature surfaces in a half-space of* $\mathbf{R}^3$ *with boundary in the boundary of the half-space*, J. Diff. Geom., **44** (1996), 807–817.

[S1] J. Serrin, *On surfaces of constant mean curvature which span a given space curve*, Math. Z., **112** (1969), 77–88.

[S2] J. Serrin, *Uniqueness of positive radial solutions of $\Delta u + f(u) = 0$ in* $\mathbf{R}^n$, Math. Z., **112** (1969), 115–145.

[W] H. Wente, *The symmetry of sessile and pendant drops*, Pacific J. Math., **88** (1980), 387–397.

School of Mathematics
Georgia Institute of Technology
Atlanta, GA 30332-0383
U. S. A.
`mccuan@math.gatech.edu`

Advanced Studies in Pure Mathematics 34, 2002
Minimal Surfaces, Geometric Analysis and Symplectic Geometry
pp. 219–229

Discrete Spectrum and Weyl's Asymptotic Formula for Incomplete Manifolds

Jun Masamune and Wayne Rossman

Abstract.

Motivated by recent interest in the spectrum of the Laplacian of incomplete surfaces with isolated conical singularities, we consider more general incomplete m-dimensional manifolds with singularities on sets of codimension at least 2. With certain restrictions on the metric, we establish that the spectrum is discrete and satisfies Weyl's asymptotic formula.

§1. Discreteness of the Spectrum

When one studies the Morse index of minimal surfaces in Euclidean 3-space $\mathbb{R}^3$ or of mean curvature 1 surfaces in hyperbolic 3-space $\mathbb{H}^3$, the problem reduces to the study of the number of eigenvalues less than 2 of the spectrum of the Laplace-Beltrami operator on Met_1 surfaces [FC], [UY], [LR]. (Met_1 surfaces are incomplete 2-dimensional manifolds with constant curvature 1 and isolated conical singularities.) Met_1 surfaces are known to have pure point spectrum and satisfy Weyl's asymptotic formula.

Here we will show that the spectrum is discrete and that Weyl's asymptotic formula holds for more general incomplete manifolds. We allow the dimension to be arbitrary; we do not make any specific assumptions about the curvature; and we allow more general singularities, of at least codimension 2 (in a sense to be made precise below). This more general setting allows us to consider singularities such as a product of an $m-n$ dimensional metric cone with a portion of $\mathbb{R}^n$ ($m \geq n+2$), one of our desired examples. In this example, the incomplete metric is singular only in the direction of the metric cone and not on the portion of $\mathbb{R}^n$ itself, so generally the incomplete manifolds and their metrics $\tilde{g}$ that we consider will not be conformally equivalent to open sets of compact Riemann manifolds, unlike the case of Met_1 surfaces. With this in

---

2000 *Mathematics Subject Classification.* Primary 58J50; Secondary 35J05, 35P10, 35P20, 53C20.

mind, we now define the types of incomplete manifolds and metrics $\tilde{g}$ that we will study here.

Let (M, g) be a compact manifold of dimension m with smooth Riemannian metric g. Let N be a compact submanifold of dimension n with codimension $m - n \geq 2$. Suppose further that in a neighborhood of N the metric g can be diagonalized; that is, there exist local coordinates $(x_1, \ldots, x_{m-n}, y_1, \ldots, y_n)$, where $(0, \ldots, 0, y_1, \ldots, y_n)$ are coordinates for N, so that $(dx_1, \ldots, dx_{m-n}, dy_1, \ldots, dy_n)$ is globally defined in some open neighborhood of N and so that the metric g is written

$$g = \begin{pmatrix} g_1 & 0 \\ 0 & g_2 \end{pmatrix},$$

where g_1 is an $m - n \times m - n$ positive definite matrix, and g_2 is an $n \times n$ positive definite matrix. (For example, such a case can occur if M has a product structure $M = M_1 \times N$ near N, where M_1 is an $m - n$ dimensional compact Riemannian manifold.)

Theorem 1.1. *Let N be an n-dimensional compact submanifold of an m-dimensional compact manifold (M, g) with $m \geq n + 2$ such that the metric g can be diagonalized near N. Choose local coordinates in a neighborhood of N so that*

$$g = \begin{pmatrix} g_1 & 0 \\ 0 & g_2 \end{pmatrix}$$

in this neighborhood. Let $\tilde{g}$ be another smooth regular metric on $M \setminus N$ so that

$$\tilde{g} = \begin{pmatrix} f^2 g_1 & 0 \\ 0 & g_2 \end{pmatrix}$$

in a neighborhood of N, where $f \in C^\infty(M \setminus N)$.

If $m = 2$, assume that $f \in L_g^{2+\epsilon}(M)$ for some $\epsilon \in (0, \infty)$.

If $m \geq 3$, assume that $\inf(f) > 0$ and $f \in L_g^{(m(m-n)/2)+\epsilon}(M)$ for some $\epsilon \in (0, \infty)$.

Then the Sobolev space $W_{\tilde{g}}^{1,2}(M \setminus N)$ with respect to $\tilde{g}$ is compactly included in $L_{\tilde{g}}^2(M \setminus N)$.

Proof. When $m \geq 3$ and $p \in (2, 2m/(m-2))$ (resp. $m = 2$ and $p \in (2, \infty)$), then the inclusion $W_g^{1,2}(M)$ into $L_g^p(M)$ is compact. When $m \geq 3$ and $f \in L^{(m(m-n)/2)+\epsilon}$ (resp. $m = 2$ and $f \in L^{2+\epsilon}$) for some positive ϵ, then the inclusion $L_g^p(M)$ into $L_{\tilde{g}}^2(M \setminus N)$ is continuous, by Hölder's inequality. For example, when $m \geq 3$, we can choose

$$p = \frac{m + (2\epsilon/(m-n))}{(m/2) + (\epsilon/(m-n)) - 1},$$

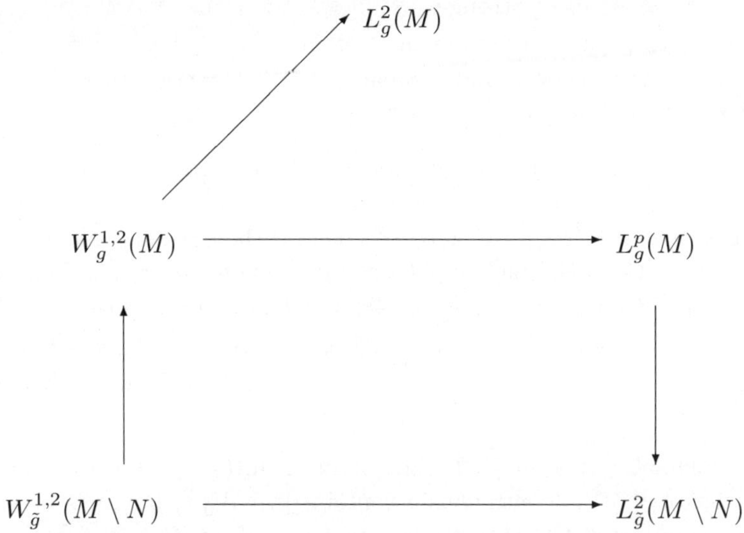

Figure 1. The compact inclusion of $W_{\tilde{g}}^{1,2}(M \setminus N)$ into $L_{\tilde{g}}^2(M \setminus N)$.

and then the Hölder inequality implies

$$\|u\|_{L_{\tilde{g}}^2} = \sqrt{\int u^2 f^{m-n} dA} \le c \cdot \|u\|_{L_g^p}$$

for

$$c = \left(\int f^{(m(m-n)/2)+\epsilon} dA \right)^{((m/2)+(\epsilon/(m-n)))^{-1}/2} < \infty .$$

So we only need to show that $W_{\tilde{g}}^{1,2}(M \setminus N)$ is continuously contained in $W_g^{1,2}(M)$ to conclude $W_{\tilde{g}}^{1,2}(M \setminus N)$ is compactly contained in $L_{\tilde{g}}^2(M \setminus N)$. When $m \ge 3$, this is clear, since $\inf(f) > 0$. When $m = 2$, then $n = 0$, and g and $\tilde{g}$ are conformally equivalent on $M \setminus N$. Suppose by way of contradiction that the inclusion is not continuous, that is, that there exists a sequence of functions u_k such that $\|u_k\|_{W_g^{1,2}} = 1$ and $\|u_k\|_{W_{\tilde{g}}^{1,2}} < 1/k$. By choosing a subsequence if necessary, we may assume the following:

(1) there exists a function u such that $u_k \to u$, $W_g^{1,2}$-weakly,
(2) there exists a function v such that $u_k \to v$, L_g^p-strongly,
(3) $u_k \to v$, $L_{\tilde{g}}^2$-strongly,

(4) $u_k \to v$, L^2_g-strongly.

The fourth item follows from the fact that $\|u_k - v\|_{L^2_g} \leq \hat{c} \cdot \|u_k - v\|_{L^p_g}$, since (M, g) is smooth and compact. As u_k converges to both u and v L^2_g-weakly, $u = v$. Also,

$$1 = \liminf_{k\to\infty} \|u_k\|_{W^{1,2}_g} \geq \|u\|_{W^{1,2}_g} .$$

Let ∇ and dA (resp. $\tilde{\nabla}$ and $d\tilde{A}$) denote the gradient and area-form with respect to the metric g (resp. $\tilde{g}$). Then, using $\int_M |\nabla u_k|^2_g dA = \int_M |\tilde{\nabla} u_k|^2_{\tilde{g}} d\tilde{A}$, we have $\int_M u_k^2 dA \to 1$ and $\int_M u^2 dA = 1$ and $\int_M |\nabla u|^2_g dA = 0$, so u is a nonzero constant. Also, $\int_M u_k^2 d\tilde{A} \to \int_M u^2 d\tilde{A} = 0$, so $\int_M d\tilde{A} = 0$. This is a contradiction, since f is smooth on $M \setminus N$ and not identically zero. □

Remark. For $m \geq 3$, the condition $\inf(f) > 0$ is a simple way to ensure $W^{1,2}_{\tilde{g}}$ is continuously contained in $W^{1,2}_g$, but it is necessary. This is not generally a continuous inclusion if $\inf(f) = 0$. For example, suppose $\inf(f) = 0$, and $n = 0$. Let $M_k = \{p \in M \setminus N \mid |f(p)| < 1/k\} \neq \emptyset$. Choose u_k so that $\mathrm{supp}(u_k) \subset M_k$ and $\|u_k\|^2_{W^{1,2}_g} = 1$. Then $\tilde{g} = f^2 g$ and g are conformally equivalent and

$$\|u_k\|^2_{W^{1,2}_{\tilde{g}}} = \int_{M_k} (u_k)^2 f^m dA + \int_{M_k} |\nabla u_k|^2_g f^{m-2} dA \leq \frac{1}{k^{m-2}} \|u_k\|^2_{W^{1,2}_g}$$
$$= \frac{1}{k^{m-2}} \to 0$$

as $k \to \infty$. Hence, we do not have continuous inclusion.

Let $\overline{\triangle}^F_{\tilde{g}}$ denote the Freidrichs' self-adjoint extension of the Laplacian with domain $C^\infty_0(M \setminus N)$, and let $W^{1,2}_{0,\tilde{g}}(M \setminus N)$ be the closure of $C^\infty_0(M \setminus N)$ in the $W^{1,2}_{\tilde{g}}(M \setminus N)$ norm. Standard arguments give the following:

Corollary 1.1. *Let $(M \setminus N, \tilde{g})$ be as in Theorem* 1.1. *The operator $\overline{\triangle}^F_{\tilde{g}}$ on $(M \setminus N, \tilde{g})$ has discrete spectrum consisting of eigenvalues $0 = \lambda_1 < \lambda_2 \leq \cdots \leq \lambda_j \leq \cdots \to +\infty$, each with multiplicity* 1. *The corresponding eigenfunctions $\phi_1, \phi_2, \ldots \in W^{1,2}_{\tilde{g}}(M \setminus N)$ can be chosen as an orthonormal basis for $L^2_{\tilde{g}}(M \setminus N)$. Furthermore, the variational characterization for the eigenvalues holds*:

$$\lambda_j = \inf_{V^j} \sup_{\phi \in V^j, \phi \neq 0} \frac{\|\nabla \phi\|^2_{L^2_{\tilde{g}}(M \setminus N)}}{\|\phi\|^2_{L^2_{\tilde{g}}(M \setminus N)}} ,$$

where V^j represents an arbitrary j-dimensional subspace of $W^{1,2}_{0,\tilde{g}}(M \setminus N)$.

Remark. When $W^{1,2}_{0,\tilde{g}}(M \setminus N) = W^{1,2}_{\tilde{g}}(M \setminus N)$, $(M \setminus N, \tilde{g})$ has negligible boundary, in Gaffney's sense [G]. Therefore, the Laplacian considered on Gaffney's domain of functions is essentially self-adjoint. Let $\overline{\triangle}^G_{\tilde{g}}$ denote the unique self-adjoint extension. One can show that the two self-adjoint operators $\overline{\triangle}^F_{\tilde{g}}$ and $\overline{\triangle}^G_{\tilde{g}}$ have equal domains, that is,

$$\triangle := \overline{\triangle}^G_{\tilde{g}} = \overline{\triangle}^F_{\tilde{g}} .$$

Since $(M \setminus N, \tilde{g})$ has negligible boundary, this operator's domain has no boundary conditions at N. So when $W^{1,2}_{0,\tilde{g}}(M \setminus N) = W^{1,2}_{\tilde{g}}(M \setminus N)$, this is the operator for which we will study the spectrum, and it is the same operator as that used in the study of Morse index of minimal surfaces in $\mathbb{R}^3$ and mean curvature 1 surfaces in $\mathbb{H}^3$.

As seen in the above remark, we would like to consider the cases where $W^{1,2}_{0,\tilde{g}}(M \setminus N) = W^{1,2}_{\tilde{g}}(M \setminus N)$. We will also need this property for establishing Weyl's asymptotic formula, so we now give a sufficient condition to imply this property [M2]. In order to state it, here we introduce the notion of capacity and Cauchy boundary [M1],[M2].

Definition 1.1. Let M be an arbitrary Riemannian manifold. We denote by $\mathcal{O}$, the family of all open subsets of the completion $\overline{M}$ of M. For $A \in \mathcal{O}$, we define the set of functions L_A by

$$L_A = \{f \in W^{1,2}(M) \,|\, f \geq 1 \text{ a.e. on } A\}.$$

We define the capacity of A, $\mathrm{Cap}(A)$, by

$$\mathrm{Cap}(A) = \begin{cases} \inf_{f \in L_A} \|f\|_{W^{1,2}}, & L_A \neq \phi, \\ \infty, & L_A = \phi. \end{cases}$$

For a Borel set $B \subset \overline{M}$, we define the capacity $\mathrm{Cap}(B)$ by

$$\mathrm{Cap}(B) = \inf_{A \in \mathcal{O},\, B \subset A} \mathrm{Cap}(A).$$

We say that a subset B of $\overline{M}$ is *almost polar* if Cap(B) = 0.

Definition 1.2. The Cauchy boundary ∂M of M is defined by

$$\partial M := \overline{M} \setminus M,$$

where $\overline{M}$ is the completion of M with respect to the Riemannian distance.

Lemma 1.1 ([M2]). *For an arbitrary Riemannian manifold M, let ∂M denote the Cauchy boundary of M. If the capacity of ∂M is finite, then the two Sobolev spaces $W_0^{1,2}(M)$ and $W^{1,2}(M)$ coincide if and only if ∂M is an almost polar set.*

In the case of Theorem 1.1, the Cauchy boundary of $M \setminus N$ is N. It is shown in [M1] that when the lower Minkowski codimension of the Cauchy boundary is not less than 2, then ∂M is almost polar, where the lower Minkowski codimension is defined as follows:

Definition 1.3. The lower Minkowski codimension of ∂M is defined to be

$$\underline{\mathrm{codim}}_{\mathcal{M}}(\partial M) := \liminf_{R \to 0} \frac{\log(\mathrm{vol}(\mathcal{N}_R))}{\log(R)} ,$$

where $\mathcal{N}_R$ is a radius R tubular neighborhood of ∂M.

We now consider some examples.

Example 1.1. Consider the "football". Set $M = \mathbb{C} \cup \{\infty\}$ and $N = \{0, \infty\}$ ($m = 2$ and $n = 0$) and set

$$g = \frac{4(dx^2 + dy^2)}{(1 + r^2)^2} , \quad f = \frac{\mu r^{\mu - 1}(1 + r^2)}{1 + r^{2\mu}} , \quad \mu \in \mathbb{R}^+ , \quad \tilde{g} = f^2 g ,$$

where $r = \sqrt{x^2 + y^2}$. Note that $f \in L_g^{2+\epsilon}(M \setminus N)$ for some $\epsilon > 0$, and $\underline{\mathrm{codim}}_{\mathcal{M}}(N) = 2$ for any μ. When $\mu < 1$, the football is an Alexandrov space and $\triangle$ has discrete spectrum, by [KMS] or by Theorem 1.1 above. When $\mu > 1$, the football is not an Alexandrov space, but the spectrum is still discrete, by Theorem 1.1 (see also Lemma 4.3 of [LR]).

Example 1.2. Consider a compact m-dimensional manifold M with metric g, and suppose M contains the unit ball B^m so that g is the standard Euclidean metric on $B^m \subset M$. Let $N = \{\vec{o}\}$ be the center point of $B^m \subset M$. Let $f = r^\ell$ on B^m and $\tilde{g} = f^2 g$ with $\ell \in (-2/m, 0)$, and extend f to be positive and smooth on $M \setminus B^m$. Thus $(M \setminus N, \tilde{g})$ is not complete, and $\underline{\mathrm{codim}}_{\mathcal{M}}(N) = m$ for any ℓ. Also, as f satisfies the conditions of Theorem 1.1, $\triangle$ on $(M \setminus N, \tilde{g})$ has discrete spectrum.

Example 1.3. As it is known that Alexandrov spaces have discrete spectrum [KMS], we are interested in finding examples that are not Alexandrov spaces and for which Theorem 1.1 can be applied. The footballs with $\mu > 1$ provide such examples in two dimensions. The following example shows that one can easily find such examples in higher dimensions as well. (We choose a slightly complicated function f in order to easily verify it will not be an Alexandrov space.)

Consider the previous example with $m = 3$; that is, M is compact, 3-dimensional, $N = \{\vec{o}\} \subset B^3 \subset M$, and g is the Euclidean metric on B^3. Set $f = \cos^2(\phi)r^\ell + \sin^2(\phi)(1 + r^{3/2}) \in L_g^{(9/2)+\epsilon}(M)$ on $B^3 \subset M$, where (r, θ, ϕ) are the spherical coordinates of B^3, and $\ell \in (-2/3, 0)$. Extend f to be positive and smooth on $M \setminus B^3$, and let $\tilde{g} = f^2 g$. Then $(M \setminus N, \tilde{g})$ is not complete, and the conditions of Theorem 1.1 are satisfied. Hence the spectrum of $\overline{\triangle}_{\tilde{g}}^F$ is discrete. The ball B^3 is invariant under the isometry $(r, \theta, \phi) \to (r, \theta, \pi - \phi)$, thus the sectional curvature in the $\{\phi = \pi/2\}$-plane is $K_{\tilde{g}} = -(\triangle \ln(f))/f^2 \to -\infty$ near N, so it is not an Alexandrov space.

Example 1.4. Consider the 3-dimensional torus $M = T^3 = \mathbb{R}^3/\mathbb{Z}^3$ with the standard Euclidean metric g, and the 1-dimensional torus $N = S^1 = (\mathbb{R}/\mathbb{Z}, 0, 0) \subset M$. We will use cylindrical coordinates (x, r, θ), where r is the radial distance to N and x is the arc-length along N. Let $f = \cos^2(\theta) + \sin^2(\theta)r^\ell$ near N with $\ell \in (-2/3, 0)$, and extend f to be positive and smooth away from N. This manifold is incomplete, and $\overline{\triangle}_{\tilde{g}}^F$ has discrete spectrum, by Theorem 1.1 and Corollary 1.1.

Remark. Suppose M is 2-dimensional and contains B^2 so that g is the standard Euclidean metric when restricted to B^2. Suppose $N = \{\vec{o}\} \subset B^2 \subset M$ and $f = -1/(r\ln(r))$ near N. Then, with respect to $\tilde{g} = f^2 g$, we have a complete end at N that is a curvature -1 psuedosphere of finite area, so the spectrum is not discrete [D], [Mu]. Since $f \in L^2(M, g)$, but $f \notin L^{2+\epsilon}(M, g)$ for all positive ϵ, we know Theorem 1.1 is sharp when $m = 2$. (If we had chosen $f = 1/r \in L^{2-\epsilon}(M, g)$ for all small positive ϵ instead, we would have produced a round cylindrical end of radius 1 which does not have discrete spectrum and does not have finite area.)

Remark. Consider $M = T^2 \times T^{m-2}$ and $N = T^{m-2}$ and $f = -1/(r\ln(r))$ near N, where r is radial distance to N. Let the diagonalized coordinates near N be $(x_1, x_2, y_1, \ldots, y_{m-2})$, inherited from the standard rectangular Euclidean coordinates of $\mathbb{R}^m$. Then $(M \setminus N, \tilde{g})$ is complete, and the sectional curvatures are

$$K_{\tilde{g}}(\partial_{x_1}, \partial_{x_2}) = -1\ , \quad K_{\tilde{g}}(\partial_{x_i}, \partial_{y_j}) = 0\ , \quad K_{\tilde{g}}(\partial_{y_i}, \partial_{y_j}) = 0\ .$$

So the Ricci curvature is bounded below, and hence the essential spectrum is not empty [D, Theorem 3.1]. So Theorem 1.1 is not true for this $f \in L^p$, $p \le 2$. Hence, for all m, the restriction on f in Theorem 1.1 cannot be weakened to $f \in L^p$ for some $p \le 2$.

Remark. Donnelly and Li [DL] have found complete examples $(M \setminus N, \tilde{g})$ where $M = \mathbb{R}^m \cup \{\infty\}$ and $N = \infty$ $(n = 0)$ and $\tilde{g}$ is rotationally invariant, so that sectional curvature converges to $-\infty$ at N

and $(M \setminus N, \tilde{g})$ has pure point spectrum. For example, let $m = 2$ and $\tilde{g} = dr^2 + \exp(-r^k)d\theta^2$, $k > 1$, in radial coordinates (r, θ) of $\mathbb{R}^2$. It is complete and its single end is conformally a punctured disk, and since the curvature converges to $-\infty$ at the end, it has pure point spectrum [DL]. Theorem 1.1 does not apply to such examples.

§2. Weyl's formula

In this section, let M be an m-dimensional Riemannian manifold with finite volume and finite diameter. M can be noncompact and incomplete.

Remark. The manifolds $(M \setminus N, \tilde{g})$ in Theorem 1.1 have finite volume, since $f \in L_g^{(m(m-n)/2)+\epsilon} \subseteq L_g^{m-n}$ implies $\mathrm{vol}(M \setminus N, \tilde{g}) = \int_M d\tilde{A} = \int_M f^{m-n} dA < \infty$.

Before stating and proving Weyl's asymptotic formula, we establish some notation. Let $\mathcal{N}_R$ be a radius R tubular neighborhood of the Cauchy boundary ∂M of M. Note that $\mathrm{vol}(M \setminus \mathcal{N}_R) + \mathrm{vol}(\mathcal{N}_R) = \mathrm{vol}(M)$ and $\mathrm{vol}(\mathcal{N}_R) \to 0$ as $R \to 0$. Define the Neumann isoparimetric constant of $\mathcal{N}_R$ by

$$C_R := \inf_{\gamma} \frac{\mathrm{vol}(\gamma)}{\min\{\mathrm{vol}(M_1), \mathrm{vol}(M_2)\}^{(m-1)/m}} \, ,$$

where the infimum is taken over all hypersurfaces γ of $\mathcal{N}_R$ which divide $\mathcal{N}_R$ into two parts M_1 and M_2, and where $\mathrm{vol}(\gamma)$ represents the $m-1$ dimensional volume of γ and $\mathrm{vol}(M_j)$ represents the m-dimensional volume of M_j.

Here, we will assume that

$$C := \inf_{R>0} C_R > 0 \, .$$

Then, since M has finite volume, one can see that M, N_R and $M \setminus N_R$ all have pure point spectra. Let $\lambda_j^{1,N}$ (resp. $\lambda_j^{2,N}$) be the Neumann eigenvalues on $\mathrm{Int}(M \setminus \mathcal{N}_R)$ (resp. $\mathrm{Int}(\mathcal{N}_R)$) counted with their multiplicities (i.e. listed in nondecreasing order, and the number of times that any eigenvalue appears in the list equals its multiplicity). Let $\lambda_j^{3,N}$ be the Neumann eigenvalues of $\mathrm{Int}(M \setminus \mathcal{N}_R) \cup \mathrm{Int}(\mathcal{N}_R)$ counted with their multiplicities. Let λ_j^D be the Dirichlet eigenvalues on $\mathrm{Int}(M \setminus \mathcal{N}_R)$ counted with their multiplicities. Here, we state Weyl's asymptotic formula for M.

Theorem 2.1. *Let M be an m-dimensional Riemannian manifold with finite volume and finite diameter. If the Cauchy boundary of M is* *an almost polar set and $C > 0$, then the eigenvalues λ_j of the Laplacian $\triangle$ satisfy Weyl's asymptotic formula*

$$\lim_{j\to\infty} \frac{\lambda_j^{m/2}\mathrm{vol}(M)}{j} = \frac{(2\pi)^m}{\mathrm{vol}(B^m)} .$$

Proof. Let $W = (2\pi)^m/\mathrm{vol}(B^m)$. Note that $\lambda_j \le \lambda_j^D$ by Dirichlet-Neumann bracketing techniques (see, for example, volume 4 of [RS]). Note also that, since M has finite diameter and therefore $M \setminus \mathcal{N}_R$ is relatively compact, the λ_j^D satisfy Weyl's asymptotic formula on $M\setminus\mathcal{N}_R$. So

$$\lambda_j \le \lambda_j^D \approx W\mathrm{vol}(M \setminus \mathcal{N}_R)^{-2/m} j^{2/m} \to W\mathrm{vol}(M)^{-2/m} j^{2/m}$$

for large j, as $R \to 0$. This implies

$$\limsup_{j\to\infty} \frac{\lambda_j^{m/2}\mathrm{vol}(M)}{j} \le W .$$

Consider the Neumann heat kernel

$$H_R(x,y,t) = \sum_{i=1}^{\infty} e^{-\lambda_i^{2,N} t}\phi_i^{2,N}(x)\phi_i^{2,N}(y)$$

on $\mathcal{N}_R$, where $\{\phi_i^{2,N}\}_{i=1}^{\infty}$ is an orthonormal basis of eigenfunctions in $L^2(\mathcal{N}_R)$ associated to the eigenvalues $\lambda_i^{2,N}$. Using the method of [LT], we know that the Neumann heat kernel on $\mathcal{N}_R$ belongs to the Sobolev space $W^{1,2}(\mathcal{N}_R)$ and has the above form. As the isoperimetric constant C_R of $\mathcal{N}_R$ is positive and the coarea formula on $\mathcal{N}_R$ holds for nonnegative functions, the associated Neumann Sobolev constant of $\mathcal{N}_R$ is also positive. Additionally, we have $H_R(x,y,t)$ in the above form, so the methods in [CL] can be applied to show

$$\lambda_j^{2,N} \ge \alpha(m)C_R^2\left(\frac{j}{\mathrm{vol}(\mathcal{N}_R)}\right)^{2/m} \ge \alpha(m)C^2\left(\frac{j}{\mathrm{vol}(\mathcal{N}_R)}\right)^{2/m} ,$$

where $\alpha(m)$ is a positive constant depending only on m.

Note that the list $\{\lambda_j^{3,N}\}$ is equal to the disjoint union of the lists $\{\lambda_j^{1,N}\}$ and $\{\lambda_j^{2,N}\}$ rearranged in increasing order. Note also that $\lambda_j \ge \lambda_j^{3,N}$, by Dirichlet-Neumann bracketing. Since $\lambda_j^{2,N} \ge \alpha(m)C^2(j/\mathrm{vol}(\mathcal{N}_R))^{2/m}$ and $\lambda_j^{1,N} \approx W\mathrm{vol}(M\setminus\mathcal{N}_R)^{-2/m}j^{2/m}$ for large j, and since $\mathrm{vol}(\mathcal{N}_R) \to 0$ and $\mathrm{vol}(M \setminus \mathcal{N}_R) \to \mathrm{vol}(M)$ as $R \to 0$, we have

$$\liminf_{j\to\infty} \frac{\lambda_j^{m/2}\mathrm{vol}(M)}{j} \geq W .$$

□

Example 2.1. Examples 1.1 and 1.2 satisfy the conditions of Theorem 2.1, hence their eigenvalues satisfy Weyl's asymptotic formula.

Remark. Using the methods of [CL], one can additionally conclude that $\lambda_j^{m/2} \geq \alpha(m)\hat{C}^{m/2}j/\mathrm{vol}(M)$ for some positive constant $\hat{C}$ depending only on the lower bound of the Sobolev constants of $\mathcal{N}_R$ for all $R > 0$.

Remark. Because the "football" in Example 1.1 satisfies the conditions of Theorem 2.1, it is clear that all Met_1 surfaces also satisfy the conditions of Theorem 2.1. The authors hope to consider the more general case where the conical singularities form a fractal set, and hope that Theorem 2.1 can be applied to such cases. As an example of such a case, since Minkowski dimension and Hausdorff dimension coincide on self-similar fractals, the Cauchy boundary of $(S^3 \setminus \mathcal{C}, g_{S^3})$ is almost polar, where $\mathcal{C}$ is a Cantor set.

Remark. The results here bear some relation to the work [KS], in which Kuwae and Shioya have recently studied the convergence of the spectra of a sequence of Riemannian manifolds (they do not assume completeness of the manifolds). Some of the results in [KS] involve the almost polarity condition.

References

[CL] S. Y. Cheng and P. Li, *Heat kernel estimates and lower bound of eigenvalues*, Comment. Math. Helv., **56** (1981), 327–338.

[D] H. Donnelly, *On the essential spectrum of a complete Riemannian manifold*, Topology, **20** (1981), 1–14.

[DL] H. Donnelly and P. Li, *Pure point spectrum and negative curvature for noncompact manifolds*, Duke Math. J., **46(3)** (1979), 497–503.

[FC] D. Fischer-Colbrie, *On complete minimal surfaces with finite Morse index in three manifolds*, Invent. Math., **82** (1985), 121–132.

[G] M. Gaffney, *A special Stoke's theorem for complete Riemannian manifolds*, Ann. of Math., **60(1)** (1954), 140–145.

[KMS] K. Kuwae, Y. Machigashira and T. Shioya, *Sobolev spaces, Laplacian, and heat kernel on Alexandrov spaces*, Math. Z., to appear.

[KS] K. Kuwae and T. Shioya, *Convergence of spectral structures: a functional analytic theory and its applications to spectral geometry*, preprint.

[L] P. Li, *On the Sobolev constant and the p-spectrum of a compact Riemannian manifold*, Ann. Sci. École Norm. Sup., **13(4)** (1980), 451–469.

[LT] P. Li and G. Tian, *On the heat kernel of Bergmann metric on algebraic varieties*, J. Amer. Math. Soc., **8(4)** (1995), 857–877.

[LR] L. L. de Lima and W. Rossman, *On the Index of Mean Curvature* 1 *Surfaces in* $\mathbb{H}^3$, Indiana Univ. Math. J., **47(2)** (1998), 685–723.

[M1] J. Masamune, *Essential self adjointness of Laplacians on Riemannian manifolds with fractal boundary*, Comm. Partial Diff. Eqs., **24(3-4)** (1999), 749–757.

[M2] J. Masamune, *On Liouville property, conservativeness and self-adjointness of the Laplacian on manifolds with fractal boundary*, preprint.

[Mu] W. Müller, *Spectral theory for Riemannian manifolds with cusps and a related trace formula*, Math. Nachr., **111** (1983), 197–288.

[RS] M. Reed and B. Simon, Methods of Modern Mathematical Physics (vol. I-IV), Academic Press, New York-London, 1979.

[UY] M. Umehara and K. Yamada, *Complete surfaces of constant mean curvature one in the hyperbolic* 3*-space*, Ann. of Math., **137** (1993), 611–638.

Jun Masamune
Assegno biennale di ricerca (1999–2001)
preso l'universita' degli Studi della Basilicata
Dipartimento della Matematica
Macchia Romana, 85100 Potenza
Italy

Wayne Rossman
Department of Mathematics
Faculty of Science
Kobe University
Rokko, Kobe 657-8501
Japan
wayne@math.kobe-u.ac.jp
http://www.math.kobe-u.ac.jp/HOME/wayne/wayne.html

Advanced Studies in Pure Mathematics 34, 2002
Minimal Surfaces, Geometric Analysis and Symplectic Geometry
pp. 231–236

Brieskorn Manifolds and Metrics of Positive Scalar Curvature

Hiroshi Ohta*

Abstract.

Using relevant properties of the Seiberg-Witten monopole invariants, a non-existence theorem of metrics of positive scalar curvature on certain series of Brieskorn 3-manifolds is proved. This gives a short proof without the help of Gromov-Lawson's theorem.

§1. Introduction

In this note, we will study a non-existence theorem of Riemannian metrics of positive scalar curvature on certain series of Brieskorn 3-manifolds by using relevant properties of the Seiberg-Witten monopole invariants. By contrast, this theorem can be regarded as a characterization of Bireskorn 3-manifolds obtained by the so-called simple singularites of type A_n, D_n, E_6, E_7 and E_8.

Let p, q and r be integers grater than or equal to 2. Let $\Sigma(p,q,r)$ be a smooth oriented compact 3-dimensional manifold defined by the intersection of the complex algebraic surface $\{(z_1,z_2,z_3) \in \mathbf{C}^3 \mid z_1^p + z_2^q + z_3^r = 0\}$ with the unit sphere $\{(z_1,z_2,z_3) \in \mathbf{C}^3 \mid |z_1|^2 + |z_2|^2 + |z_3|^2 = 1\}$. We call $\Sigma(p,q,r)$ the Brieskorn 3-manifold of type (p,q,r). These manifolds have a non-trivial S^1-action. Roughly speaking, the family of Brieskorn 3-manifolds can be devided geometrically into three classes, according as the rational number $1/p + 1/q + 1/r - 1$ is positve, zero or negative. Milnor [Mi] showed that in the positive case, $\Sigma(p,q,r)$ can be described as a quotient space of S^3 by a certain finite subgroup Γ of $SU(2)$. It is well-known that finite subgroups of $SU(2)$ can be classified by the Dynkin diagrams of type A_n, D_n, E_6, E_7 and E_8. Of course, since Γ acts on S^3 isometrically, $\Sigma(p,q,r)$ actually carries a metric of positive

*Partly supported by the Grant in-Aid for Scientific Research (C) No. 10640070, the Ministry of Education, Science, Sports and Culture, Japan.

2000 *Mathematics Subject Classification.* Primary 57R57; Secondary 32Q05, 32S25, 57R17.

scalar curvature in these cases. In [O-O], in collaboration with Ono the present author studied this positive case from the point of view of a relation with singularity theory and contact/symplectic geometry. In this note, we consider the negative case. By making use of the same technique as that employed in the paper [O-O], we can prove the following theorem.

Theorem. *Let $\Sigma(p,q,r)$ be a Brieskorn rational homology 3-sphere. If $1/p+1/q+1/r<1$, then $\Sigma(p,q,r)$ carries no Riemannian metric of positive scalar curvature.*

Remark. (1) The assumption that $\Sigma(p,q,r)$ is a rational homology 3-sphere is not necessary. In the proof of Theorem, we will use a vanishing theorem on the Seiberg-Witten invariant. Although it is known as folklore that the vanishing theorem can be applied to the case when $\Sigma(p,q,r)$ is not a rational homology 3-sphere, there is no published proof of the vanishing theorem for general cases. We have a proof for the ratioanl homology 3-shpere case in [O-O]. If we admit the vanishing theorem for general cases, we might be able to prove Theorem without the assuming that $\Sigma(p,q,r)$ is a rational homology 3-sphere, by slightly modifing the argument below. It should be also remarked that when $\Sigma(p,q,r)$ is a rational homology 3-sphere, it is a total space of an S^1-bundle over S^2 with three singular fibres. Of course, a total space of an S^1-bundle over S^2 with no singular fibres admits a Riemannian metric of positive scalar curvature. Thus this theorem implies that singular fibres of the S^1-bundle is related to an obstruction for the total space to admit a metric of positive scalar curvature. The similar phenomenon can be seen in the 4-dimensional manifolds which have an elliptic fibration structure.

(2) A celebrated theorem deu to Gromov and Lawson [G-L] shows that any 3-dimensional $K(\pi,1)$-space does not admit a metric of positive scalar curvature. On the other hand, Milnor showed in [Mi] that in the case $1/p+1/q+1/r<1$, the universal covering space of the Brieskorn 3-manifold $\Sigma(p,q,r)$ is the universal cover of $SL(2,\mathbf{R})$. His proof is based on a study of automorphic forms of fractional degree. Combing these two deep theorems, we can show Theorem. In this sense, this theorem is not new. But we will give an independent short proof of Theorem without the help of Gromov-Lanson's theorem and Milnor's theorem. We use relevant properties of the Seiberg-Witten monopole invariants.

(3) For the case $1/p+1/q+1/r=1$, Milnor showed that $\Sigma(p,q,r)$ is a nil-manifold and a circle bundle over a torus. (The triple (p,q,r) must be either $(2,3,6)$, $(2,4,4)$ or $(3,3,3)$.) By a result of [G-L] again, we find that $\Sigma(p,q,r)$ carries no metric of positive scalar curvature in

this case. In this note we do not discuss this case from through the monopole invariants.

This work was first (and last) presented at a workshop at Hakone, Japan around 1995–96, as a part of author's talk entitled "Bakusho! Koneta Shu" ("Collection of short farces"). The author would like to thank K. Ono who was interested in this work and encouraged him to publish it.

§2. Proof of Theorem

We put the polynomial $f_{p,q,r}(z_1, z_2, z_3) = z_1^p + z_2^q + z_3^r$. Let $M(p,q,r)$ be the Milnor fibre defined by $\{(z_1, z_2, z_3) \in \mathbf{C}^3 \mid f_{p,q,r}(z_1, z_2, z_3) = \epsilon\}$, where ϵ is a non-zero number. By the argument similar to that in [O-O], we use the compactification of the Milnor fibre in a certain weighted projective space. (In this case we put $\epsilon = 1$.) To do this, we regard $f_{p,q,r}(z_1, z_2, z_3)$ as a weighted homogeneous polynomial. We denote by $l(p,q,r)$ the least common multiple of p, q and r, which gives the degree of the weighted homogenous polynomial. The weights are given by $l(p,q,r)/p$, $l(p,q,r)/q$ and $l(p,q,r)/r$, respectively. Let us consider a weighted projective space

$$\mathbf{P}(l(p,q,r)/p, l(p,q,r)/q, l(p,q,r)/r, 1) = (\mathbf{C}^4 \setminus 0)/\sim,$$

where $(z_1, z_2, z_3, z_4) \sim (t^{l(p,q,r)/p} z_1, t^{l(p,q,r)/q} z_2, t^{l(p,q,r)/r} z_3, t z_4)$ for $t \in \mathbf{C}^*$. By a map ι from $\mathbf{C}^3$ to $\mathbf{P}(l(p,q,r)/p, l(p,q,r)/q, l(p,q,r)/r, 1)$ defined by $\iota(z_1, z_2, z_3) = [z_1 : z_2 : z_3 : 1]$, $\mathbf{C}^3$ can be embedded as an open dense subset such that

$$\begin{aligned}&\mathbf{P}(l(p,q,r)/p, l(p,q,r)/q, l(p,q,r)/r, 1)\\ &\quad = \mathbf{C}^3 \cup \mathbf{P}(l(p,q,r)/p, l(p,q,r)/q, l(p,q,r)/r).\end{aligned}$$

Here

$$\begin{aligned}&\mathbf{P}(l(p,q,r)/p, l(p,q,r)/q, l(p,q,r)/r)\\ &\quad = \{[z_1 : z_2 : z_3 : 0] \in \mathbf{P}(l(p,q,r)/p, l(p,q,r)/q, l(p,q,r)/r, 1)\}.\end{aligned}$$

We denote by $\overline{M(p,q,r)}$ the closure of $\iota(M(p,q,r))$ in

$$\mathbf{P}(l(p,q,r)/p, l(p,q,r)/q, l(p,q,r)/r, 1).$$

It can be identified with a hypersurface $f_{p,q,r}(z_1, z_2, z_3) = z_4^{l(p,q,r)}$ in the weighted projective space $\mathbf{P}(l(p,q,r)/p, l(p,q,r)/q, l(p,q,r)/r, 1)$. The singular points on $\overline{M(p,q,r)}$ lie on the complement:

$$\begin{aligned}&\overline{M(p,q,r)} \setminus \iota(M(p,q,r))\\&\quad = \overline{M(p,q,r)} \cap \mathbf{P}(l(p,q,r)/p, l(p,q,r)/q, l(p,q,r)/r).\end{aligned}$$

Taking the minimal resolution $\widetilde{M(p,q,r)} \to \overline{M(p,q,r)}$, we have a compact smooth projective complex surface $\widetilde{M(p,q,r)}$. (See, for example, [S]). This contains the Milnor fibre $M(p,q,r)$ as an open dense subset.

We put $M(p,q,r)^{\text{out}} = \widetilde{M(p,q,r)} \setminus \iota(M(p,q,r) \cap B^6)$, where we denote by B^6 the unit open ball around the origin in $\mathbf{C}^3$. We denote by $b_2^+(M(p,q,r)^{\text{out}})$ the number of positive eigenvalues of the intersection form of $M(p,q,r)^{\text{out}}$, which is a non-degenerate quadratic form because the boundary of $M(p,q,r)^{\text{out}}$ is a rational homology 3-sphere. Then we can show the following.

Lemma. $b_2^+(M(p,q,r)^{\text{out}}) \geq 1$.

Proof. We put

$$B(p,q,r) := \{(z_1,z_2,z_3) \in \mathbf{C}^3 \mid f_{p,q,r}(z_1,z_2,z_3) = 0\} \cap \overline{B^6}.$$

We have an isolated normal singularity at the origin. Taking the minimal resolution $\widetilde{B(p,q,r)} \to B(p,q,r)$, we have a non-singular Kähler surface with boundary $\Sigma(p,q,r)$. The intersection form of $\widetilde{B(p,q,r)}$ is negative definite [Mu]. Note that we have a symplectically fillable contact structure ξ on the boundary $\Sigma(p,q,r)$, for which $\widetilde{B(p,q,r)}$ is a symplectically filling 4-manifold. On the other hand, we put

$$M(p,q,r)^{\text{in}} := M(p,q,r) \cap \overline{B^6}.$$

By the map ι, we can embed $M(p,q,r)^{\text{in}}$ into $\widetilde{M(p,q,r)}$ so that

$$\widetilde{M(p,q,r)} = M(p,q,r)^{\text{in}} \cup_{\Sigma(p,q,r)} M(p,q,r)^{\text{out}}.$$

Here we identify $\iota(M(p,q,r)^{\text{in}})$ with $M(p,q,r)^{\text{in}}$. Note that $M(p,q,r)^{\text{in}}$ is also a non-singular Kähler surface with boundary $\Sigma(p,q,r)$. We have a symplectically fillable contact structure ξ_ϵ on the boundary, for which $M(p,q,r)^{\text{in}}$ is a symplectically filling 4-manifold. The contact structure ξ is nothing but ξ_0, so we have a smooth family of contact structures on $\Sigma(p,q,r)$. Hence Gray's stability theorem [G] implies that ξ and ξ_ϵ with $\epsilon \neq 0$ is isotopic. So we have two symplectically filling 4-manifolds $\widetilde{B(p,q,r)}$ and $M(p,q,r)^{\text{in}}$ of $\Sigma(p,q,r)$ with the specific fillable contact structure. Then by the argument similar to Lemma 2.1 in [O-O], we can glue $\widetilde{B(p,q,r)}$ and $M(p,q,r)^{\text{out}}$ to obtain a closed symplectic 4-manifold

$$Z(p,q,r) = \widetilde{B(p,q,r)} \cup_{\Sigma(p,q,r)} M(p,q,r)^{\text{out}},$$

by replacing $M(p,q,r)^{\text{in}}$ in $\widetilde{M(p,q,r)}$ by $\widetilde{B(p,q,r)}$. Since we have a symplectic form on it, we have $b_2^+(Z(p,q,r)) \geq 1$. We recall that $b_2^+(\widetilde{B(p,q,r)}) = 0$. Therefore we have $b_2^+(M(p,q,r)^{\text{out}}) \geq 1$, which proves Lemma. □

Now, let us consider the compact non-singular Kähler surface with the following decomposition:

$$\widetilde{M(p,q,r)} = M(p,q,r)^{\text{in}} \cup_{\Sigma(p,q,r)} M(p,q,r)^{\text{out}}.$$

The signature of $M(p,q,r)^{\text{in}}$ is computed by E. Brieskorn [B]. According to his result, $b_2^+(M(p,q,r)^{\text{in}})$ can be calculated by counting the number of triples of integers (k,l,m) satisfying the follwing inequality:

$$0 < \frac{k}{p} + \frac{l}{q} + \frac{m}{r} < 1 \pmod 2.$$

Therefore, if $1/p+1/q+1/r < 1$, we have $b_2^+(M(p,q,r)^{\text{in}}) \geq 1$ obviously. (Note that if $1/p+1/q+1/r > 1$, we have $b_2^+(M(p,q,r)^{\text{in}}) = 0$). Thus, by taking Lemma into account, we have $b_2^+(\widetilde{M(p,q,r)}) \geq 2$. Hence, by Taubes' theorem [T], we find that the Siberg-Witten monopole invariant for the canonical Spin^c structure of $\widetilde{M(p,q,r)}$ is non-trivial. Now suppose that $\Sigma(p,q,r)$ carries a Riemannian metric of positive scalar curvature. Then we have the following vanishing theorem on the Seiberg-Witten monopole invariants.

Proposition. *Let X be a closed oriented connected smooth 4-manifold. Suppose that X has a decomposition $X = X_1 \cup_Y X_2$ such that $b_2^+(X_i) \geq 1$, $(i = 1,2)$ and Y is a closed oriented 3-manifold with a metric of positive scalar curvature. Then all the Seiberg-Witten invariants of X vanish.*

See [O-O] for the proof in the case when Y is a rational homology 3-sphere. Since we have $b_2^+(M(p,q,r)^{\text{in}}) \geq 1$ and $b_2^+(M(p,q,r)^{\text{out}}) \geq 1$ by Lemma, we can apply the proposition to the case $X = \widetilde{M(p,q,r)}$. Hence we have a contradiction to Taubes' theorem. Thus $\Sigma(p,q,r)$ carries no Riemannian metric of positive scalar curvature, whenever $1/p + 1/q + 1/r < 1$.

References

[B] E. Brieskorn, *Beispiele zur Differentialtopologie von Singularitäten*, Invent. Math., **2** (1966), 1–14.

[G] J. W. Gray, *Some global properties of contact structures*, Ann. of Math., **69** (1959), 421–450.

[G-L] M. Gromov and B. Lawson, *Positive scalar curvature and the Dirac operator on complete Riemannian manifolds*, Publ. Math. I. H. E. S., **58** (1983), 295–408.

[Mi] J. Milnor, *On the 3-dimensional Brieskorn manifolds* $M(p,q,r)$, Ann. of Math. Studies, **84**, 175–225, *"Knots, Groups and 3-Manifolds"*, Princeton Univ. Press, Princeton, 1975.

[Mu] D. Mumford, *The topology of normal singularities of an algebraic surface and a criterion for simplicity*, Publ. Math. I. H. E. S., **9** (1961), 229–246.

[O-O] H. Ohta and K. Ono, *Simple singularities and topology of symplectically filling 4-manifold*, Comment. Math. Helv., **74** (1999), 575–590.

[S] K. Saito, *A new relation among Cartan matrix and Coxeter matrix*, J. Algebra, **105** (1987), 149–158.

[T] C. H. Taubes, *The Seiberg-Witten invariants and symplectic forms*, Math. Res. Lett., **1** (1994), 809–822.

Graduate School of Mathematics
Nagoya University
Nagoya, Chikusa, 464-8602
Japan

Advanced Studies in Pure Mathematics 34, 2002
Minimal Surfaces, Geometric Analysis and Symplectic Geometry
pp. 237–243

Space of Geodesics on Zoll Three-Spheres

Kaoru Ono

Abstract.

The space of geodesics on a Zoll manifold, i.e., a Riemannian manifold all of whose geodesics are closed with the same minimal period, carries a natural symplectic structure. In this note, it is shown that the space of geodesics on a Zoll three-sphere is symplectomorphic to the product of two copies of two-spheres with the same area.

§1. Introduction

The geodesic flow of a Riemannian manifold (M, g) is a Hamiltonian system on its cotangent bundle T^*M. More precisely, we identify the tangent bundle TM and the cotangent bundle T^*M via the Riemannian metric and the Hamiltonian function for the geodesic flow is given by half of the square of the fiber norm on T^*M. Note that the unit cotangent bundle U^*M is an energy hypersurface, i.e., a level set of the Hamiltonian function. The trajectories of the geodesic flow $\{\Psi_t\}$ on the unit tangent bundle of (M, g) are characteristic curves on U^*M with respect to the standard symplectic structure on T^*M.

The space of geodesics, $\mathrm{Geod}(M, g)$, is the quotient space of U^*M by the characteristic flow, which may not be a nice space, in general. If all the geodesics are closed with the same minimal period, the quotient space becomes a manifold. We call a metric g enjoying this property a Zoll metric. In such a case, the space of geodesics is considered as the symplectic reduction of the energy hypersurface U^*M by the circle action and carries the natural symplectic structure. The cohomology class of the symplectic form is the negative of the first Chern class of the S^1-bundle $U^*M \to \mathrm{Geod}(M, g)$. A special feature of the symplectic

2000 *Mathematical Subject Classification.* Primary 53D05, Secondary 53C22.

Partly supported by the Grant-in-Aid for Scientific Research, Nos. 09640095 and 11440015, the Ministry of Education, Science, Sports and Culture, Japan.

manifold Geod(M, g) is that it contains Lagrangian spheres which sweep up the whole space. Namely, the set L_p of geodesics passing through a point $p \in M$ is the image of a fiber of U^*M at p by the projection $U^*M \to$ Geod(M, g), and hence a Lagrangian sphere. It is obvious that Geod(M, g) is swept by these Lagrangian spheres.

It is known that the space of geodesics on an n-dimensional Zoll sphere is simply connected and has the same cohomology ring as that of the complex hyperquadric of complex dimension $n-1$ (see [Y]). In the case of low dimensional spheres, Hajime Sato proved that the Chern classes of the space of geodesics of a Zoll 3-sphere has the same form as in the case of the standard 3-sphere in their cohomology rings, which are isomorphic, and the space of geodesics of a Zoll 4-sphere is diffeomorphic to the complex hyperquadric [S].

In this short note, we shall prove the following

Theorem. *The space of geodesics on any Zoll 3-sphere is symplectomorphic to the product of two copies of symplectic 2-spheres with the same area.*

Remark. All known examples of Zoll metrics, up to now, are deformations of the standard metric through Zoll metrics. (see [B], [K] for examples of Zoll metrics.) In such a case, the result of our Theorem follows from Moser's stability theorem [M]. Note also that Theorem implies the existence of homogeneous symplectomorphism between the cotangent bundles with the zero section removed of any Zoll three-sphere and the round three-sphere, which intertwines their geodesic flows.

§2. Preliminaries

In this section, we collect several results, which are necessary for our argument.

A symplectic manifold (X, ω) is called monotone[1], if the symplectic class $[\omega]$ and the first Chern class $c_1(X)$ are positively proportional:

$$c_1(X) = \lambda[\omega] \text{ for some } \lambda > 0.$$

For 4-dimensional monotone symplectic manifolds, we have the following

Theorem 2.1 (cf. [O-O]). *A monotone symplectic 4-manifold (X, ω) is diffeomorphic to one of Del-Pezzo surfaces. If X is minimal,*

---

[1]Monotonicity is usually defined to be the positive proportionality of $[\omega]$ and $c_1(M)$, as homomorphisms from $\pi_2(M)$ to $\mathbf{R}$. Our assumption here is stronger than this "spherical monotonicity", in case that M is not simply connected.

i.e., X does not contain (-1) symplectic sphere, then it must be symplectomorphic to either $\mathbf{C}P^2$ with a multiple of the Fubini-Study form or the product of two symplectic 2-spheres with the same area.

The result is obtained by combining a consequence of Taubes' theorem "SW=GW" and McDuff's theorem [MD] concerning rational or ruled symplectic 4-manifolds. In [O-O], we used Taubes' theorem and found a rational curve with non-negative intersection number. Then McDuff's theorem implies that it must be a rational or ruled surface, and we get classification up to diffeomorphism. The statement about symplectic structure follows from the uniqueness result [MD] (see also [L-MD]).

Remark 2.2. As we mentioned, the space of geodesics on a Zoll 3-sphere is a cohomology $S^2 \times S^2$. So it is sufficient to show that $\mathrm{Geod}(M, g)$ is a monotone symplectic manifold.

Remark 2.3. It is possible to prove our theorem by combining Theorem 2.1 with Sato's result mentioned above, which implies the monotonicity. Here we shall give another method of computing the Chern class.

The unit cotangent bundle U^*M carries a contact structure $\xi \subset TU^*M$, which inherits the structure of symplectic vector bundle from the canonical symplectic form on T^*M. The tangent bundle along fibers of $U^*M \to M$ is a Lagrangian subbundle, which we denote by Ver. Associated to a closed geodesic $\gamma : [0, L] \to M$, we consider the loop $\mathcal{L}_\gamma$ of Lagrangian subspaces in ξ:

$$\mathcal{L}_{\gamma(t)} = d\Psi_t(\mathrm{Ver}_{\gamma(0)}).$$

We denote by $\mu(\gamma)$ the Maslov index of the loop $\mathcal{L}_\gamma$ with respect to the vertical Lagrangian distribution Ver. The relation between the Maslov index $\mu(\gamma)$ and the Morse index of the geodesic γ is established in [D]. Here we recall the following fact.

Lemma 2.4 (cf. [B, Theorm 7.23]). *$\mu(\gamma)$ must be 4. (the Morse index $= 2$, the nullity $= 2$)*

§3. Proof of Theorem

Let g_0 be the standard metric on $M = S^3$ and g a metric, all of whose geodesics are closed with the common minimal period. By multiplying a suitable real number to g, we may assume that the unit cotangent ball bundle $B^{g_0}M$ with respect to g_0 is strictly contained in the unit cotangent ball bundle $B^g M$ with respect to g. By the assumption that characteristics on unit cotangent bundles $U^*_{g_0}M$ and $U^*_g M$ are periodic (with different periods), we can apply symplectic cutting construction [L] along boundaries to the compact symplectic manifold with boundary $B^g M \setminus Int(B^{g_0}M)$ and get a closed symplectic 6-manifold (Z, ω). Namely, we consider the orbit spaces of boundaries by the circle actions and replace the boundaries of the compact manifold by the orbit spaces. It turns out that the space obtained is a smooth manifold, if the circle actions are free, and carries a natural symplectic structure. This is the symplectic cutting construction. It is clear that $\mathrm{Geod}(M, g_0)$ and $\mathrm{Geod}(M, g)$ are symplectically embedded in Z.

For the standard metric g_0, the space of geodesics $\mathrm{Geod}(M, g_0)$ is diffeomorphic to the complex hyperquadric of complex dimension 2, i.e., the product of two copies of S^2. We denote by α_1 and β_1 the Poincaré duals of $S^2 \times \mathrm{pt}$ and $\mathrm{pt} \times S^2$, respectively.

Since the cohomology ring of $\mathrm{Geod}(M, g)$ is also isomorphic to that of $S^2 \times S^2$, we denote by α_2 and β_2 the corresponding generators of the second cohomology group so that $\alpha_2 \cup \beta_2[\mathrm{Geod}(M, g)] = 1$ and $\alpha_2^2[\mathrm{Geod}(M, g)] = \beta_2^2[\mathrm{Geod}(M, g)] = 0$.

Lemma 3.1. *The first Chern class of* $\mathrm{Geod}(M, g)$ *is* $\pm 2(\alpha_2 + \beta_2)$.

Proof. Note that the signature of $\mathrm{Geod}(M, g)$ is zero. Hence the first Pontrjagin class must be zero. Since $p_1 = c_1^2 - 2c_2$ and c_2 is identified with the Euler number, $c_1^2[\mathrm{Geod}(M, g)] = 8$. For a simply connected 4-manifold, the fact that the intersection form is of even type implies the vanishing of the second Stiefel-Whitney class w_2. Combining the fact that c_1 modulo 2 equals w_2, we have $c_1(M) = \pm 2(\alpha_2 + \beta_2)$. □

Remark 3.2. The symplectic manifold, which has the same cohomology ring as $S^2 \times S^2$, contains an embedded Lagrangian sphere, only if the symplectic class is proportional to $\alpha_2 + \beta_2$. The reason is the following. The self intersection number of a Lagrangian two-sphere must be -2 and its homology class must be $\pm(\alpha_2 - \beta_2)$. Then the Lagrangian condition implies that the integration of the symplectic form over α_2 and β_2 are the same, which implies the proportionality, although it is not neccessarily positive proportionality.

We may change, if necessary, the orientation on both factors of $S^2 \times S^2$ simultaneously and assume that the symplectic class is a positive multiple of $\alpha_2 + \beta_2$. Then, once we know that $c_1(\mathrm{Geod}(M, g)) = 2(\alpha_2 + \beta_2)$, our theorem follows from Theorem 2.1. The rest of the argument is devoted to computing the first Chern class.

Let γ_1 and γ_2 be orbits of the geodesic flow on $U^*_{g_0}M$ and U^*_gM, respectively. We can regard them as transversal loops in the projective cotangent bundle (PT^*M, ξ), by identifying U^*_gM and $U^*_{g_0}M$ with projective cotangent bundle using the projection

$$T^*M \setminus O_M \to PT^*M = (T^*M \setminus O_M)/\mathbf{R}^+.$$

Since PT^*M is diffeomorphic to $S^3 \times S^2$, it is a 5-dimensional simply connected manifold. Then we can find an isotopy of smooth embeddings between loops γ_1 and γ_2. By a C^0-small perturbation of the isotopy, we get an isotopy through transversally embedded loops. Hence we obtain the following

Lemma 3.3. *Two transversal loops γ_1 and γ_2 above are transversally isotopic.*

Using this lemma, we can construct an symplectically embedded 2-sphere as follows. Let $\{\gamma_t\}$ denote the transversal isotopy between γ_1 and γ_2. We assume that it is independent of t near $t = 1, 2$, respectively. Take a family of metrics $g_t = (2-t)g_0 + (t-1)g$. Using the identification between $U^{g_t}M$ and PT^*M, we consider that γ_t is a loop in $U^{g_t}M$. Note that the unit cotangent bundles with respect to g_t are mutually disjoint. We may rescale the metric the metric small real number so that the unit cotangent ball bundle $B^{g_0}M$ is sufficiently small and the t-direction is stretched out and the map

$$F : S^1 \times [1, 2] \to B^gM \setminus \mathrm{Int}(B^{g_0}M)$$

given by $(s, t) \mapsto \gamma_t(s)$ yields a symplectically embedded cylinder. It is easy to see that after symplectic cutting construction, we get an symplectically embedded sphere C in Z, which intersects $\mathrm{Geod}(M, g_0)$ and $\mathrm{Geod}(M, g)$ with intersection index 1. We denote by N and S the points in C corresponding to γ_1 and γ_2, respectively.

By the Mayer-Vietris exact sequence, we have

Lemma 3.4. *The second homology group $H_2(Z; \mathbf{Z})$ is generated by $H_2(\mathrm{Geod}(M, g_0; \mathbf{Z}))$ and C.*

We identify $\mathrm{Geod}(M, g_0)$ with $S^2 \times S^2$ and denote by A and B the generators $[S^2 \times \mathrm{pt}]$ and $[\mathrm{pt} \times S^2]$ of $H_2(S^2 \times S^2; \mathbf{Z})$.

Then the image of $H_2(\mathrm{Geod}(M, g); \mathbf{Z})$ in $H_2(Z; \mathbf{Z})$ is the submodule spanned by $A+C$ and $B+C$. Note that $\omega\cdot(A+C) = \omega\cdot(B+C) > 0$. Since Lemma 3.1 implies that $c_1(\mathrm{Geod}(M, g))[A+C] = c_1(\mathrm{Geod}(M, g))[B+C]$, it suffices to prove that $c_1(\mathrm{Geod}(M, g))[A + C] > 0$. $\mathrm{Geod}(M, g)$ is a symplectic submanifold, so that an almost complex submanifold with respect to a compatible almost complex structure and we have

$$c_1(\mathrm{Geod}(M,g)) = c_1(TZ|_{\mathrm{Geod}(M,g)}) - PD([\mathrm{Geod}(M,g)]),$$

where $PD([\mathrm{Geod}(M,g)])$ is the Poincaré dual of the fundamental class of $\mathrm{Geod}(M,g)$. Hence we have $c_1(\mathrm{Geod}(M,g))[A+C] = c_1(Z)[A+C]-1$ and $c_1(Z)[A+C] = c_1(Z)[A] + c_1(Z)[C]$. From the explicit description of a neighborhood of $\mathrm{Geod}(M,g_0)$, we have $c_1(Z)[A] = 1$.

We have the following

Lemma 3.5. $c_1(Z)[C] = 2$.

Proof. Since C is a symplectically embedded sphere, we may assume that TC is a complex subbundle of a complex vector bundle $TZ|_C$. We denote by N_C its normal bundle. Then we have

$$c_1(Z)[C] = c_1(TC)[C] + c_1(N_C)[C] = 2 + c_1(N_C)[C].$$

Note that N_C restricted to $C\backslash\{N,S\}$ is identified as the restriction of ξ and contains a Lagrangian subbundle given by the vertical distribution. Around N and S, they may have non-trivial Maslov index and may not extend over C. But by Lemma 2.4, these Maslov indices are the same. Since the bundle N_C is the quotient of $\xi|_{F(S^1\times[1,2]}$ by the circle actions along boundaries, i.e., the differential of the geodesic flows $d\Psi_t$ with respect to the Zoll metrics g_0 and g, this implies that the symplectic vector bundle N_C is trivial and has vanishing first Chern class. Therefore we have $c_1(Z)[C] = 2$. □

Combining our computation, we have

$$c_1(\mathrm{Geod}(M,g))[A+C] = c_1(Z)[A] + c_1(Z)[C] - 1 = 1 + 2 - 1 = 2 > 0,$$

which completes the proof of the Theorem.

Acknowledgments. This work is an outcome of the discussion with Juan Carlos Alvarez Paiva duirng my stay at IHES in summer 1998. I would like to thank him and the institute. Thanks are also due to the referee for comments.

References

[B] A. Besse, Manifolds all of whose geodesics are closed, Ergeb. Math. Grenzgeb, **93** (1978), Springer-Verlag, Berlin-Heidelberg-New York.

[D] J. J. Duistermaat, *On the Morse index of variational calculus*, Adv. Math., **21** (1976), 173–195.

[K] K. Kiyohara, *Two classes of Riemannian manifolds whose geodesic flows are integrable*, Mem. Amer. Math. Soc., **130** (1997).

[L-MD] F. Lalonde and D. McDuff, J-curves and the classification of rational and ruled symplectic 4-manifolds, Contact and Symplectic Geometry (C. B. Thomas ed.), Publ. Newton Inst., **8**, Cambridge Univ. Press, Cambridge, 1996, 3–42.

[L] E. Lerman, *Symplectic cuts*, Math. Res. Letters, **2** (1995), 247–258.

[MD] D. McDuff, *Rational and ruled symplectic 4-manifolds*, J. Amer. Math. Soc., **3** (1990), 679–712, Errata, ibid., **5** (1992), 987–988.

[M] J. Moser, On the volume elements of a manifold, Trans. Amer. Math. Soc., **120** (1965), 286–294.

[O-O] H. Ohta and K. Ono, *Note on symplectic 4-manifolds with* $b_2^+ = 1$, Intern. J. Math., **7** (1996), 755–770.

[S] H. Sato, *On the manifold of periodic geodesics*, Geometry of Geodesics and Related Topics (K. Shiohama ed.), Adv. Stud. Pure Math., **3**, North-Holland, Amsterdam-New York, 1984, 213–230.

[Y] C. T. Yang, *Odd dimensional Wiedersehen manifolds are spheres*, J. Diff. Geom., **15** (1980), 91–96.

Department of Mathematics
Hokkaido University
Sapporo, 060-0810
Japan
ono@math.sci.hokudai.ac.jp

Advanced Studies in Pure Mathematics 34, 2002
Minimal Surfaces, Geometric Analysis and Symplectic Geometry
pp. 245–253

Constant Mean Curvature 1 Surfaces with Low Total Curvature in Hyperbolic 3-Space

Wayne Rossman, Masaaki Umehara and Kotaro Yamada

Abstract.

Surfaces of constant mean curvature one in hyperbolic 3-space have quite similar properties to minimal surfaces in Euclidean 3-space. We shall list the possibilities of constant mean curvature one surfaces in hyperbolic 3-space with low total absolute curvature, or low dual total absolute curvature, and compare them with the known classification of minimal surfaces with low total curvature. Complete proofs of the new results will be published in two forthcoming papers (listed in the bibliography).

§ Introduction

Recent developments in the study of constant mean curvature 1 (CMC-1) surfaces in hyperbolic 3-space H^3 (of constant sectional curvature -1) have led to many recently-discovered examples of such surfaces, and it is now well-known that CMC-1 surfaces in H^3 share quite similar properties with minimal surfaces in Euclidean 3-space $\boldsymbol{R}^3$. (See [1], [3], [7], [8], [11], [12], [13], [14], [15] and [16].)

The total absolute curvature of a complete minimal surface in $\boldsymbol{R}^3$ is a 4π-multiple of a nonnegative integer and is equal to the area of the Gauss image of the surface. All such surfaces with finite total absolute curvature less than or equal to 8π have been classified by Lopez [4]. Here we consider the corresponding problem for CMC-1 surfaces in H^3.

Classifying CMC-1 surfaces in H^3 with low total absolute curvature turns out to be more difficult and subtle than Lopez's classification, for the following reasons: Unlike the case of minimal surfaces in $\boldsymbol{R}^3$, the Bryant representation formula, which is an analogy of the Weierstrass representation formula, is not formulated by using line integration, but rather uses parallel transport along a path in the non-commutative group

2000 *Mathematics Subject Classification.* Primary 53A10; Secondary 53A35, 53A42.

$\mathrm{SL}(2, \boldsymbol{C})$. Moreover, also unlike the case of minimal surfaces in $\boldsymbol{R}^3$, CMC-1 surfaces in H^3 have two Gauss maps, the hyperbolic Gauss map G and the secondary Gauss map g. The total absolute curvature of CMC-1 surfaces in H^3 is equal to the area of the image of the secondary Gauss map g, but since g might not be single-valued on the surface, the total absolute curvature might not be a 4π-multiple of an integer. The hyperbolic Gauss map G, on the other hand, does not relate to the total absolute curvature of the surface directly, but it has much clearer geometric meaning, namely the image $G(p)$ lies in the ideal boundary S^2 of the hyperbolic space at the point corresponding to the end of the normal geodesic emanating from the point p on the surface. Therefore, unlike the secondary Gauss map g, the hyperbolic Gauss map G is single-valued on the surface, but it may have essential singularities, even when the total absolute curvature is finite.

There is a natural notion of *dual* total absolute curvature for CMC-1 surfaces in H^3. A duality for CMC-1 surfaces is introduced in [13, Remark 1.8], [15], and Yu [17] (called *inverse surfaces* in [17]), which interchanges the role of the hyperbolic Gauss map and the secondary Gauss map (see Section 1.2). The total absolute curvature of the dual CMC-1 surfaces, i.e., the dual total absolute curvature, is equal to the area of the image of the hyperbolic Gauss map G. In particular, the dual total absolute curvature is always a 4π-multiple of an integer. Though the total absolute curvature of CMC-1 surfaces satisfies only the Cohn-Vossen inequality, the dual total absolute curvature has a much stronger lower bound, which is an analogue of the Osserman inequality for minimal surfaces (cf. [15], [18]).

The purpose of this note is to list the possibilities of CMC-1 surfaces in H^3 with low total absolute curvature, or low dual total absolute curvature, and compare them with Lopez's classification. Complete proofs of the new results will be published in forthcoming papers [9], [10]. Though the results at present do not achieve a full classification of CMC-1 surfaces with total absolute curvature or dual total absolute curvature less than or equal to 8π, the authors hope the results might be of help to readers interested in this subject.

§1. Preliminaries

1.1. Total absolute curvature

Let $f\colon M \to H^3$ be a conformal immersion with CMC-1 of a Riemann surface M into H^3. Denote the Gaussian curvature, the induced metric, and the induced area element by K, ds^2, and dA, respectively.

Then K is non-positive and $d\sigma^2 := (-K)ds^2$ is a conformal pseudometric of constant curvature 1 on M. We call the developing map $g\colon \widetilde{M} \to \boldsymbol{CP}^1$ the *secondary Gauss map* of f, where $\widetilde{M}$ is the universal cover of M. Namely, g is a conformal map such that the pull-back of the Fubini-Study metric of $\boldsymbol{CP}^1$ coincides with $d\sigma^2$.

In addition to the secondary Gauss map, the following two holomorphic invariants G and Q are closely related to the geometric properties of CMC-1 surfaces. The *hyperbolic Gauss map* $G\colon M \to \boldsymbol{CP}^1$ is defined as a holomorphic map on M as follows: Identifying the ideal boundary of H^3 with $\boldsymbol{CP}^1$, $G(p)$ is the asymptotic class of the normal geodesic of $f(M)$ starting at $f(p)$ and oriented in the mean curvature vector's direction. The *Hopf differential* Q is the $(2,0)$-part of the complexified second fundamental form, and is a symmetric holomorphic 2-differential on the Riemann surface M.

As K is a non-positive number, we can define the *total absolute curvature*

$$\mathrm{TA} := \int_M (-K)\, dA \in [0, +\infty].$$

Then TA is the area of the image in $\boldsymbol{CP}^1$ of the secondary Gauss map. The value of TA might not be an integral multiple of 4π — for example, the total curvature of the catenoid cousins [1, Example 2] admits *any* positive real number except 4π.

If the induced metric ds^2 is complete and of finite total absolute curvature (i.e., $\mathrm{TA} < +\infty$), then there exists a compact Riemann surface $\overline{M}$ and a finite set of points $\{p_1, \dots, p_n\} \subset \overline{M}$ such that M is biholomorphic to $\overline{M} \setminus \{p_1, \dots, p_n\}$. We call the p_j's the *ends* of f.

For CMC-1 surfaces, equality never holds in the Cohn-Vossen inequality [11]:

$$\frac{\mathrm{TA}}{2\pi} > -\chi(M) = n - 2 + 2\gamma\,, \tag{1.1}$$

where $\chi(M)$ denotes the Euler characteristic of M, and γ is the genus of $\overline{M}$.

1.2. Dual total absolute curvature

The *dual* CMC-1 immersion of a conformal CMC-1 immersion is defined as follows ([15], [17]): For a conformal CMC-1 immersion $f\colon M \to H^3$, there exists a holomorphic null immersion $F\colon \widetilde{M} \to \mathrm{SL}(2, \boldsymbol{C})$ such that $f = FF^*$, where $\widetilde{M}$ is the universal cover of M and $F^* = {}^t\overline{F}$. Here, we consider $H^3 = \mathrm{SL}(2,\boldsymbol{C})/\mathrm{SU}(2) = \{aa^* \mid a \in \mathrm{SL}(2,\boldsymbol{C})\}$ in the Hermitian model. We call F the *lift* of f. Then, the inverse matrix F^{-1} is also a holomorphic null immersion, and hence we have a new CMC-1 immersion $f^{\#} = F^{-1}(F^{-1})^* : \widetilde{M} \to H^3$, which is called the *dual* of f. The hyperbolic Gauss map (resp. secondary Gauss map, Hopf differential) of the dual immersion $f^{\#}$ is the secondary Gauss map g (resp. hyperbolic Gauss map G, sign-changed Hopf differential $-Q$) of f.

Although the dual immersion might only be defined on the universal cover $\widetilde{M}$ of M, the induced metric $ds^{\#2}$ and the Gaussian curvature $K^{\#}$ are well-defined on M itself. Hence we can define the *dual total absolute curvature* as

$$\mathrm{TA}^{\#} := \int_M (-K^{\#})\, dA^{\#} ,$$

where $dA^{\#}$ is the area element induced by $ds^{\#2}$. Since the secondary Gauss map of $f^{\#}$ is the hyperbolic Gauss map G of f, $d\sigma^{\#2} := (-K^{\#})ds^{\#2}$ is a pseudo-metric of constant curvature 1 with developing map G. Hence $\mathrm{TA}^{\#}$ is the area of the image of G on $\boldsymbol{CP}^1$.

As shown in [15], [17], the induced metric ds^2 of f is complete if and only if the dual metric $ds^{\#2}$ is complete. If we assume the immersion f is complete and of finite dual total absolute curvature (i.e., $\mathrm{TA}^{\#} < +\infty$), then, as in the finite total curvature case, M is biholomorphic to a finitely punctured compact Riemann surface: $M = \overline{M} \setminus \{p_1, \dots, p_n\}$. Unlike the minimal surface case, the hyperbolic Gauss map might not extend to a meromorphic function on $\overline{M}$. The dual total absolute curvature $\mathrm{TA}^{\#}$ is finite if and only if the hyperbolic Gauss map can be extended to a meromorphic function on $\overline{M}$, and in this case, $\mathrm{TA}^{\#} = 4\pi \deg G$. In particular, $\mathrm{TA}^{\#}$ is an integral multiple of 4π.

For $\mathrm{TA}^{\#}$, a hyperbolic analogue of the Osserman inequality holds [15], namely

$$\frac{\mathrm{TA}^{\#}}{2\pi} \geq 2n - 2 + 2\gamma . \tag{1.2}$$

1.3. Notation

Assume f is complete with $\mathrm{TA} < \infty$ or $\mathrm{TA}^{\#} < \infty$, and let $M = \overline{M} \setminus \{p_1, \dots, p_n\}$, where $\overline{M}$ is a compact Riemann surface. Then Q extends to a meromorphic differential on $\overline{M}$ [1]. We say an end p_j ($j = 1, \dots, n$) of a CMC-1 immersion is *regular* if the hyperbolic Gauss map is holomorphic at p_j. When $\mathrm{TA} < \infty$, an end is regular if and only if the order of the Hopf differential Q at p_j is at least -2. Otherwise, the hyperbolic Gauss map has an essential singularity at the end [1].

In this way, the orders of the Hopf differential at the ends are closely related to properties of the surface, so we now introduce a notation for these orders. In the following discussion, we say a surface

Type	TA	The surface	cf.
$\mathbf{O}(0)$	0	Plane	
$\mathbf{O}(-4)$	4π	Enneper's surface	
$\mathbf{O}(-5)$	8π		[4, Theorem 6]
$\mathbf{O}(-6)$	8π		[4, Theorem 6]
$\mathbf{O}(-2,-2)$	4π	Catenoid	
	8π	Double cover of the catenoid	
$\mathbf{O}(-1,-3)$	8π		[4, Theorem 5]
$\mathbf{O}(-2,-3)$	8π		[4, Theorem 4, 5]
$\mathbf{O}(-2,-4)$	8π		[4, Theorem 5]
$\mathbf{O}(-3,-3)$	8π		[4, Theorem 4]
$\mathbf{O}(-1,-2,-2)$	8π		[4, Theorem 5]
$\mathbf{O}(-2,-2,-2)$	8π		[4, Theorem 5]
$\mathbf{I}(-4)$	8π	Chen-Gackstatter surface	[4, Theorem 5], [2]

Table 1. Classification of minimal surfaces in $\boldsymbol{R}^3$ with TA $\leq 8\pi$ [4].

is of *type* $\mathbf{\Gamma}(d_1,\dots,d_n)$ if the surface is given as an immersion $f\colon \overline{M}\setminus\{p_1,\dots,p_n\}\to H^3$, where the order of the Hopf differential at p_j is d_j for each $j=1,\dots,n$. We use $\mathbf{\Gamma}$ because it is the capitalized letter corresponding to γ, the genus of $\overline{M}$. For instance, the class $\mathbf{I}(-4)$ means the class of surfaces of genus 1 with 1 end so that Q has a pole of order 4 at the end, and the class $\mathbf{O}(-2,-3)$ is the class of surfaces of genus 0 with two ends so that Q has a pole of order 2 at one end and a pole of order 3 at the other.

1.4. Minimal surfaces with TA $\leq 8\pi$

Using the above notation, the classification of complete minimal surfaces in $\boldsymbol{R}^3$ with TA $\leq 8\pi$ (Lopez [4]) is listed in Table 1.

§2. Complete CMC-1 surfaces with TA $\leq 4\pi$

It is well-known that the only complete minimal surfaces in $\boldsymbol{R}^3$ of total curvature less than or equal to 4π are the plane, the Enneper surface, and the catenoid. In this section, we shall introduce a complete classification of CMC-1 surfaces in H^3 with TA $\leq 4\pi$.

Assume $f\colon M\to H^3$ is a complete conformal immersion of TA $\leq 4\pi$. Then, by the Cohn-Vossen inequality (1.1), the genus γ and the number

Type	TA	The surface	cf.
$\mathbf{O}(0)$	0	Horosphere	
$\mathbf{O}(-4)$	4π	Enneper cousins	[1, Example 1]
$\mathbf{O}(-2,-2)$	$4\pi\mu$ $(0<\mu<1)$	Catenoid cousins and their m-fold covers	[1, Example 2] [11, Theorem 6.2]
$\mathbf{O}(-2,-2)$	4π	$\star$	[11, Theorem 6.2]

Table 2. Classification of CMC-1 surfaces in H^3 with TA $\leq 4\pi$.

of ends n are restricted to the following cases:

$$(\gamma, n) = (0,1), \quad (0,2), \quad (0,3), \quad (1,1).$$

However, the cases $(\gamma, n) = (0,3)$ and $(1,1)$ do not occur. More precisely, the following theorem holds:

Theorem 1 ([10]). *Any complete* CMC-1 *surface in* H^3 *with* TA $\leq 4\pi$ *is one of those in Table* 2.

The case marked $\star$ is the class of immersions $f\colon \boldsymbol{C}\setminus\{0\} \to H^3$ given by the Weierstrass data

$$\left(g, \omega := \frac{Q}{dg}\right) = \left(az^l + b, \frac{m^2 - l^2}{4a}\frac{dz}{z^2}\right), \quad a \in \boldsymbol{C}\setminus\{0\},\ b \in \boldsymbol{C}, \tag{2.1}$$

for $l = 1$ and $m = 2, 3, \ldots$ as in the equation (6.5) in [11]. When $b = 0$, the surface is a catenoid cousin. However, if $b \neq 0$, the surface is not rotationally symmetric.

Though we do not give the details of the proof here, we remark that the proof is more difficult than for the corresponding case of minimal surfaces in $\boldsymbol{R}^3$. For example, the nonexistence of CMC-1 surfaces in H^3 with $(\gamma, n) = (1,1)$ is shown by applying a flux formula in [8]. The nonexistence of CMC-1 surfaces with $(\gamma, n) = (0,3)$ is shown by first applying the classification of irreducible CMC-1 surfaces of type $\mathbf{O}(-2,-2,-2)$ in [16], and then one can show that TA $\geq 4\pi$ for such surfaces. In [10], we will show the stronger inequality TA $> 4\pi$ for CMC-1 surfaces with $(\gamma, n) = (0,3)$. However, the proof is not simple. In [10], some other results for surfaces with TA $\leq 8\pi$ will also be discussed.

§3. Complete CMC-1 surfaces with TA$^{\#} \leq 8\pi$

In this section, we introduce a partial result on classification of CMC-1 surfaces in H^3 with TA$^{\#} \leq 8\pi$. Note that TA may take the

Type	$\mathrm{TA}^{\#}$	Status	cf.
$\mathbf{O}(0)$	0	classified^0	Horosphere
$\mathbf{O}(-4)$	4π	classified	Dual Enneper cousins [7, Example 5.4]
$\mathbf{O}(-2,-2)$	4π	classified	Catenoid cousins, and ⋆⋆ [1, Example 2], [11, Theorem 6.2]
$\mathbf{O}(-5)$	8π	classified	[9]
$\mathbf{O}(-6)$	8π	classified	[9]
$\mathbf{O}(-2,-2)$	8π	classified	Double covers of catenoid cousins, and ⋆⋆⋆ [11, Theorem 6.2]
$\mathbf{O}(-1,-4)$	8π	classified^0	[9]
$\mathbf{O}(-2,-3)$	8π	classified	[9]
$\mathbf{O}(-2,-4)$	8π	classified	[9]
$\mathbf{O}(-3,-3)$	8π	existence^d	[9]
$\mathbf{O}(-1,-1,-2)$	8π	classified^0	[9]
$\mathbf{O}(-1,-2,-2)$	8π	classified	[9]
$\mathbf{O}(-2,-2,-2)$	8π	existence	Classified for the irreducible case [16]
$\mathbf{I}(-3)$	8π	unknown	
$\mathbf{I}(-4)$	8π	existence^d	[9]
$\mathbf{I}(-1,-1)$	8π	unknown	
$\mathbf{I}(-2,-2)$	8π	existence^1	[6]

Table 3. Classification of CMC-1 surfaces in H^3 with $\mathrm{TA}^{\#} \leq 8\pi$.

value $+\infty$ even if $\mathrm{TA}^{\#}$ is finite. By (1.2), the genus γ and the number n of ends of such surfaces are restricted to the following cases:

$$(\gamma, n) = (0,1), \quad (0,2), \quad (0,3), \quad (1,1), \quad (1,2), \quad (2,1).$$

However, the case $(\gamma, n) = (2,1)$ does not occur, which is a consequence of the flux formula in [8]. A list of possible surfaces with $\mathrm{TA}^{\#} \leq 8\pi$ is shown in Table 3 (for the proof, see [9]). In this table,

- *classified* means the complete list of the surfaces in such a class is known,
- *classified*^{0} means there exists a unique surface (up to isometries of H^3 and deformations that come from its reducibility [7, Theorem 3.2]),
- *existence* means that examples of such surfaces are known to exist, but they are not yet classified,

- *existence*^{d} means that examples can be obtained by deforming from a minimal surface in $\boldsymbol{R}^3$, using the method in [7],
- *existence*^{1} means there exists a 1 parameter family of examples, which is not deformations coming from reducibility,
- *unknown* means that neither existence nor nonexistence is known yet.

The case marked $\star\star$ (resp. $\star\star\star$) is the class of surfaces given by the Weierstrass data (2.1) for $m = 1$ and $l = 2, 3, \dots$ (resp. $m = 2$ and $l = 1, 3, 4, \dots$).

It is interesting to compare Table 3 with Table 1. In the case of minimal surfaces in $\boldsymbol{R}^3$ with TA $\le 8\pi$, the cases

$$\mathbf{O}(-1,-4), \quad \mathbf{O}(-1,-1,-2), \quad \mathbf{I}(-2,-2)$$

do not occur, whereas these cases really do occur for CMC-1 surfaces in H^3. On the other hand, there is no CMC-1 surface in H^3 of type $\mathbf{O}(-1,-3)$, in spite of the fact that such minimal surfaces exist in $\boldsymbol{R}^3$. Although the existence of CMC-1 surfaces in H^3 of type $\mathbf{I}(-3)$ and $\mathbf{I}(-1,-1)$ is still unknown, Table 3 shows the existence of CMC-1 surfaces in H^3 for which the corresponding minimal surfaces in $\boldsymbol{R}^3$ cannot exist.

References

[1] R. Bryant, *Surfaces of mean curvature one in hyperbolic space*, Astérisque, **154–155** (1987), 321–347.

[2] C. C. Chen and F. Gackstatter, *Elliptische und hyperelliptische Funktionen und vollständige Minimalflächen vom Enneperschen Typ*, Math. Ann., **259** (1982), 359–369.

[3] P. Collin, L. Hauswirth and H. Rosenberg, *The geometry of finite topology surfaces properly embedded in hyperbolic space with constant mean curvature one*, Ann. of Math., to appear.

[4] F. J. Lopez, *The classification of complete minimal surfaces with total curvature greater than* -12π, Trans. Amer. Math. Soc., **334** (1992), 49–74.

[5] R. Osserman, A Survey of Minimal Surfaces, 2nd ed., Dover, 1986.

[6] W. Rossman and K. Sato, *Constant mean curvature surfaces with two ends in hyperbolic space*, Experimental Math., **7(2)** (1998), 101–119.

[7] W. Rossman, M. Umehara and K. Yamada, *Irreducible constant mean curvature* 1 *surfaces in hyperbolic space with positive genus*, Tôhoku Math. J., **49** (1997), 449–484.

[8] ———, *A new flux for mean curvature* 1 *surfaces in hyperbolic* 3*-space, and applications*, Proc. Amer. Math. Soc., **127** (1999), 2147–2154.

[9] ______, *Mean curvature* 1 *surfaces in hyperbolic* 3*-space with low total curvature* I, preprint, `math. DG/0008015`.

[10] ______, *Mean curvature* 1 *surfaces in hyperbolic* 3*-space with low total curvature* II, preprint.

[11] M. Umehara and K. Yamada, *Complete surfaces of constant mean curvature-*1 *in the hyperbolic* 3*-space*, Ann. of Math., **137** (1993), 611–638.

[12] ______, *A parameterization of Weierstrass formulae and perturbation of some complete minimal surfaces of* $\boldsymbol{R}^3$ *into the hyperbolic* 3*-space*, J. Reine Angew. Math., **432** (1992), 93–116.

[13] ______, *Surfaces of constant mean curvature-c in* $H^3(-c^2)$ *with prescribed hyperbolic Gauss map*, Math. Ann., **304** (1996), 203–224.

[14] ______, *Another construction of a* CMC-1 *surface in* H^3, Kyungpook Math. J., **35** (1996), 831–849.

[15] ______, *A duality on* CMC-1 *surface in the hyperbolic* 3*-space and a hyperbolic analogue of the Osserman Inequality*, Tsukuba J. Math., **21** (1997), 229–237.

[16] ______, *Metric of constant curvature one with three conical singularities on* 2*-sphere*, Illinois J. Math., **44(1)** (2000), 72–94.

[17] Z. Yu, *Value distribution of hyperbolic Gauss maps*, Proc. Amer. Math. Soc., **125** (1997), 2997–3001.

[18] ______, *The inverse surface and the Osserman Inequality*, Tsukuba J. Math., **22** (1998), 575–588.

Wayne Rossman
Department of Mathematics
Faculty of Science
Kobe University
Kobe 657-8501
Japan
`wayne@math.kobe-u.ac.jp`

Masaaki Umehara
Department of Mathematics
Faculty of Science
Hiroshima University
Higashi-Hiroshima 739-8526
Japan
`umehara@math.sci.hiroshima-u.ac.jp`

Kotaro Yamada
Faculty of Mathematics
Kyushu University 36, 6-10-1
Hakozaki, Higashi-ku, Fukuoka 812-8185
Japan
`kotaro@math.kyushu-u.ac.jp`

Advanced Studies in Pure Mathematics 34, 2002
Minimal Surfaces, Geometric Analysis and Symplectic Geometry
pp. 255–259

A Note on the Symplectic Volume of the Moduli Space of Spatial Polygons

Tatsuru Takakura*

Abstract.

We present an altenative proof of the volume formula of the moduli space of spatial polygons, which was given by Kamiyama-Tezuka. Our method is based on the commutativity of geometric quantization and symplectic reduction, originating from a conjecture of Guillemin-Sternberg.

§1. Introduction

Consider the following space $\mathcal{M}_n$, often called the moduli space of spatial polygons;

$$\mathcal{M}_n = \{(a_1, \ldots, a_n) \in (S^2)^n \mid a_1 + \cdots + a_n = 0\}/SO_3,$$

where $n \geq 3$ and each S^2 is the unit sphere in $\mathbb{R}^3$ with the standard SO_3-action. For simplicity, we asseme that n is odd. In this case $\mathcal{M}_n$ is a compact connected smooth manifold of dimension $2(n-3)$. The topology and geometry of this space has been studied from various points of view (see, e.g., [K-T] and references cited there). For example, it is well-known that $\mathcal{M}_n$ admits a natural symplectic (in fact, Kähler) form ω_n. Concerning symplectic properties of $\mathcal{M}_n$, Kamiyama and Tezuka [K-T] proved, among other things, the following formula[1].

2000 *Mathematics Subject Classification.* Primary 53D20; Secondary 53D50, 14L24.

*Partly supported by the Grant-in-Aid for Encouragement of Young Scientists (No. 11740048), The Ministry of Education, Science and Culture, Japan.

[1] *Added in Proof.* After this article was submitted, the author was informed that S. Martin (*Transversality theory, cobordisms, and invariants of symplectic quotients*, preprint) had also obtained this formula. His method is different from those of Kamiyama-Tezuka and us.

Theorem 1.1 ([K-T, Theorem C]). *The symplectic volume of the space $\mathcal{M}_n$ is given as follows.*

$$\mathrm{Vol}(\mathcal{M}_n) = \frac{-1}{2(n-3)!} \sum_{r=0}^{[n/2]} (-1)^r \binom{n}{r} (n-2r)^{n-3}.$$

(Strictly speaking, the volume in [K-T] is $(n-3)!$ times our $\mathrm{Vol}(\mathcal{M}_n)$. Besides, their formula is written in a slightly different form, but is indeed equivalent to ours (see [K-T, §5]).)

The aim of this paper is to give an alternative proof of the above, as well as to prove the following more general result.

Theorem 1.2. *Let $\mathcal{L}_n$ be the complex line bundle over $\mathcal{M}_n$ with $c_1(\mathcal{L}_n) = [\omega_n]$. Then for each non-negative integer k, we have*

$$\int_{\mathcal{M}_n} \mathrm{Td}(\mathcal{M}_n)\mathrm{Ch}(\mathcal{L}_n^{\otimes k}) = \dim \left(V_k^{\otimes n}\right)^{SO_3}$$
$$= -\frac{1}{2} \sum_{r=0}^{N(n,k)} (-1)^r \binom{n}{r} \binom{(n-2r)k+n-r-2}{n-3}.$$

Here, V_k is the irreducible representation of SO_3 of dimension $2k+1$, $\left(V_k^{\otimes n}\right)^{SO_3}$ is the invariant subspace of $V_k^{\otimes n}$, and $N(n,k) = \left[\dfrac{kn+1}{2k+1}\right]$.

Note that a similar formula for $k=1$ is given by Kamiyama [Ka], where the results in [K-T] on the intersection pairings $\int_{\mathcal{M}_n} \alpha \cdot \beta$ ($\alpha, \beta \in H^*(\mathcal{M}_n)$) are essential. On the other hand, we do not use such information in our proof. (In fact, our approach in this paper is able to be applied to derive general intersection pairings [T].) Actually, the proof of Theorem 1.2 is a simple application of the "quantization commutes with reduction" theorem, originally due to Guillemin-Sternberg.

Acknowledgement. I am grateful to Yasuhiko Kamiyama for explaining his work and for useful discussions.

§2. Preliminaries

2.1. First, we need to specify the symplectic structure ω_n. For this purpose, let us describe the symplectic manifold $(\mathcal{M}_n, \omega_n)$ as a symplectic (Kähler) quotient, or a reduced phase space due to Marsden-Weinstein. (See, e.g., [Ki], for this notion, as well as the relation to the geometric invariant theory.) Let us consider the symplectic manifold

$(\mathbb{P}^1, 2\omega_{\mathrm{FS}})^n$, that is, the product of n-copies of complex projective line with twice the Fubini-Study Kähler form. (The Fubini-Study Kähler form ω_{FS} is supposed to be normalized by $[\omega_{\mathrm{FS}}] = c_1(H) \in H^2(\mathbb{P}^1)$, where H is the hyperplane bundle on $\mathbb{P}^1$.) The diagonal action of $SO_3 = SU_2/\{\pm 1\}$ on $(\mathbb{P}^1, 2\omega_{\mathrm{FS}})^n$ is Hamiltonian. In fact, if we identify each $\mathbb{P}^1$ with a coadjoint orbit of SO_3 in $\mathfrak{so}_3^*$ (or a round 2-sphere in $\mathbb{R}^3$), then the moment map $\Phi : (\mathbb{P}^1)^n \to \mathfrak{so}_3^* \cong \mathbb{R}^3$ is given by $\Phi(a_1, \ldots, a_n) = a_1 + \cdots + a_n$. Now our symplectic manifold $(\mathcal{M}_n, \omega_n)$ is defined as the symplectic quotient $\Phi^{-1}(0)/SO_3$ with the reduced symplectic form.

Remark. In [K-T], the symplectic structure of $\mathcal{M}_n$ is defined differently. But it does coincide with ours.

Moreover, $\mathcal{M}_n$ inherits a complex structure J_n from the standard one on $(\mathbb{P}^1)^n$, so that $(\mathcal{M}_n, \omega_n, J_n)$ is a Kähler manifold (called the Kähler quotient).

Let $\mathbb{L} = H^{\otimes 2}$. It is an SO_3-equivariant holomorphic line bundle over $\mathbb{P}^1$. Then the outer tensor product $\mathbb{L}^{\boxtimes n}$ over $(\mathbb{P}^1)^n$ naturally defines a holomorphic line bundle $\mathcal{L}_n$ over $\mathcal{M}_n$ such that $c_1(\mathcal{L}_n) = [\omega_n]$. In particular, $\mathcal{M}_n$ is projective.

2.2. Now let us recall the "quantization commutes with reduction" theorem. This theorem arose from a conjecture of Guillemin-Sternberg [G-S], and has been proved (and improved) by several people and by various methods. See, e.g., [S] for a survey and references on this topic. We do not intend to state it in full generality, but restrict ourselves only to the following special situation.

For a compact complex manifold X and a holomorphic line bundle L over X, define $\chi(X, L) := \sum(-1)^i H^i(X, \mathcal{O}(L))$, as a virtual vector space. By the Hirzebruch-Riemann-Roch theorem (or Atiyah-Singer index theorem), we have $\dim \chi(X, L) = \displaystyle\int_X \mathrm{Td}(X)\mathrm{Ch}(L)$. If a compact group G acts holomorphically on X and L is G-equivariant, we can regard $\chi(X, L)$ as a virtual representation of G, i.e., an element of the representation ring $R(G)$ of G.

Suppose in addition that X admits an G-invariant Kähler form ω and L admits an G-invariant hermitian connection ∇ such that $c_1(\nabla) = \omega$. Then G-action on X is Hamiltonian. Let $\Phi : X \to \mathfrak{g}^*$ be the moment map. For simplicity, we assume 0 is a regular value of Φ and G acts freely on $\Phi^{-1}(0)$, so that the Kähler quotient $X_G = \Phi^{-1}(0)/G$ is a smooth Kähler manifold and we have the reduced holomorphic line bundle L_G over X_G. The theorem we need is the following.

Theorem 2.1. *Under the assumption as above, we have $\chi(X,L)^G = \chi(X_G, L_G)$, where the left hand side is the G-invariant part of the virtual representation $\chi(X,L)$ of G.*

Remarks. (1) By the argument in §2.1, we are able to apply this theorem to the case when $G = SO_3$, $(X,L) = ((\mathbb{P}^1)^n, \mathbb{L}^{\boxtimes n})$ (with an appropriate invariant hermitian connection) and $(X_G, L_G) = (\mathcal{M}_n, \mathcal{L}_n)$.

(2) The original result in [G-S] states that $H^0(X, \mathcal{O}(L))^G = H^0(X_G, \mathcal{O}(L_G))$ for the spaces of global holomorphic sections instead of χ. Since we can show that the higer cohomologies vanish for $((\mathbb{P}^1)^n, \mathbb{L}^{\boxtimes n})$ and $(\mathcal{M}_n, \mathcal{L}_n)$, this may be enough for our purpose.

(3) If we replace L (resp. L_G) to $L^{\otimes k}$ (resp. $L_G^{\otimes k}$) for a non-negative integer k, the same formula holds. It is obvious when $k \geq 1$. See [M-S] for the case $k = 0$.

§3. Proofs

Theorem 1.1 follows from Theorem 1.2, since

$$\begin{aligned}\mathrm{Vol}(\mathcal{M}_n) &= \int_{\mathcal{M}_n} \exp(\omega_n) = \int_{\mathcal{M}_n} \mathrm{Ch}(\mathcal{L}_n) \\ &= \frac{1}{k^{n-3}} \int_{\mathcal{M}_n} \mathrm{Ch}(\mathcal{L}_n^{\otimes k}) = \lim_{k\to\infty} \frac{1}{k^{n-3}} \int_{\mathcal{M}_n} \mathrm{Td}(\mathcal{M}_n)\mathrm{Ch}(\mathcal{L}_n^{\otimes k}).\end{aligned}$$

In order to prove Theorem 1.2, note that, for each $k(\geq 0)$,

$$\begin{aligned}\int_{\mathcal{M}_n} \mathrm{Td}(\mathcal{M}_n)\mathrm{Ch}(\mathcal{L}_n^{\otimes k}) = \chi(\mathcal{M}_n, \mathcal{L}_n^{\otimes k}) &= \chi((\mathbb{P}^1)^n, (\mathbb{L}^{\otimes k})^{\boxtimes n})^{SO_3} \\ &= \left(\chi(\mathbb{P}^1, \mathbb{L}^{\otimes k})^{\otimes n}\right)^{SO_3} \\ &= \left(V_k^{\otimes n}\right)^{SO_3}.\end{aligned}$$

Indeed, the first equality is a direct consequence of Theorem 2.1, the second one is due to the multiplicative property for χ, and the third one is an elementary fact about the representations of SO_3 (which is a typical example of the Borel-Weil theorem).

Now, the invariant part of the representation is computed by the integration of its character. Namely,

$$\begin{aligned}\dim \left(V_k^{\otimes n}\right)^{SO_3} &= \frac{1}{\pi}\int_0^{2\pi}\left(\frac{\sin(2k+1)\theta}{\sin\theta}\right)^n \sin^2\theta d\theta \\ &= -\frac{1}{2}\mathrm{Res}_{z=0}\frac{1}{z}\left(\frac{z^{2k+1}-z^{-(2k+1)}}{z-z^{-1}}\right)^n (z-z^{-1})^2 \\ &= -\frac{1}{2}\mathrm{Res}_{z=0}\frac{1}{z}z^{-2(kn+1)}(z^{2(2k+1)}-1)^n(z^2-1)^{-(n-2)}.\end{aligned}$$

By considering the Laurent expansion, we obtain Theorem 1.2.

Remark. When n is even(≥ 4), the space $\mathcal{M}_n$ has singularities. Nevertheless, Theorem 1.1 holds also in this case (as proved in [K-T]). So does Theorem 1.2, after modifying the definition of $\chi(\mathcal{M}_n, \mathcal{L}_n^{\otimes k})$. These follow from a generalization of Theorem 2.1 to singular quotients (see [M-S]).

References

[G-S] V. Guillemin and S. Sternberg, *Geometric quantization and multiplicities of group representations*, Invent. Math., **67** (1982), 515–538.

[K-T] Y. Kamiyama and M. Tezuka, *Symplectic volume of the moduli space of spatial polygons*, J. Math. Kyoto Univ., **39** (1999), 557–575.

[Ka] Y. Kamiyama, *Sheaf cohomology of the moduli space of spatial polygons and lattice points*, Tokyo J. Math., **24** (2001), 205–209.

[Ki] F. C. Kirwan, *Cohomology of quotients in symplectic and algebraic geometry*, Mathematical Notes, vol. 31, Princeton Univ. Press, Princeton, NJ, 1984.

[M-S] E. Meinrenken and R. Sjamaar, *Singular reduction and quantization*, Topology, **38** (1999), 699–762.

[S] R. Sjmaar, *Symplectic reduction and Riemann-Roch formulas for multiplicities*, Bull. Amer. Math. Soc., **33** (1996), 327–338.

[T] T. Takakura, *Intesection theory on symplectic quotients of products of spheres*, Internat. J. Math., **12** (2001), 97–111.

Department of Mathematics
Chuo University
1-13-27 Kasuga, Bunkyo-ku
Tokyo 112-8551
Japan
takakura@math.chuo-u.ac.jp

Advanced Studies in Pure Mathematics

A SERIES OF UP-TO-DATE GUIDES OF LASTING INTEREST TO ADVANCED MATHEMATICS

Volume 1	Algebraic Varieties and Analytic Varieties. Edited by S. Iitaka. February, 1983	
Volume 2	Galois Groups and their Representations. Edited by Y. Ihara. December, 1983	
Volume 3	Geometry of Geodesics and Related Topics. Edited by K. Shiohama. June, 1984	
Volume 4	Group Representations and Systems of Differential Equations. Edited by K. Okamoto. March, 1985	
Volume 5	Foliations. Edited by I. Tamura. February, 1986	
Volume 6	Algebraic Groups and Related Topics. Edited by R. Hotta. March, 1985	
Volume 7	Automorphic Forms and Number Theory. Edited by I. Satake. February, 1986	
Volume 8	Complex Analytic Singularities. Edited by T. Suwa and P. Wagreich. February, 1987	
Volume 9	Homotopy Theory and Related Topics. Edited by H. Toda. February, 1987	
Volume 10	Algebraic Geometry, Sendai, 1985. Edited by T. Oda. July, 1987	
Volume 11	Commutative Algebra and Combinatorics. Edited by M. Nagata and H. Matsumura. October, 1987	
Volume 12	Galois Representations and Arithmetic Algebraic Geometry. Edited by Y. Ihara. November, 1987	
Volume 13	Investigations in Number Theory. Edited by T. Kubota. March, 1988	
Volume 14	Representations of Lie Groups, Kyoto, Hiroshima, 1986. Edited by K. Okamoto and T. Oshima. February, 1989	
Volume 15	Automorphic Forms and Geometry of Arithmetic Varieties. Edited by K. Hashimoto and Y. Namikawa. July, 1989	

Volume 16 Conformal Field Theory and Solvable Lattice Models.
Edited by M. Jimbo, T. Miwa and A. Tsuchiya.
June, 1988

Volume 17 Algebraic Number Theory—in honor of K. Iwasawa.
Edited by J. Coates, R. Greenberg, B. Mazur and I. Satake.
August, 1989

Volume 18 I—Recent Topics in Differential and Analytic Geometry.
Edited by T. Ochiai. November, 1990

II—Kähler Metric and Moduli Spaces.
Edited by T. Ochiai. November, 1990

Volume 19 Integrable Systems in Quantum Field Theory and Statistical Mechanics.
Edited by M. Jimbo, T. Miwa and A. Tsuchiya.
November, 1989

Volume 20 Aspects of Low Dimensional Manifolds.
Edited by Y. Matsumoto and S. Morita.
December, 1992

Volume 21 Zeta Functions in Geometry.
Edited by N. Kurokawa and T. Sunada.
December, 1992

Volume 22 Progress in Differential Geometry.
Edited by K. Shiohama. September, 1993

Volume 23 Spectral and Scattering Theory and Applications.
Edited by K. Yajima. February, 1994

Volume 24 Progress in Algebraic Combinatorics.
Edited by E. Bannai and A. Munemasa. May, 1996

Volume 25 CR-Geometry and Overdetermined Systems.
Edited by T. Akahori, G. Komatsu, K. Miyajima, M. Namba and K. Yamaguchi. July, 1997

Volume 26 Analysis on Homogeneous Spaces and Representation Theory of Lie Groups, Okayama-Kyoto.
Edited by M. Kashiwara, T. Kobayashi, T. Matsuki, K. Nishiyama and T. Oshima. April, 2000

Volume 27 Arrangements – Tokyo 1998.
Edited by M. Falk and H. Terao. August, 2000

Volume 28 Combinatorial Methods in Representation Theory.
Edited by M. Kashiwara, K. Koike, S. Okada, I. Terada, and H.-F. Yamada. December, 2000

Volume 29 Singularities – Sapporo 1998.
Edited by J.-P. Brasselet and T. Suwa.
December, 2000

Volume 30 Class Field Theory – Its Centenary and Prospect.
Edited by K. Miyake. April, 2001

Volume 31 Taniguchi Conference on Mathematics Nara '98.
Edited by M. Maruyama and T. Sunada. May, 2001

Volume 32 Groups and Combinatorics — in memory of Michio Suzuki.
Edited by E. Bannai, H. Suzuki, H. Yamaki and T. Yoshida.
October, 2001

Volume 33 Computational Commutative Algebra and Combinatorics.
Edited by T. Hibi. February, 2002

Volume 34 Minimal Surfaces, Geometric Analysis and Symplectic Geometry.
Edited by K. Fukaya, S. Nishikawa and J. Spruck.
June, 2002

TO BE CONTINUED